Handbook of Measurement Science

Science

Volume 3 Elements of Change

WILEY SERIES IN MEASUREMENT SCIENCE AND TECHNOLOGY

Chief Editor

Peter H. Sydenham
*University of
South Australia*

Advisory Editor

L. Finkelstein
*City University
London
UK*

Instruments and Experiences: Papers on Measurement and Instrument Design
R. V. Jones

Temperature Measurement
L. Michalski, K. Eckersdorf and J. McGhee,

Handbook of Measurement Science, Volume 1
Edited by P. H. Sydenham

Handbook of Measurement Science, Volume 2
Edited by P. H. Sydenham

Handbook of Measurement Science, Volume 3
Edited by P. H. Sydenham and R. Thorn

Introduction to Measurement Science & Engineering
P. H. Sydenham, N. H. Hancock and R. Thorn

Handbook of Measurement Science

Volume 3   Elements of Change

Edited by
P. H. Sydenham and R. Thorn
University of South Australia

JOHN WILEY & SONS
Chichester · New York · Brisbane · Toronto · Singapore

Copyright © 1992 by John Wiley & Sons Ltd.
Baffins Lane, Chichester
West Sussex PO19 1UD, England

Other Wiley Editorial Offices

John Wiley & Sons Inc., 605 Third Avenue,
New York, NY 10158-0012, USA

Jacaranda Wiley Ltd, G.P.O. Box 859, Brisbane,
Queensland 4001, Australia

John Wiley & Sons (Canada) Ltd, 22 Worcester Road,
Rexdale, Ontario M9W 1L1, Canada

John Wiley & Sons (SEA) Pte Ltd, 37 Jalan Pemimpin #05-04,
Block B, Union Industrial Building, Singapore 2057

Library of Congress Cataloging-in-Publication Data
(Revised for volume 3)

Handbook of measurement science.
  'A Wiley-Interscience publication.'
  Includes bibliographies and indexes.
  Contents: v. 1. Theoretical fundamental / edited by P. H. Sydenham —   —
v. 3. Elements of change / edited by P. H. Sydenham and R. Thorn.
  1. Mensuration.   I. Sydenham, P. H.
T50.H26   1982   620'.0044   81–14628
ISBN 0 471 10037 4 (v. 1)
ISBN 0 471 92219 6 (v. 3)

British Library Cataloging in Publication Data
A catalogue record for this book is available
from the British Library
ISBN 0 471 92219 6

Typeset by Pure Tech Corporation India
Printed in Great Britain by Biddles Ltd, Guildford and King's Lynn

To Bernard Veltmann
for his initial encouragement and support
that led to these Handbooks

Contributing Authors

A. E. Baldin   *Studio Baldin Consulenze Di Direzione, Milan, Italy*

W. E. Duckworth   Formerly *Fulmer Ltd, Stoke Poges, Berkshire, UK*

L. Finkelstein   *School of Engineering, The City University, London, UK*

S. Hamilton   *Department of Physics, University of Manchester, UK*

N. H. Hancock   *School of Engineering, University of Southern Queensland, Australia*

D. D. Harris   *Sensor Science and Engineering Group, University of South Australia, Australia*

M. R. Haskard   *Microelectronics Centre, University of South Australia, Australia*

D. Hofmann   *Department of Measurement Engineering, Friedrich—Schiller University, Jena, Germany*

S. Howell   *Intelligent Solutions Partnership, Manchester, UK*

J. R. Jordan   *Department of Electrical Engineering, University of Edinburgh, UK*

I. KARUBE   *Research Centre for Advanced Science and
                     Technology, University of Tokyo, Japan*

L. C. LYNNWORTH   *Panametrics Inc., Waltham, Massachusetts,
                     USA*

J. V. NICHOLAS   *DSIR Physical Sciences, Lower Hutt,
                     New Zealand*

R. C. SPOONCER   *The Brunel Centre for Manufacturing
                     Metrology, Brunel University of West London, UK*

P. H. SYDENHAM   *Sensor Science and Engineering Group, University
                     of South Australia, Australia*

R. THORN   *Sensor Science and Engineering Group, University of
                     South Australia, Australia*

F. W. UMBACH   *Department of Electrical Engineering, Twente
                     University of Technology, The Netherlands*

M. J. WILLISON   *Environmental Technology Centre, UMIST,
                     Manchester, UK*

Contents

34. Relationship of Legal Issues to Measurement
J. V. Nicholas

35. Condition monitoring
A. E. Baldin

36. Environmental Monitoring

M. J. Willison

PART 2—DEVELOPING TECHNOLOGIES

37. Microelectronics in Instrumentation
M. R. Haskard

38. Ultrasonics in instrumentation
L. C. Lynnworth

39. Fibre Optics in instrumentation
R. C. Spooncer

40. Biosensors
I. Karube

PART 3—ADVANCES IN DESIGN AND MANUFACTURING TECHNIQUES

41. Computer aided engineering of instrumentation
N. H. Hancock

42. Communication standards for measurement and control
J. R. Jordan

43. Active and passive role of materials in measurement systems
W. E. Duckworth, D. D. Harris and P. H. Sydenham

44. Quality control through measurement

D. Hofmann

45. Intelligent instruments

S. Howell and S. Hamilton

Series Editor's Preface

Reader interest in the first two volumes of the *Handbook of Measurement Science* was a key factor that encouraged the birth of this series.

The series provides authorative books, written by international experts, on the many topics that constitute measurement science (today also known as sensing) and its engineering.

As well as commissioning books on its teaching, such as *Introduction to Measurement Science and Engineering* and on specific measurands, such as *Temperature Measurement*, there is a continuing need to provide timely reference works that address the overall issues in a comprehensive way. Volume 3 of the *Handbook of Measurement Science* provides material on the changing aspects of fundamental issues that form the now accepted structure of the discipline of measurement science and its engineering.

PETER SYDENHAM
Editor in Chief

Editors' Preface

This book is Volume 3 in the The Handbook of Measurement Science series. The series began with the *Handbook of Measurement Science, Volume 1—Theoretical Fundamentals*, which was first published in 1982, and continued with the *Handbook of Measurement Science, Volume 2—Practical Fundamentals*, which was first published in 1983.

The *Handbook of Measurement Science, Volume 3* deals with the factors that are responsible for the emergence of new measurement systems.

As was clearly shown in Volume 1, measurement systems while appearing at first sight to be many and various, are really no more than information machines which can be broken down into a finite set of building blocks. The fundamental analysis and design techniques that can be applied to each block is the same. It is the interconnection between them, the choice of implementation, and thus the imagination of the designer, that results in a different measurement system being produced.

Since fundamental principles rarely change (although our understanding of them may improve), the concepts in Volume 1 are unlikely to require updating very often.

Similarly Volume 2, which was concerned with fundamentals of practice, covers techniques rather than specific implementations.

So if the first two volumes are concerned with the relatively static fundamentals of theory and practice, what does change? Something must, because new (or apparently new) measuring instruments are constantly appearing on the market. This volume examines key factors responsible for change. These factors around which the book is designed may be classified into three groups:

1. pressures for change—the needs;
2. emergence of applicable new technologies—the means;
3. advances in design and manufacturing techniques—the methods.

Chapter 33, the first of Volume 3, deals with the formation of the engineers and scientists who bring about change in measurement science. As people become increasingly concerned about the environment they live and work in, then more advanced measurement systems are required to monitor such parameters. Chapters 34–6, deal with the legal aspects of measurement and the use of measurement systems for industrial plant health monitoring and environmental monitoring.

Chapters 37–40 are concerned with technologies which are playing an important part in the development of new measurement systems. Some of these, such as microelectronic, ultrasonic and fibre optics, are already well established but are finding increased usage as academic and industrial development of their capabilities continues. The newest of the technologies considered, biosensors, are still yet to fully emerge but they have enormous potential in many areas of measurement.

Chapters 41–6 are concerned with the way in which advances in design and manufacturing techniques are affecting the development of measurement systems. These chapters highlight the influence of computers for the design and operation of measurement systems, the increasing importance of materials and quality control, and the design and communications problems facing those implementing large measurement systems.

There are many people we wish to thank for their help in this project. Firstly, the authors themselves, who although located in many different parts of the world, all did their best to respond quickly to our requests and occasional changes of mind.

Mrs Wendy Johns and Mrs Isla Gordon, School of Electronic Engineering, University of South Australia, are thanked for typing all correspondence and retyping texts where needed. Even though we often placed unfair deadlines on the work, it was always accomplished with good humour and on time.

Dr Denis Mulcahy, School of Chemical Technology, University of South Australia, is thanked for his comments and sound advice.

We thank the production team at John Wiley & Sons, UK for transforming a large but varying quality of text and diagrams into a well-produced Handbook. Any errors that remain are our responsibility.

Finally, most of all we thank our wives, Patricia and Danusia, for their continued encouragement and support, and that is something that cannot be measured.

PETER H. SYDENHAM
RICHARD THORN

PART 1

Pressures for Change

Chapter

33 L. FINKELSTEIN

General Principles of Formation in Measurement Science and Technology

Editorial introduction

The theme of this volume is the changes taking place in measurement science and technology. Many of these changes are significant and affect almost every aspect of the discipline's theory and practice. It is all too easy to regard change as being the result of the changes we see in technological artefacts and practices. The engine of change is, of course, really the people who bring about the changes. It is fitting, therefore, that the first chapter of this volume addresses the formation of the engineers and scientists for this activity.

33.1 INTRODUCTION

It is appropriate that a handbook which presents and analyses the general and fundamental principles of measurement science and technology should consider the principles underlying education and training in the field. This is firstly because measurement science and technology is made effective through people who are able to create and apply measurement instruments and techniques, and education and training is the process by which people acquire that competence. Secondly, one of the most essential motivations for the organization of measurement principles into a systematic science is the need to teach it effectively.

Handbook of Measurement Science, Volume 3
Edited by P. H. Sydenham and R. Thorn
© 1992 John Wiley & Sons Ltd

Thus, the present chapter considers education and training in measurement science and technology. As part of a work which deals essentially with fundamentals, the chapter will not consider details of the present state of the art, but will rather be a survey and analysis of the general and fundamental principles.

Measurement science is so fundamental that its basic principles form a core topic, or rather should form a core topic, of all education. The present consideration will, however, focus on the education, training and professional development of engineers and scientists, while giving some limited thought to wider problems.

Again, as part of a work which treats measurement science and technology systematically the present chapter will attempt to consider education and training in the subject as part of the wider system of formation, examining briefly the underlying concepts.

While examining education and training in measurement science and technology in the widest context, the chapter will concentrate on the inculcation of the basic, theoretical principles of the discipline and address in particular the teaching of these principles at the most advanced and specialist level. This is not because this is the most important practical task; indeed it is not. It is rather because it is at this level that the systematic organization of the principles of measurement science and technology is most important and because it is from this level that methods and approaches trickle down to all levels of formation.

33.2   MEASUREMENT SCIENCE AND TECHNOLOGY

It is proposed at this stage to define the terms measurement science and technology in the sense in which they are going to be used in this chapter. Of course this definition is to a certain extent superfluous since the whole of this handbook is essentially devoted to this task. Nevertheless the definition is useful to maintain the internal consistency and completeness of this presentation.

Science will be defined, for the purpose of this chapter, as 'a body of systematically organized objective knowledge, possessing a framework of concepts, principles, theories and models of a domain'.

Technology will, for the same purpose, be defined as 'the science and art of the application of the materials and forces of nature for the use of man'. The term science has been defined above and its use here indicates that the application of the materials and forces of nature has, in various domains, a body of systematically organized objective knowledge. The term 'art' is used to describe the fact that in addition to such a systematic body of knowledge, the various domains of technology possess a body of individual experiential, empirical or heuristic items.

The term discipline is commonly used in connection with discussions of formation and so it is useful to define it. Discipline is basically an area of knowledge organized in such a way that it can be taught to disciples. It will be so used in this presentation.

Measurement can, in the most general terms, be described as 'the empirical acquisition and symbolic representation of objective knowledge about objects and events of the real world'.

Measurement instrumentation will be defined here as 'a system of machines and processes which acquire, transform, communicate and output measurement knowledge'.

Measurement science will be the term used to 'describe the systematically organized principles of measurement and instrumentation'.

Measurement technology will be the term used for 'the design and application of measurement and instrumentation'.

The terms measurement science and technology will be used to describe both the science and the art of measurement and instrumentation. The term measurement and instrumentation is sometimes used to describe this subject.

33.3 FORMATION OF ENGINEERS AND SCIENTISTS

It seems appropriate at the outset to consider the basic nature of the system of formation of which the development of competence in measurement science and technology forms part.

The term 'formation' is used in this chapter to denote the aggregate of all the processes of developing in a person knowledge, skills and attitudes. For professional engineers and scientists, with which this account is mainly concerned, the process of formation may be seen as consisting of three major components: education, training and continuing professional development. Before discussing these components it is necessary to recall briefly what is meant by the terms knowledge, skills and attitudes in the literature of education.

Knowledge is a difficult and complex concept, but will be defined, in this context, as the 'totality of facts, beliefs and conventions to which the human mind has access'. It may be useful to distinguish between *descriptive knowledge,* concerned with what, how and why things are, and *procedural knowledge* concerned with how to do things. An important aspect of knowledge is the capability to acquire and organize new knowledge. Further it is helpful to distinguish here between *deep knowledge* consisting of theories and models of a domain, and *surface knowledge*, represented by individual experiential, empirical or heuristic items.

A *skill* is anything that the individual has learned to do with ease and accuracy and may be either a physical or mental performance or a combination of both.

Finally, an *attitude* is a predisposition to react specifically towards an object, situation or value usually accompanied by feeling or emotions.

It is the combination of all three, knowledge, skills and attitudes, that forms professional competence, that is the ability to function effectively as a professional in a particular field, with which this chapter is mainly concerned.

It is now time to return to the three components of formation: education, training and experience.

The term 'education' is frequently used to denote the whole of the process which is here termed *formation*. However, it is more convenient for the present purpose to confine the term education to denoting the process of schooling, a more or less formal and structured process, which is mainly concerned with the imparting and acquisition of general, fundamental and transferable knowledge, and skills as well as general attitudes. Concepts of generality, breadth and transferability will be taken later.

It is useful to consider the education of professionals as being divided into a general and a professional phase. The general phase consists of what is usually termed primary, secondary and the like school education. Such general education is basic and not directed towards a profession or vocation.

Professional *education* is substantially concerned with preparation for professional competence and takes place, in general, in an institution of higher education. It is only one component of professional formation, concerned, as explained above, with general, fundamental and transferable knowledge, broad, transferable skills and general attitudes. In the formation of engineers and scientists the knowledge is basically a knowledge of fundamental principles but education must develop some skill of application and a motivation to do so.

Training is instruction and learning with clearly determined goals. In professional technological formation it is mainly oriented towards the development of skills which are practical, specific and detailed, and generally concerned with a narrow range of technology and with an ability to perform practical tasks here and now. Training is commonly undertaken outside educational institutions and closely linked with employment. In some systems of formation, however, education and training are closely integrated.

The knowledge, skills and attitudes of the professional are acquired to a significant extent by an intensive phase of initial formation which constitutes preparation for entry to the profession. However they continue to develop throughout active professional life. This professional development has a number of objectives: maintenance of theoretical competence, updating, enhancement of capability and adaptation of the professional to new tasks. This development may take place by formal education and training, informal methods of which individual study forms the essential component, and most importantly, by experience gained in the performance of professional tasks of appropriate depth and breadth.

Finally a systematic consideration of the formation of professionals must consider two aspects. Firstly, professionals are not merely that. Their formation must be concerned with the development of them as individuals with personal psychological and spiritual needs and with their wider role in society. Secondly, their formation must be for their complete life taking into account, among others, their possible, indeed probable, future employment in other functions such as management.

33.4 MEASUREMENT SCIENCE IN GENERAL EDUCATION

Competence in measurement is essential for humans to function effectively in modern civilization. It must be developed by general education. Education in measurement in the general phase does not, and should not, take the form of explicit presentation of principles of measurement. Rather competence in measurement is developed by the inculcation of these principles in various parts of the curriculum, particularly science and mathematics, but also through social studies. It would seem desirable that curriculum developers should have a clear and explicit grasp of the principles of measurement.

It is suggested that the general phase of education should at least develop a familiarity with some simple forms of measurement observation, and some skill in their performance, an understanding of the basic concepts of errors and accuracy and some competence in the interpretation of observations.

The general phase must, of course, also lay down the foundations for those who will proceed to further formation in which development of competence in measurement is involved.

33.5 MEASUREMENT SCIENCE AS A GENERAL INTELLECTUAL DISCIPLINE BEYOND SCHOOL

This presentation must make a claim for the general intellectual significance of the principles of measurement and the place which this significance should earn for these principles in general education beyond the school stage.

It is suggested that it is desirable that a sound, advanced, intellectual education should provide an understanding of at least the following: the basic principles which underlie the quantification of qualities; the limitations of measurement; accuracy, errors and estimation in their appropriate statistical context; and the interpretation of relatively complex measurement data. The claim is that this understanding should go beyond merely day-to-day practical competence and form part of a general intellectual equipment and the basis of an approach to problems.

It is not suggested that there should be explicit teaching of general principles of measurement to all or even to much wider circles in higher education. Rather it is suggested that such principles should be incorporated and integrated in curricula by teachers who should take steps to understand them clearly, and in an organized and systematic way, themselves.

33.6 MEASUREMENT SCIENCE IN THE FORMATION OF SOCIAL AND BEHAVIOURAL SCIENTISTS

Having mentioned the place of measurement science in general education and before entering on the discussion of education of natural scientists and engineers it is important to consider the formation of a group to whom the principles of measurement science are of great practical significance and whose perspectives are special, namely behavioural and social scientists.

Behavioural and social scientists encounter particularly difficult measurement problems and these lie at the very core of their sciences. They require a very clear and systematic understanding of the foundational concepts of measurement and scaling, of the problems of statistical treatment of observational data, of model building, validation and identification, and of the interpretation of measurement data.

Education and training in these topics takes place to some extent explicitly and systematically through courses on methodology, to some extent implicitly through other subjects and finally through practical training.

It is important to realize that behavioural and social scientists, on the one hand, and natural scientists and engineers, on the other, share many common problems and solutions in the field of measurement and that more could be done to pool and share experience.

33.7 THE FORMATION OF ENGINEERS AND SCIENTISTS AND THE PLACE OF MEASUREMENT SCIENCE AND TECHNOLOGY IN IT

Passing now to the main area of concern of this chapter, the formation of scientists (meaning now natural scientists) and engineers in measurement science and technology, it is necessary to begin by recognizing the different levels of scientific and technical cadres and different degrees of special expertise in measurement that they may need to possess.

It is possible, at the first level, to distinguish engineers and scientists whose formation has equipped them with a mastery of the science of their discipline, in the sense of an organized body of principles and methods. Their principal distinguishing characteristic is their ability to transfer their knowledge and skills to new problems, not only to apply their branch of science or technology but to extend it.

At the second level there are engineers and scientists whose formation has has been oriented towards established technology and its application, giving them a mastery of such technology and ability to apply it with competence to standard problems in situations of some difficulty and complexity.

Finally, for this purpose, we may recognize technicians whose formation has equipped them with a competence to apply established technology to routine problems and with a high degree of the technical skills of implementation.

It is now possible to recognize three degrees of expertise in measurement science and technology that the scientific and technical cadres at any level may need to possess.

At the generalist degree of expertise all scientific and technical cadres need to possess a familiarity with general principles of measurement and the basic principles underlying measurement instruments and techniques in their discipline as well as having some skill in their practical application.

At the specialist level there is a restricted, but nevertheless significant, group of engineers, who might be called measurement and instrumentation engineers. They are concerned as their principal professional task with the design, operation and maintenance of measurement and instrumentation equipment and systems and with research, development and design in the field. Their formation will be given special consideration.

Finally, there is a degree of expertise intermediate between the two which might be called ancillary expertise. It is the expertise required by those who, say, incorporate measurement and instrumentation in systems, such as control or aerospace engineers, or those whose technology has substantial application to measurement and instrumentation.

The generalist measurement and instrumentation expertise required by all engineers and scientists may be imparted in their education either by a special introductory course, or laboratory teaching of fundamental scientific subjects. For the latter method, it is essential that the teaching of general principles of measurement is a central objective and a distinctive constituent of such laboratory teaching.

The development of ancillary expertise in the professional education process requires distinct formal courses in measurement science and technology embracing the general principles of the discipline, any specially relevant basic science not otherwise provided and the principal aspects of the application of measurement science and technology in the main discipline of specialization.

As to the education of specialists in measurement science and technology, there are divergent views about whether there is room for such a specialty in the education process at all, or whether such specialists should be formed by training and continuing professional development following a professional education in some other fundamental discipline.

However, no single established discipline provides all the coverage of basic knowledge which a measurement specialist requires. Electrical and electronic engineering does not provide an adequate coverage of physics or mechanical technology. Information technology is to an increasing extent divorced from physical realisation. Applied physics is the traditional and perhaps the most obvious path. However, it is not really flourishing as a discipline and where it is, its treatment of information technology is superficial, it does not often provide a good education for design and it is divorced from control with which measurement and instrumentation is most closely technically connected.

A specialist curriculum in measurement science and technology forms an excellent education. Curricula in the discipline should be based on a broad coverage of basic physical science, as well as mechanical, electrical, electronic, optical and information technology. The core of the curriculum should be formed by measurement science and technology, in particular at the latter stages. The subject should embrace both general principles and a broad overage of application. A major project commonly forms a key part of professional engineering curricula. In specialist measurement science and technology education the major project should of course lie in the specialist field. Finally, all engineering and applied science education should have design as a central core. As will be discussed later measurement science and technology as a discipline is particularly design-orientated and this linkage must be closely maintained in specialist courses.

Notwithstanding the fact that an educational curriculum of the kind outlined above has cohesiveness, rigour, breadth and transferability as well as excellent employment prospects for graduates, it is widely perceived as too narrow and specialized for entrants to professional engineering education. Specialists in measurement science and technology are most commonly formed by a specialist option in an applied physics or electrical and electronic course. Specialist education in the option must remedy deficiencies in the core as best as it can.

With reference to training in measurement science and technology, it should not be merely unstructured experience. It must, for those who receive it, be organized on the basis of general principles of the discipline, which should provide the framework for the art and practice.

With regard to continuing professional development some special aspects of measurement science and technology should be mentioned. Firstly, given the relative scarcity of good initial professional education and training in measurement science and technology there is much room for courses in the discipline as part of continuing professional development. Secondly, any instruction on advances in the technology of measurement and instrumentation provided by such means as taught courses, tutorial seminars and the like should be firmly linked to the organized scientific concepts and principles of the subject.

33.8 SYSTEMATIC PRINCIPLES IN THE EDUCATION OF ENGINEERS AND SCIENTISTS IN MEASUREMENT SCIENCE AND TECHNOLOGY

Discussion now passes to the nature of education in measurement science and technology in whatever curriculum it is embedded.

This exposition argues that measurement science and technology should be taught to engineers and scientists as an organized body of principles. In other words it should, in terms of the definition of knowledge given above, be taught as deep knowledge, rather than as a surface one.

It is taken for granted in this handbook, and in particular in this chapter, that there is such an organized body of principles underlying measurement science and technology. The nature and scope of these principles and their relation to the principles of other fundamental and general disciplines is briefly discussed below.

33.9 THE NATURE AND SCOPE OF THE SYSTEMATIC PRINCIPLES OF MEASUREMENT SCIENCE AND TECHNOLOGY

The fundamental systematic principles of measurement science and technology are basically the science which forms the subject of this handbook. It is thus superfluous and inappropriate to discuss these principles at length in this chapter. However, a brief summary of the principles will be presented in order both to provide the special view of the author about what they constitute and also to discuss how they relate to other disciplines.

Measurement is the assignment of numbers or other symbols by an objective, empirical process to attributes of objects or events of the real world in such a way as to describe them. Information may be viewed by what is carried by a symbol about a referent by virtue of a defined relation the symbol bears to the referent. Measurement may thus be viewed as an information process.

Instruments are information machines which sense a power or material flow from an object under measurement at the input, assign to it a symbol and carry out operations on the symbol, providing at the output either a display symbol to a human operator, a symbol which is processed further by other processes or information machines or finally effectuate the information by operating actuators or similar machines.

It is convenient to discuss the fundamental principles of instruments and of the measurement process in terms of the architecture of a measuring instrument system. The system consists of a number of subsystems. There is firstly the system under measurement. This is connected to a sensor system which interacts with it by a flow of matter or energy. The sensor converts this flow into a signal, maintaining a functional relation between the input flow and the information carrying characteristics of the signal. There is usually a signal-conditioning block which converts this signal into a symbol which may be conveniently handled by the following block, which performs any required functions of information transformation and communication. This system passes the information to the effector block, to further processing or to the human operator. The measuring instrument system operates under the control of a control block. An important part of the system is the human-machine interface. Through this interface the operator effects supervisory control of the measurement process. The interface also embodies any displays.

In the great majority of modern systems, and to a rapidly increasing extent, once information has been acquired by a sensor and conditioned, it is processed and effectuated by standard computing equipment. The control of the measurement process and the display of information to the operator also takes place through standard human computer interfaces. Thus much of instrumentation is implemented by modern computer equipment.

Thus, measurement being an information process, and instruments being information machines, realized substantially by standard computer technology, it is argued that systematic principles of measurement science and technology form part of the wider science and technology of information and knowledge. There are, however, specific aspects and problems of measurement science and technology.

It is now possible to review in summary, in terms of this architecture and in the light of advances of information technology, the principles underlying measurement science and technology as they are generally agreed and as they are set out in, for example, this handbook.

Underlying measurement science and technology are the basic principles of representation of the attributes of objects or events of the real world by symbols which are based on the representational theory of measurement and its extensions. In the light of recent advances in knowledge engineering this can now be seen as an aspect of the general principles of knowledge representation.

The principles underlying the analysis and design of signals and of signal processes are based on signal theory and so-called information theory.

The basis of any systematic organization of principles of measurement science and technology is the fact that information machines are most effectively viewed using the methods of systems theory and technology. The essence of these methods is, firstly, analysis and synthesis of such machines as systems of simpler blocks. Complex and diverse systems are built up from a limited set of simpler components and connection architectures. Secondly, systems are viewed holistically, and their analysis and synthesis must consider any super system in which the information system is embedded and also the environments in which it exists in its life cycle. The realization of instrument systems substantially by computer hardware and software has reinforced the use of the systems approach. In particular, measuring instrumentation is considered as systems built up of a limited variety of building blocks organized in a limited number of possible architectures.

A systematic organization of knowledge such as is involved in the systematic principles of measurement science and technology requires a good scheme of knowledge representation. Information machines are most usefully described by abstract models. Such models make clear the isomorphisms that exist among systems of different physical nature. There are a number of levels or perspectives of abstraction which may be employed: the models may be, at the lowest level of abstraction, power or matter flow models; they may be, at higher abstraction, signal flow models which consider the information-carrying variable only; they may be symbol flow models; and at higher levels of abstraction information and knowledge flow models and the like.

Based on systemic concepts and using models described above, the general principles of instrumentation involve a functional classification of the principal building blocks of instrumentation and a consideration of architectures for use in measurement and instrumentation, including in particular feedback and push–pull.

Recent advances in the science underlying information technology have led to the development of the understanding of the process of abstraction and of methods of abstract modelling in particular in relation to data and knowledge representation. Only active quantities, that is those which characterize the flow of power and energy, can be input to a sensor. Those quantities which are passive, that is quantities which characterize storage, transformation or transmission of energy or matter, can only be observed by interrogating the system under observation by exciting it and estimating the quantities under observation as parameters of a model of the system. The principles of system identification are one of the bases of the general principles of measurement and instrumentation.

The theory of errors and uncertainty in measurement forms a central part of measurement science and technology as a discipline and its most distinctive component. The problems of error and uncertainty are in many ways general to all information machines and processes, being the problem of distortion of signals by transformations, and the estimation of signals in the presence of noise. They are specially important to measurement processes and instruments presenting specific aspects, and demand special treatment in their context. There is the well-known body of random-error theory based on the assumption of Gaussian error distribution and the statistical methods derived from it. Further, however, there are the principles of treatment of systematic errors based on the analysis of instrument models and the effects of sensitivity to parameter variations and to external influences. Following from error theory there is a systematic body of principles of error avoidance and compensation, based on the theory of invariance and employing systematically such methods as disturbance suppression, disturbance feedforward, feedback and finally filtering. The latter presents, as already mentioned, a link to signal theory.

The above discussion has been concerned with the systems and information principles of the measurement process and of measuring instrumentation. These principles are, in general, abstract and are not concerned with physical realization.

However, the measurement process is concerned substantially with two components where such abstract treatment is inadequate. They are the system under observation and the sensor-measured system interaction. Their treatment represents special problems of measurement science and technology as a discipline. As already stated above the system under observation and the sensor-measured system interaction are considered in the discipline of measurement science and technology in terms of abstract mathematical models at the power or material flow levels. Such models are, however, founded not just on the principles of mathematical modelling but also on the physical theory of the relevant systems and phenomena. The discipline of measurement science and technology is thus linked with the discipline of physics. The organization of the latter and its interface with measurement science and technology is thus particularly important.

The discipline of measurement science and technology is concerned not only with the general principles outlined above but also with the technology of their implementation. Here there have been very major changes as a result of the advances of information technology. Since all the information processing and the human–machine interface are realized by standard computing techniques the discipline of measurement science and technology is no longer concerned with the detailed technology of realization of information processing or of the human–machine interface which once constituted its very core. However, the technology of sensors and of signal conditioning remains an essential part of the discipline.

The principles outlined above are orientated towards description and analysis. However the essential task of measurement science and technology is the design and realisation of appropriate measuring systems and processes for particular needs. The needs and the available solution as already stated are too diverse and over a spectrum which is too extensive for the teaching of all the solutions to all the problems. Indeed a catalogue of all known solutions to all known measurement problems could not be effectively constructed or employed. In any case many measurement tasks and most of the really important ones require fundamental innovation. Measurement and instrumentation technology as a discipline must be based on the teaching of general principles of design applied to its particular area. Such general principles of design have been developed and embrace an understanding of the nature and structure of the process of design, and also systematic methodologies for the generation of design concepts.

The above account, and more generally this handbook, should adequately demonstrate the existence of general systematic principles of measurement science and technology. The question that is legitimately asked, however, is: are these principles unique to the subject or are they a collection, in another framework, of principles of other subjects, such as information technology or applied physics? In a discussion on formation it should suffice to say that no other discipline provides an adequate formation for professional competence in the area.

33.10 VALUE OF MEASUREMENT SCIENCE AND TECHNOLOGY AS A DISCIPLINE IN THE FORMATION OF ENGINEERS AND SCIENTISTS

The value of measurement science and technology as a discipline in the formation of engineers and scientists has two aspects.

Firstly, it is of great practical significance. A high proportion of engineers and scientists requires expertise in the field at the ancillary level and there is a significant demand for specialist experts.

Secondly, the principles of the discipline have a significant intrinsic value because of their breadth and transferability. The significance of measurement science as an intellectual discipline has already been argued. One may add that it is an excellent vehicle for the teaching of high technology design, as well as being an excellent example of interdisciplinarity and of the systems approach to engineering.

33.11 TEACHING METHODS IN MEASUREMENT SCIENCE AND TECHNOLOGY

It is necessary to say something about teaching methods in measurement science and technology.

Formal teaching of fundamental concepts and principles by way of lectures is the spine of professional education. It is a presentation of what there is to be known in the form of a framework, which the student must build on by personal study and experience. Informal methods make the acquisition of a systematic overview of the subject a difficult and haphazard affair.

The usual way of testing the students knowledge and understanding of the subject in technical disciplines is by means of problems. Because they are a means of examination, such problems often determine the real, as distinct from the ostensible, syllabus. This is unfortunate because the practical constraints of problem setting, marking and so on means that the problems are generally artificially simple and analytically orientated. This is particularly undesirable in a design-oriented discipline like measurement science and technology. In this discipline analytical, closed problems must be augmented by open ended, discussion-oriented assignments.

Laboratory teaching by way of prescribed experiments is the classical core component of the teaching of measurement science and technology. It performs the essential task of familiarization with equipment and with the practical performance of measurement processes. It should be linked in a structured way with the teaching of general principles which they should explicitly illustrate. However, such experiments are limited in their capability. The practice of measurement science and technology does not involve, in general, the performance of well-defined tasks with prescribed equipment.

The practical teaching of measurement science and technology must be undertaken by design-oriented methods. The standard way of teaching design is by way of the design–make–test project. This has very great value, but because it is expensive in terms of time and resources, it can be used to a very limited extent only.

Design–evaluate projects have very much applicability. Finally, design–analysis case studies in which an equipment is critically analysed, with tests as appropriate, can form a useful part of the teaching arsenal.

This chapter is a systematic presentation of the general principles of formation in measurement science and technology as the discipline stands at present rather than an analysis of different views or a history of the development of the subject.

As already stated this handbook itself is perhaps the best view of the nature and scope of the discipline of measurement science and technology. For some views of the development of the discipline see Finkelstein (1983, 1989.) and the literature cited therein.

33.12   SOURCES OF FURTHER ADVICE

The reader is referred to the following textbooks, journals and organizations for further information on education and training in measurement science.

Books

Feikema, H. and Finkelstein, L. (1979) *A World Directory of Institutes Providing Higher Education in Measurement and Instrumentation* IMEKO, Budapest.

Finkelstein, L. and Williams, J. C. (eds) (1976) *The Nature and Scope of Measurement Science* IMEKO, Budapest.

IMEKO Secretariat (1984) *Advances in Measurement and Instrumentation Education* IMEKO, Budapest.

Linkens, D. A. and Atherton, D. P. (eds) (1989) *Trends in Control and Measurement Education* Pergamon, New York.

Journals

Engineering Science and Education Journal IEE, UK.

IEEE Transactions on Education IEEE, USA.

ISA Transactions ISA, USA.

Measurement and Control Institute of Measurement and Control, UK.

Measurement Science and Technology (formerly *Journal of Physics E: Scientific Instruments*) Institute of Physics, UK.

Measurement IMEKO, UK.

Professional societies

International Measurement Confederation (IMEKO)
1371 Budapest
POB 457
Hungary

Instrument Society of America (ISA)
PO Box 12277
Research Triangle Park
NC 27709
USA

The Institute of Measurement and Control
87 Gower Street
London
WCIE 6AA

REFERENCES

Finkelstein L. (1983), 'Education and training of engineers and scientists in measurement and instrumentation', *Measurement*, **1**, 7–13

Finkelstein L. (1989), 'Formation of engineers and scientists in measurement and instrumentation' *Acta IMEKO XI, Instrumentation for the 21st Century*, ISA, 25–32.

Chapter

34 J. V. NICHOLAS

Relationship of Legal Issues to Measurement

Editorial introduction

Legal requirements of measurement in trade can be traced back to ancient times. As the world seeks to ensure equity and fairness in daily living, in trade and in the delivery of services it has become inevitable that many more aspects of measurement have taken on heightened legal dimensions with which scientists and engineers need to be better acquainted. This account provides an insight into the issues involved as they stand today.

34.1 INTRODUCTION

Almost since the beginning of recorded history trade measurements have had a legal basis. Trusted weights and scales are obviously needed for the exchange of goods. Today a wide variety of measurements require legal sanction for trade to proceed smoothly and with quality becoming marketable, all measurements are opened up to possible legal scrutiny. Trade is, of course, not the only area of legal interest in measurements. Concern over the environment, health and safety in an increasingly technological world has given rise to a host of regulations which require measurements in their enforcement or compliance.

A consequence of increasing legal activity is that virtually all measurements made have a related legal aspect. As legal matters can involve people personally, scientists, engineers and technicians should be well aware of the potential legal

Handbook of Measurement Science, Volume 3
Edited by P. H. Sydenham and R. Thorn
© 1992 John Wiley & Sons Ltd

implications of their measurements. For example, consider a process engineer who has installed thermocouple thermometers to monitor a manufacturing process. Originally, to cope with the thermocouple drift, he merely adjusted the thermometers when the product was starting to be unsatisfactory. Once his firm adopted a quality assurance programme in order to obtain contracts to sell their product, the engineer is required to meet the legal obligations of the contract. In practice this means keeping records of the temperatures and ensuring the thermocouples are in calibration in such a way that an audit could be made by a third party. Not only in the commercial world do legal obligations exist. Scientists involved in nuclear or particle physics, for example, need to carry out measurements to show no harm has come to any personnel or to the local environment from any radiation produced by their experiments.

As the measurement scientist or engineer becomes more involved with legal considerations he or she will need a reorientation in their outlook to their science. Scientists like to think of themselves as objective in their measurement of the physical world. From the legal point of view the opposite is assumed; people are subjective in their approach to the world and they make mistakes, have a lack of understanding, and are likely to be devious if not dishonest, to name a few of the irrational aspects of human behaviour. However, the law still requires the facts to be discovered and has developed its own techniques to minimize or remove the subjective nature of any investigation involving people. The removal of personal bias from measurements is a strong feature of the trends outlined in this chapter. Quality assurance systems are designed largely to ensure that consistency of some output can be achieved, even if people vary. Rules of evidence for expert witnesses are under review to improve the confidence courts can place in the wide variety of science and engineering topics being introduced to support legal cases. Some scientists may feel affronted that their personal integrity is being questioned. Unfortunately an unquestioned reliance on the integrity of technical experts has led to miscarriages of justice in law courts and shabby goods in the market place. A challenge for the measurement scientist is to take as much care in removing the subjective influences on the measurement as he or she puts into removing objective influences.

This chapter outlines legal concerns which impinge on measurements. These concerns are taken from a measurement science viewpoint rather than a legal viewpoint, and apply generally to English-speaking countries. Many of the issues are international with relevance to all countries. Allowance needs to be made for local nomenclature.

We firstly consider the meaning of legal units and how they are disseminated. The next section takes up measurements affected directly by the law and how they can form barriers to trade. While these two topics may have a long history there is still a pressing need for more accurate measurements to be readily available. A third section deals with quality systems which are undergoing a growth phase. Quality systems represent both a solution and a challenge to measurement scientists as they tackle the problem of putting their measurements on a legal basis. An overview of the growth of new laws and regulations arising from people's concern about their welfare shows that there is scope for better and new measurement techniques. A brief look is taken at patent laws, since the measurement scientist may need to patent new devices and processes which could arise from meeting some of the challenges of the legal issues. We also consider what happens to the measurement expert who ends up in court. The role of the forensic expert is under close scrutiny at present. This raises the question of the responsibility and liability of the science professional and in doing so the essential features of this chapter are highlighted. Finally we look at where a measurement scientist or engineer can obtain the information he or she needs to meet legal requirements.

34.2   UNITS OF MEASUREMENT

34.2.1   Metric Treaty

To provide for uniform measures throughout a country, a legal definition of the units of measurement is required. These definitions need to align with those of other nations. One difficulty is that measurement units are more properly a technical than a legal matter. Technological advances can result in changes to the definition of units, which need to be adopted simultaneously by all users. Changes in the law are slow to occur and difficult to keep in line with other countries, but fortunately there is an effective solution. A nation can enter into a diplomatic treaty with France to adhere to the Convention du Mètre 1875 (BIPM, 1987) and approximately 47 nations have already done so. The organizational infrastructure associated with the Treaty is given in Figure 34.1 (see HBMS[†], Chapter 3, Volume 1 for further information). The effect of the Treaty is to pass the responsibility of an adequate technical meaning of the units to the CGPM (Conférence Générale des Poids et Mesures), to which the member nations can contribute. The Treaty infrastructure is designed to keep the units up to date with technological and scientific advances.

National legislation is needed to refer the base units to the Treaty, i.e. the units for length, time, mass, electrical current, temperature and luminous intensity. The meaning of the multitude of other derived units can be made by reference to the International System of Units (SI), which is also decided on by the CGPM. Reference can be made directly to the recommendations of the International Standards Organization (ISO) (ISO, 1982), or through the adoption of these recommendations by the national organization for specification standards. If

---

[†] HBMS = Handbook of Measurement Science

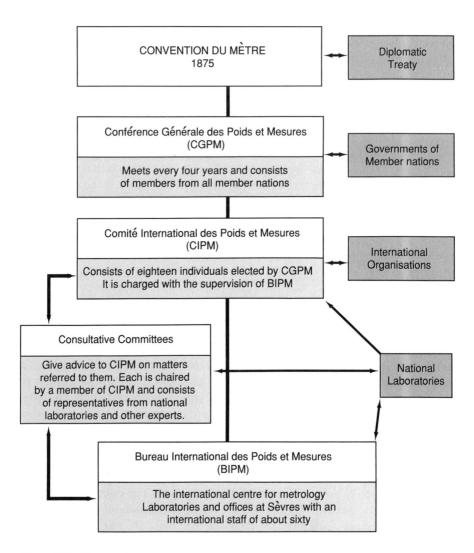

Figure 34.1   The metric treaty

units other than metric ones are in common use, then the ISO conversion factors can be used or they could be specified directly in legislation. Correctly worded legislation would require no change when technological advances resulted in new definitions, because they would have an immediate legal basis through the review made by the CGPM or ISO.

The changes occurring in the definition of units (see Section 34.2.3 on Consultative Committees) are not merely about increasing accuracy, they also involve the legal aspects of measurement traceability. Historically attempts were made to define measurement units around common physical properties, e.g. the foot was meant to be the length of a person's foot. However, these proved unsatisfactory even when attempts were made to link them to less variable physical properties than those based on the human body. Definitions in terms of arbitrary artefacts proved to be most reliable, e.g. when the artefacts for the British pound and yard were destroyed by fire, reconstruction of the artefacts was effected more readily from the copies of the artefacts than from physical properties (Glazebrook, 1931). Under the original treaty the artefacts were known as the prototype metre, prototype kilogram, etc. Only one such prototype can exist for each unit since exact copies are not physically possible. Obviously extreme care is needed in preserving the artefact and access to it needs to be strictly limited. Copies of the prototype can be made and calibrated against it. These copies are suitable for national laboratories who carefully guard them and transfer the unit to other working reference standards used for routine calibrations.

There are at least three main problems with using artefacts;

- In the long term the single prototype is not safe even if it remained stable.

- The best accuracy is not available to the instrument user because high-precision instruments will be at least 3 steps down the calibration chain.

- There is no way of checking if human error has biased the standard. Only a very few people will have the skills necessary and the opportunity to transfer the unit from the prototype. Sources of systematic errors are notoriously difficult to eliminate and there is no way of checking for possible errors that may have been overlooked or not recognized.

Advances have allowed all but the kilogram to be based on fundamental physical properties rather than artefacts. The fundamental physical properties have been shown to be more repeatable than the original definitions. For example, the distance between the two scratches which was used to define the metre is dependent on the choice of where the scratch is considered to be located, thus limiting the accuracy to around 0.5 μm (Glazebrook, 1931). A tacit assumption in using defined physical properties is that the unit is constant in all the universe.

Thus, any competent person can construct a primary physical standard for any of the units so defined. Immediately the concern over the uniqueness of the defining prototype is eliminated. Precision instruments can be compared directly with the primary standard, thus ensuring that state-of-the-art accuracy is readily available. Advances in modern technology often depend on this direct access to a standard (see Section 34.2.3). Initially a primary physical standard is costly to develop and time consuming in its use. After a period of instrumental development, a useful and reliable primary physical standard may result, within the reach of laboratories who can utilize the accuracy. Note that this desirable state of affairs is presently not available for all units.

The problem of eliminating systematic errors due to oversight is also greatly reduced. With many primary physical standards in existence, intercomparisons will test the competency of the construction and operation of the standard. Genuine differences will help to determine the accuracy of the method, as well as pointing to overlooked systematic errors. The existence of multiple primary standards does raise a legal problem as to which is the legal standard and Section 34.2.2 addresses this problem.

Adoption of physically based units with the resulting increase in accuracy has resulted in a new approach to standards. In well-defined situations units are usually related to each other by a fundamental constant of nature. This is, of course, how fundamental constants can be measured. However, the process can just as easily be reversed. Provided that at least one unit is defined, then a defined fundamental constant can be used to fix the value of other units.

Some units have already been defined in this manner, e.g. the metre is related to the second by defining the speed of light. The role the BIPM now serves is in determining what physical systems are of sufficient stability to serve as primary physical standards.

34.2.2   National Standards Laboratories

Legislation, which provides for the meaning of measurement units, is not sufficient for a nation to ensure the orderly application of uniform measures. Disputes can arise over measurements. The existence of more than one primary physical standard for a unit requires someone to adjudicate over differences. Therefore, legislation is required to establish a means by which measured units can be declared legal. Usually a national standards laboratory (NSL is used here as a generic acronym) is established and given the responsibility of holding the national reference standards (perhaps as primary physical standards). The NSL can be located inside the scientific and technological structure of a nation either as an independent laboratory or attached to other government research organizations. Besides the legal requirements of providing for uniform measures and certifying reference standards, the NSL can be charged with meeting other national priorities. Figure 34.2 outlines the various functions a NSL may perform as it contributes to the economic well being of a developing nation. For developed countries such a figure can be very tangled.

The laws on units are relatively simple and do not have the accompanying legal problems as, say, the laws on contracts, in spite of measurements being so pervasive in our societies that they are taken for granted like the air we breathe or the water we drink. This lack of legal complication needs further comment. For measurements, the NSL provides authoritative standards and leadership which assures that units do not drift or vary. Thus, the reliance is not on legal cases and legal proof, but on scientific merit and scientific proof. Legal proof and scientific proof are not necessarily the same and this is a major concern for

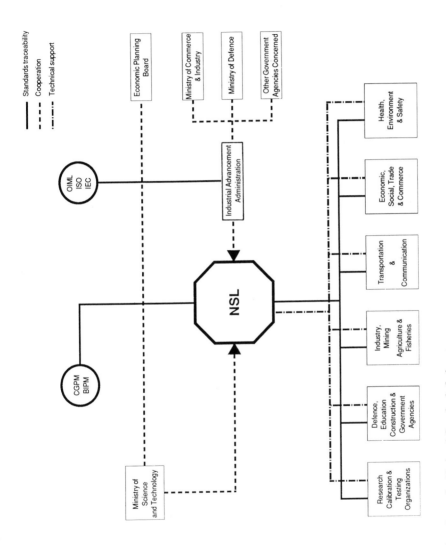

Figure 34.2 The functions of a National Standards Laboratory

rules of evidence (see Section 34.7). Therefore the NSL needs to operate so that it can meet the requirements of both legal proof and scientific proof.

In general the NSL is not involved in enforcement, though there is a function for enforcement (see Section 34.3.1) where trade is concerned. Because of the pervasive nature of units, measurements have some form of traceability back to an NSL somewhere even if there is no documentary proof. It is important to remember that the value of units would drift if there were nothing to underpin them. For instance, when instruments differ in their readings, a common procedure, where no reliable calibration is conveniently available, is to adopt the average of the readings. Over time drift must occur if the variations are random. (Experience indicates that about 20% of all instruments submitted for calibration are out of specification (Moss, 1978).) With deceit the variation would be even greater. Importation of new instruments may not help as the tendency will be to adjust these to the local average. Thus 'local' units can easily arise.

A vital role for the NSL is in the development of the definitions of the base units and their dissemination (Christmas, 1984). Primary physical standards are developed, constructed and operated by the NSL and hence they cannot, in principle, have traceability back to other NSLs. Instead intercomparisons are carried out to establish to what level of accuracy the units can be considered equivalent. Differences in units can and do arise, requiring research to determine if the difference lies in an unknown property of nature or in the procedure used by any of the NSLs. A general observation is that intercomparison accuracies are worse than the accuracy claims of the individual NSL. A trend is for an NSL to mutually recognize the equivalence of units held by another NSL, if there are political reasons to do so and equivalence can be demonstrated scientifically. Of course, some NSLs will hold secondary reference standards which will be traceable back to another NSL or BIPM. Nations who wish to develop high-technology industries are likely to need their NSL to hold primary physical standards for maximum credibility.

34.2.3 Consultative Committees

A measurement scientist who wishes to keep in touch with changes in definitions and also in improvements to the transfer of units needs to be aware of the work of the Consultative Committees of the CIPM. At present there are eight committees (BIPM, 1987):

- The Comité Consultatif pour la Définition de la Seconde (CCDS).

- The Comité Consultatif pour la Définition du Métre (CCDM).

- The Comité Consultatif d'Electricité (CCE).

- The Comité Consultatif pour la Masse et les grandeurs apparentées (CCM).

• The Comité Consultatif de Thermométrie (CCT).

• The Comité Consultatif de Photométrie et Radiométrie (CCPR).

• The Comité Consultatif pour les Étalons de Mesure des Rayonnements Ionisants (CCEMRI).

• The Comité Consultatif des Unités (CCU).

-- Some information on the work of these committees can be obtained from the scientific journal *Metrologia* edited and published by the BIPM but most of the information is obtained through the BIPM membership of a country. Outlined here are trends occurring for the base units for they illustrate points already raised.

(1) *CCDS*. The unit 'second' is the most accurately (1 in 10^{13}) known unit as it is defined in terms of a highly stable frequency of an atomic transition. Physical standards of the defining system are readily available commercially. Frequency and time standards are usually readily disseminated by an NSL. Time of day is now derived through the second to give International Atomic Time (TAI). TAI is based on an international ensemble of clocks kept in touch by satellite communication to an accuracy of around 100 ns, though resolution to 1 ns is possible. Time of day is a global example of measurement drift as it represents the average of many measurements which have no external reference. The drift is somewhat smaller than the drift due to the Earth's rotation which used to define the time of day. One motivation for obtaining even higher accuracies is to lower this drift and provide navigation systems with a positioning accuracy better than 1 m.

(2) *CCDM*. The metre is related directly to the unit 'second' through the definition of the speed of light. A frequency measurement can then give the wavelength of an electromagnetic radiation. In practice the primary physical standard is an iodine-stabilized laser. The laser can be used to directly transfer the length unit to line and end standards, such as tapes and gauge blocks. Lasers have made the dissemination of high-accuracy length measurements relatively easy.

(3) *CCE*. In spite of the defined unit being the ampère in practice it is the ohm and the volt which are realized. Both have recently, in 1990, been defined in terms of fundamental constants, the volt through the Josephson effect and the ohm by the quantum Hall effect. The development of large-scale Josephson arrays has made this primary standard more directly available to users at a time when digital voltmeters are requiring high accuracy. On the other hand the quantum Hall effect is expensive to realize with only a few people skilled in its use because it is a recent development. The next few years should see the quantum Hall effect become more accessible.

(4) *CCM*. The international prototype of the kilogram has only been used three times in 115 years of its existence. Masses are disseminated with a high accuracy and this has made it difficult to find a physical definition of higher accuracy. The mole has been related to the kilogram by defining atomic weights and while attempts have been made to reverse this definition they have fallen short of the accuracy required (1 in 10^9). The possibility of relating mass to electrical units is being explored.

The CCM also considers gravity, pressure and force measurements.

(5) *CCT*. Unlike the other base units, the kelvin is not an additive quantity so the main concern is with the recommended scale that is used. As the thermodynamic scale is largely impractical a defined scale is used. A new definition of this scale was adopted in 1990. At high and low temperatures the scale is based on thermodynamics but over most of the common range it is based on platinum resistance thermometry which is readily disseminated. The physical standard of the defining unit is relatively inexpensive and components for realizing parts of the scale are available.

(6) *CCPR*. Radiation measurements are related directly back to electrical power units. The photometric units are obtained by using a defined conversion constant. Intercomparison accuracies have an uncertainty of around 1% even though 0.1% should be possible. Unlike other base units, radiation units are not readily disseminated and care is needed in finding an authoritative traceability path. Measurement drift can easily occur with these measurements. Meteorologists, for example, had determined the solar constant to 0.1% using internally consistent calibration methods. However when finally compared to the defined radiation units they were out by 2%.

34.3 LEGAL METROLOGY

34.3.1 Weights and Measures

Legal control of weights and measures is one of the oldest forms of consumer protection provided by nations (Becker, 1975). Laws and regulations are required to provide equity in commercial transactions not only for the consumer but also for the merchant to be assured of fair competition. Weights and measures officials are charged with the enforcement of these laws. Note that here we are taking a more restricted meaning of the term 'weights and measures' and in particular it refers to mass, length and volume. A better term would be 'trade measures'.

The location of the office for weights and measures inside the national measurement system varies. In the US the weights and measures legislation is handled at the state or city level with the NSL having the responsibility to ensure uniformity amongst the states. In the UK the local authorities have Trading Standards Officers for the enforcement of the national weights and measures legislation, and the connection to their NSL is through traceable standards. The role of the officer increasingly includes other consumer-related activities. In the US the officer is also concerned with electricity and gas meters whereas in some countries this is the responsibility of other departments of state. As the officers are usually only concerned with enforcement they do not carry out research and development, and often changes to weights and measures regulations must be initiated elsewhere.

Weights and measures activities have gone through stages of development or transition. As the aim is to ensure trade measurements are correct, the most direct approach is to inspect and adjust all measuring equipment periodically to keep it inside legal tolerances. No responsibility is placed on the merchant or equipment manufacturer. At the next level of operation, inspection is used only to determine if the equipment is accurate. Use of declared inaccurate equipment would be illegal, until repaired and retested. Here the merchant requires the equipment supplier to provide repair and adjustment services. Measuring equipment can commonly be out of calibration without the user being aware and hence there may be reluctance to take legal action unless deliberate fraud is involved. However, this can leave the consumer with short measure. A stricter regime is to hold the merchant responsible and accountable for the weighing and measuring equipment used. To comply the merchant has to attend to two important issues. Firstly, staff training will be needed to ensure correct use of equipment. Secondly, the choice of the equipment supplier will be based on the supplier's ability to keep the equipment in calibration at all times. This really calls for the obvious, that the supplier and manufacturer of measuring equipment need to be experts in the relevant measurement science.

To ensure manufacturers do make suitable equipment, weights and measures legislation often requires both a *type approval* of the equipment and its regular verification. A trend is for legislation to give detailed technical specifications directly, or indirectly, by reference to a standard (see Section 34.3.2). Formal type approval can then be given so that the merchant knows which equipment is most likely to fulfil his or her obligations.

In spite of the predominantly legal aspect of weights and measures it is a technological exercise. Legal convenience has to give way to technical capability. Different measures have different technologies and, therefore, there is variety in the means of enforcement. Many of the measurement areas still require improvements in accuracy, higher reliability or more convenience in use. For example, liquid fuels are normally sold by volume even though the volume is temperature dependent. The quantity of interest to the purchaser is the energy content, which is more related to the fuel's mass. Improvements in both the method and accuracy of fuel delivery could be made.

34.3.2   International Legal Metrology

Legal metrology is a term with a wider use than weights and measures. The formal definition is (ISO, 1984) 'that part of metrology which treats units of measurement, methods of measurement, and of measuring instruments, in relation to the mandatory technical and legal requirements which have the object of ensuring a public guarantee from the point of view of the security and of the appropriate accuracy of measurements'.

Because trade is a major application of legal metrology and is international in scope, international agreements on measurement methods are required (Andrus, 1975). To promote intergovernmental cooperation the International Organization of Legal Metrology (OIML) was founded in 1955. Organizationally the OIML consists of

- the International Conference of Legal Metrology,

- the International Committee of Legal Metrology, and

- the International Bureau of Legal Metrology.

The structure is similar to that of the Metric Treaty (Figure 34.1) except that the Bureau is purely a secretariat and does not have laboratories or technical equipment.

The objectives of OIML are to develop model laws and regulations pertaining to legal metrology and to establish uniform requirements for scientific and measurement instruments used in industry and commerce. Member nations assume a moral, but not a binding, obligation to implement OIML International Recommendations. Documents are also published to facilitate the harmonization of national regulations. Note that the word 'harmonize' does not mean 'make identical' but rather 'have the same effect as', e.g. a manufacturer who meets the regulations of one nation should be able to use the same procedures in another nation and meet all their legal requirements.

The emphasis is more on the 'fair competition' rather than the 'consumer protection' aspect of legal metrology, as technical recommendations do form a technical barrier to trade. Trade can be affected directly as in the recommendation for measuring the water content in grain, or indirectly, through the trade in instruments which is affected by the instrument recommendations. Consumers derive benefits through a nation adopting good current measurement practices and through the use of readily available equipment to enforce consumer legislation.

The measurement scientist may be more concerned with the OIML recommendations on instruments. These usually specify accuracy classes based upon how the instruments are to be used. Initial and subsequent verifications are required to assure compliance with metrological and technical specifications. Guidance may also be given on the optimal frequency of inspection. Nations need to operate a national pattern or prototype approval programme for instruments covered by the recommendations. One problem is that many of the technical specifications are design based and not performance based so that technologies become locked into the design requirements and stifle innovation. Performance-based specifications often require a more developed measurement theory and practice than that available. Design specifications can therefore be seen as a legal expediency to provide control now. Future developments need to be allowed for.

With the increasing growth of legal metrology into areas other than consumer affairs the instrument designer will need to follow and become involved in the setting of International Recommendations. The OIML publish the *Bulletin OIML* which outlines its current and previous work as well as publishing technical articles related to legal metrology.

34.3.3 Certification

An end point of many legal metrology processes is certification. Certification may be exhibited in a variety of forms—a mark, a label, a seal, a certificate, a release note, published list of accepted goods, approval or licensing of manufacturers. Certification can apply to wider fields than legal metrology. To avoid disaster, instrument designers need to know what form of certification may be required before they start on a project.

Methods of achieving certification vary widely and as such they represent a strong technical barrier to trade. International bodies such as ISO (ISO, 1980) are trying to harmonize the certification process amongst nations to remove the technical barriers. For legal force a certifying authority will need to be established in law, otherwise it will need its own status for acceptance, e.g. insurance underwriters. We consider here five procedures that could be used for certification as examples of the variety of methods and their likely acceptance. As an illustrative example we consider equipment made by a manufacturer.

(1) Legislation setting up an inspector (or other named person) to give approvals may give no specifications or technical detail. For instance the inspector may only have to declare that something was 'safe'. The quality of this certification will depend on the professional integrity of the inspector. The system is clearly open to abuse and usually unacceptable for commercial transactions. This method arises presumably because of the difficulty in giving adequate specification and hence indicates a measurement challenge.

(2) Adequate technical specifications are given but not a detailed certification method. Acceptance is then likely to depend on a test of only one sample or even a prototype. Where there is a policy of always accepting the lowest tender this certification method will flourish, because of the lower initial cost. There is, of course, no guarantee that subsequent products would meet the technical specification. While this method may have once been the most

common certification method, it is no longer. The method is technically orientated in that there is an assumption that if a manufacturer has the technical competence to make one sample correctly then all will be made correctly. A major difficulty with this approach is that many manufacturers do not have the measurement skills to evaluate their products and therefore have no idea what effects modifications will have. The original approval of the product may be due more to technical advice from the test house employed than the technical skill of the manufacturer.

(3) One way around the above problems is to test 100% of the product to the technical specification. While ideal for weights and measures equipment, in general this method is too expensive and certainly not applicable to products where destructive testing is involved. Between methods (2) and (3) there is obviously a range of choices which may be more appropriate. The methods are still technically oriented in that they concentrate on a technical outcome.

(4) Acceptable certification methods now require some knowledge of the input as well as the outcome, e.g. the management practices. A suitable method involves type testing and assessment of factory quality control, and once accepted, surveillance follows that takes into account the audit of factory quality control and the testing of samples from the factory and the open market. This method provides for good industrial efficiency and obviously calls for a developed measurement technology. For instance, this method should apply to good-quality measuring instruments, where checks on the instrument accuracy will have been made but not a full calibration.

(5) A step from (4) is to have the factory fully audited under a quality assurance scheme. In effect the certification is carried out by the quality auditors. All the testing and conformity declaration may be carried out by the manufacturer (see Section 34.4 and Chapter 44).

The quality assurance route for certification of legal metrology is being taken up by the European Community (Cammell, 1990). It is also needed as a consequence of the product liability laws (see Section 34.7.3). Details are still being worked out. As it involves possible derogation of the legislature there will be a bureaucratic structure to implement it. For example, half of a technical specification might be concerned with applying the certification mark. In one respect the scheme is a barrier to trade in Europe, but once compliance is achieved, the whole market is available. Contrast the situation in the US. Outside of weights and measures, legal metrology may be handled by city ordinances which do not have to harmonize with state or federal laws or even underwriter's specifications. Thus it is possible that a manufacturer will have to meet four different certification requirements to supply goods to a single market.

34.4 QUALITY SYSTEMS

34.4.1 Quality Assurance

As explained in Section 34.3.3, Quality Assurance (QA) is one answer to meeting legal requirements. QA was originally developed as an industrial tool to give a manufacturer a competitive edge over rivals. Here the legal aspects of QA are examined. A more balanced approach to QA in general should be sought elsewhere (see HBMS Volume 2, Chapter 29, and this volume, Chapter 44, and Roberts, 1983). Many QA systems exist, some of which are proprietary. A representative generic QA system is now considered.

Quality control (QC) is a technical method which uses measurements and statistics to control the quality of a product. However, QC will fail if the two main causes of poor quality (Roberts, 1983), poor raw material and lack of calibration, are not also under control, let alone where management is not firmly committed to quality. Achieving quality involves most of the operation of a factory, especially the people. The growth of QA has arisen because it successfully takes these factors into account. People can have a dramatic effect on quality; they can be forgetful, make mistakes, even commit sabotage, to name a few causes of poor quality. QA attempts to minimize these effects while allowing the correct procedures to be carried out in order to achieve the required specification. Assurance is needed that actions were correct and timely. If not, then the time and place of the failures needs to be known as well as who is responsible, so that problems can be fixed to ensure they do not recur. QA, in tracking the people-related factors, parallels the normal requirements for legal proof (see Section 34.7). This is perhaps not too surprising as QA is often used to satisfy contractual obligations.

The attraction of accepting QA evaluation of a product for legal purposes is twofold. Firstly, duplication of the manufacturer's tests is not needed. Secondly, the industrial efficiency of the country will be improved. However, an appropriate QA system needs to be used.

Currently the ISO 9000 series cover the QA systems under consideration to meet European legal requirements. There are three systems (Telarc, 1991).

- ISO 9003 defines very basic quality systems suitable when a product's quality can be ascertained by final inspection (method (3), Section 34.3.3). This gives a supplier who is unable or unwilling to adopt a full QA system the means to interact with other QA systems.

- ISO 9002 builds upon ISO 9003 and introduces additional requirements to ensure defects are caught before they pass too far through the system. It also helps the manufacturer to do it right first time and to learn from mistakes that are made (methods (4) and (5) of Section 34.3.3). This system is the one most commonly adopted.

- ISO 9001 builds upon ISO 9002 by extending the quality management system into the design process. The intention is to ensure that products are designed properly in the first place—both to meet the customer's needs and to be able to be manufactured in an efficient manner.

Note that all the systems are primarily aimed at manufacturers but are worded generally so they can be adapted for service organizations.

Inefficient organizations normally gain considerable benefit by adopting a QA system as the increased efficiency can improve profitability. Widespread acceptance of QA has come from the fact that many organizations are inefficient and stand to gain from QA. On the other hand, a well-managed and efficient organization may find that there is a nett cost in adopting QA to meet the legal requirements of continuing in their traditional markets.

Here legal and technical requirements are distinguished even though they may overlap. An example illustrates the difference. A technical proof that a television set works is for the seller to demonstrate it to a purchaser. A legal proof would be the seller having sufficient evidence to convince a purchaser that a set works when neither has access to the set. Another way of looking at it is that technical proof is between peers who understand the subject and have the means to duplicate the experiment. Legal proof is proving to all by the way of witnesses or properly attested documents.

When applied to measurement, QA should only be formalizing good measurement practice. Additionally QA helps to define the meaning of terms, in particular 'traceability'. Traceability can have a variety of meanings and, as with many words, its meaning has to be taken from the context. In the QA context traceability is the assurance of the quality of a measurement. Legal proof needs to be available to demonstrate that the proper technical and managerial procedures have been followed, from the primary physical standard to the measurement made. Calibration certificates kept in a filing cabinet are not sufficient proof of traceability. Someone needs to be appointed responsible for the care and maintenance of the equipment and evidence kept that the maintenance was carried out to appropriate procedures. One reason for the variability in the meaning of traceability is that few traceability chains have the documentary evidence implied. In principle, though, the NSL is responsible for uniform measures and should be able to say what form of traceability is acceptable for different purposes (Christmas, 1984).

34.4.2   Specification Standards

Before measurements or tests can be made, appropriate procedures and specifications are needed. Because results can vary, depending on the procedure adopted, there must be agreement amongst the interested parties on a procedure to limit the variability. Most countries and also some interest groups have organizations to formulate voluntary consensus standards (see HBMS, Volume 1, Chapter 3). These standards will cover test methods, measurement practice, technical specifications, codes of practice etc. Internationally these standards are represented by the ISO and the IEC (the International Electrotechnical Commission) who work closely with each other and also with other interested organizations.

Different aspects of standardization follow the separate stages of the innovation and product development cycle. In the first stage there is a need to harmonize the language and terminology of a new technology. The second stage involves the standardization of measuring instrumentation and its transformation into process and production technology. Finally the concern is with product standardization to achieve rationalization, compatibility and reliability (Strawbridge, 1991).

Voluntary standards can become mandatory when they are referred to in legislation. This enables complex technical matter to be removed from legislation where it would be difficult to update. Also the use of consensus standards may give the legislation better acceptance, thus enforcement is minimal. Standards can be referred to in legislation by a variety of means, for example:

(1) The legislation requires compliance with a quoted standard.

(2) A general requirement is given and reference made to a standard, e.g. a 'safe device', with a standard given as an example.

(3) Specific requirements quoted with the means of determining them covered by a standard, e.g. boiler pressures quoted with specific reference to a boiler design and test standard.

With method (1) any unsafe device which complies could not be excluded until the standard was updated. However, with (2) the device could be excluded if there were good technical reasons for the device to be declared unsafe. Also new safe devices which do not comply could be allowed if there were good technical reasons to do so, e.g. technical equivalence with the standard demonstrated. Method (3) is useful where multiple uses of the standard are required. Again the choice will be largely determined by the available technology (often measurement based).

There is a strong onus on the standards organization to keep the mandatory standards technically sound. Not only should they be up to date but also there should be no ambiguities arising when complying with the standard. Consensus standards can be very good when the issue is not contentious. Otherwise, unfortunately, interest groups may influence the committee and controversial requirements may be weakened or even omitted through lack of consensus. A well-designed certification scheme should pick this up.

Most standards organizations run certification programmes along the lines of method (4) of Section 34.3.3. The certification function should be as separate as possible from the specification function. This eases any possible conflicts of interest when the certifiers need to notify the specifiers that the standard is defective. In order to do this the certification authority needs to have the technical expertise to evaluate the standard. Recognition of certification authorities (or indeed test laboratories) depends on their technical professionalism in the identification of inadequate standards. The credibility of the standards organization will depend on its ability to respond when inadequate standards are found. It should be emphasized that one of the main difficulties in obtaining international acceptance of measurement practices is this question of technical competence. There is no point to the paperwork if the technical side is inappropriate.

Overcoming technical barriers to trade requires careful attention to standards. The General Agreements on Tariffs and Trade (GATT) have established international standards, rules and procedures to reduce non-tariff barriers (Willingmyre, 1984):

- National standards should not create obstacles to international trade.

- Where possible international standards (e.g. ISO, IEC) should be adopted or used as the basis for new standards.

In general, national standards have followed these trends. However, international standards can also be barriers to trade if they make unreasonable demands, e.g. call on sophisticated test equipment of very limited availability.

Voluntary standards often become compulsory standards. This depends on the acceptability of the standards and the laws relating to contracts and liability. Good standards may become the minimum expected in any related transaction. In legal terms they become voluntary law. For example the Institute of Petroleum publish test standards that are used worldwide for commercial dealings in petroleum products.

34.4.3 Accreditation

Accreditation is a term usually applied to the certification of testing laboratories (see HBMS, Volume 2, Section 31.3). Laboratory accreditation is not new but what is new is the growth in the number of national accreditation bodies as a result of GATT agreements. Tests and measurements are at the heart of ensuring the quality of a product. Unfortunately these are often poorly done even in the physical sciences. With tests, especially those which rely on a recipe approach, the need is for the method to be followed without technical innovation. After a measurement is reported there is difficulty in assessing its quality because variations may have occurred. Laboratory accreditation allows the quality to be assessed since the test laboratory is investigated for all the factors, both objective

Guide 2:   definitions (1986)

Guide 25:   requirements for technical competence of testing laboratories (1990)

Guide 38:   requirements for the acceptance of testing laboratories (1983)

Guide 40:   requirements for the acceptance of certification bodies (1983)

Guide 43:   proficiency testing (1984)

Guide 45:   presentation of test results (1985)

Guide 49:   quality manual for a testing laboratory (1986)

Guide 54:   recommendations for the acceptance of accreditation bodies (1988)

Guide 55:   recommendations for the operation of testing laboratory accreditation systems (1988)

Figure 34.3   International ISO-IEC guides concerned with testing laboratories

and subjective, that influence measurements. However, then the question arises: how can one be sure the quality of the accreditation procedure is adequate?

The international interaction amongst accreditation bodies is through the International Laboratory Accreditation Conference (ILAC). The ILAC philosophy is (Gilmour, 1988):

● to build a worldwide network of national accreditation systems, which meet internationally recognized criteria, and

● to tie these systems together by mutual recognition agreements.

Several international criteria have been published as a guide by ISO (see Figure 34.3) and some form the basis for new European Standards. The guides establish most of the elements of a QA system for laboratories and can be fitted into the ISO 9002 QA system (Bell, 1988).

Mutual recognition of accreditation bodies appears to be the most likely route to satisfy legal requirements. Periodic audits of the accreditation bodies will be needed to see that the international criteria are being followed. Testing laboratories who joined accreditation schemes for the recognition of technical competence to operate in a national market will now find that they are being assessed as an international laboratory with an emphasis on legal credibility.

Mutual recognition is an appealing concept and where it is established it works well. Where the recognition is new, difficulties are likely, until the authorities in the partner countries learn to trust test certificates issued under the agreement (Gallagher, 1988). Users may need to be well aware of their rights and fight for acceptance.

QA auditing also needs some system of recognition. Unlike national accreditation bodies, QA auditing schemes are in direct competition with each other, therefore some form of a government-sponsored accreditation board for auditors and auditing schemes may be needed. ISO Guide 40 provides the means of providing these, by laying down rules for the organization of the auditing body, including legal identity, independence, organization and the qualification of the auditors involved in supplier audit. These rules need to be rigidly imposed and compliance with them regularly checked by means of systematic audit. The qualification and experience requirements for auditors must ensure that an auditor has the necessary skills in both the quality and technical aspects of his or her job. At present there are very few mutual recognition schemes for QA accreditation bodies between nations and while mutual recognition is desirable, the means to achieve this is not yet clear (Smith, 1990).

34.5 REGULATIONS

34.5.1 Delegated Legislation

Statutes passed by governments do not usually give detailed and specialized requirements (Montague, 1987). Instead, enabling Acts are passed that lay down the framework of the law and delegate power to appropriate bodies to make detailed regulations at some later date. Unlike the enabling Act, this delegated legislation can be challenged in the courts on the grounds that it is outside the power of the body concerned, or that it is unreasonable in scope or effect. Usually departments of state formulate the regulations and provide for their enforcement. Local bodies are another major group with delegated power to make by-laws or ordinances.

There has been a vast increase in the number and scope of regulations. Many regulations arise from people's concerns over their welfare. Historically regulations were generally introduced in response to specific problems as they arose. A modern trend is a desire to have regulatory control over technology before anything goes wrong. This puts measurement technology under pressure to provide the means for adequate enforcement of such regulations. Regulations can become technology forcing, in that they require technological solutions that are not in general use or are yet to be realized. There is a very fine line between technology forcing which is reasonable and hence enforceable, and that which is unreasonable and hence goes beyond the delegated powers (see Section 34.5.3). Some regulations may be considered to be technology hindering if only one type of technology is allowed for (see Section 34.3.2, International Legal Metrology).

The enabling Act normally uses a general word, such as 'deleterious', which the regulations interpret into the intended meaning. Interpreting concepts into measurable quantities is often difficult, e.g. how do you measure a clean environment? Reference may be made to a specification standard (see Section 34.4.2) for technical details. Full details of the enforcement rests in the procedures adopted by the officials enforcing the regulations. These could be published as a less rigorous code to be followed, and here such a code is taken as being a part of the regulation. Regulations need to be readily enforceable and also compliance should be feasible. The values chosen in the specification need to be achievable in a cost-effective manner. While small improvements in values may appear reasonable on paper, they may involve adopting a new technology which is too expensive. Any intention for technology forcing needs to be apparent in the discussion leading to the passing of the enabling Act.

Regulations often arise from international agreements such as those sponsored through the United Nations (Figure 34.4) e.g. the monitoring of nuclear power plants for possible weapon production. International aspects of any regulation

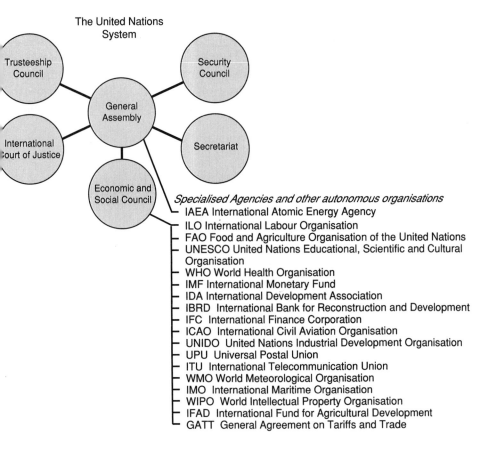

Figure 34.4   Specialized agencies of the United Nations are often involved in recommending regulations

need to be considered over and above that of meeting convention requirements or those of legal metrology (Section 34.3.2). If a nation fails to adopt regulations in line with other nations and chooses to have no or minimal control then the nation will become a target for dumping or experimentation. For example in manufacturing the yield of some products is variable and so a manufacturer will grade them according to what country will accept the various quality levels. Automobile manufacturers will field trial experimental components in countries which have no safety design rules for vehicles.

The following subsections cover several major concerns of regulations where measurement plays a vital role. They are given to illustrate trends or points raised in this chapter, and are not meant to be comprehensive.

34.5.2 Transport and Communication

The focus of transport and communication regulations is usage. Waterways, roads, railways, air lanes, and radio frequencies are all basically limited resources and their users need to consider others in order to obtain efficient and effective usage. Safety and environment are also a consideration. In these areas several international bodies have active roles in developing regulations for international conventions to provide consistency amongst nations.

Marine transport regulations are formulated through the Inter-governmental Maritime Organization (IMO). IMO covers all aspects of shipping from tonnage measurements to oil pollution at sea.

Air transport regulations are covered by the International Civil Aviation Organization (ICAO). Increasing volume of traffic in the air requires increasingly sophisticated navigation and flight control requirements. Aircraft are subject to considerable monitoring and measurement during production and throughout their lifetime with the consequence that air transport has a low accident rate. Concerns have been expressed that, in times of economic difficulty, air operators will cut back on the expense of safety requirements.

While road and rail traffic are also covered by international organizations, most regulations are still largely national in character. The European Community experience is more influential in harmonizing national regulations. Road traffic regulations come under three aspects: the driver, the vehicle and the road. Driver-related measurements can be on vision, driving ability or blood alcohol levels. Blood alcohol tests are sometimes given regulatory sanctions. Vehicles are produced for an international market and they need to meet the various national design rules which, while similar, have important differences. Safety devices require extensive measurements to ensure they will perform, e.g. safety belts. Vehicles also have increasingly strict requirements over the emission of pollution. The increasing use of computer control and sensing in vehicles could be used to reduce accidents providing appropriate sensors are used.

The International Telecommunication Union (ITU) is the overall body concerned with radio regulations (ITU, 1990) which are essential for the orderly use of radio communications. Provision is also needed for safety related aspects, such as distress signals, meteorological information and navigation systems. Like any rapidly developing area of high technology, there is considerable scope for good measurement techniques. Basic problems, such as the field evaluation of equipment, need research to determine appropriate procedures.

34.5.3 Environment

Environmental concerns come from a wide variety of causes, such as noise, air pollution, water quality or even littering. The concerns range from possible health problems to that of a nuisance or a degradation of the expected standard of living. Regulations on the environment can be quite specific, e.g. air pollution, or can be more general in nature. For example many countries require an Environmental Impact Report, covering any possible environmental degradation, before any major development proceeds. Also Town Planning regulations often contain very general requirements on preserving the environment.

Monitoring of the environment has increased (see Chapter 36), especially on a global basis, with the view of implementing possible international regulations, e.g. greenhouse effect, ozone depletion. Environmental properties are highly variable, yet small trends in the average are very important to detect. Often the trends expected can be easily masked by calibration drifts, e.g. monitoring UV-B radiation to see the effects of the change in ozone concentration suffers from instrument problems and calibration problems far greater than the 2% per year drift expected.

Noise is a very variable quantity but is measured in logarithmic units, dBA, and thus can be more readily measured and specified. Good procedural techniques have been developed for field measurements, thus enabling noise control to be extended to all environments, e.g. the factory as well as home. Technology forcing is often a feature of noise-control regulations as design procedures and materials are available that can achieve requirements with a low marginal cost. For example, airport noise control has resulted in new quieter aircraft which are just as economical to operate. Note that, in the US, municipalities cannot regulate for airport noise unless they are the airport proprietor, and airport noise control is a responsibility of the Federal Government (Berrens *et al* 1988).

The success of noise control regulations is founded on three factors; firstly the measurement and calibration technology are well established, secondly the effect of sound levels on the human ear are known, and thirdly design techniques and materials are available to achieve compliance.

Air-pollution studies are not as well established as noise studies, but public concern may force more than is feasible. As an example consider the application of the US Clean Air Act when revised regulations were proposed in 1984 for the national ambient air-quality standards (Stein, 1985). The Act provides for both primary and secondary standards ('standard' here applies to the mandatory concentration values and the secondary standard is stricter than the primary). Secondary standards should 'accurately reflect the latest scientific knowledge useful in indicating the kind and extent of all identifiable effects on public. . .welfare which may be expected from the presence of such pollution in the ambient air, in varying quantities'. Clearly the intention of the Act is for change in the standard to arise from scientific advances rather than emotive concerns. However, the proposed changes of 1984 called for stricter controls to be imposed without supporting scientific evidence since there was a lack of research. This proposed standard would have been technology forcing and would be very expensive to monitor and to comply with, thus going beyond the intention of the Act.

Scientific research had been carried out into other aspects of pollution and should have influenced the revision. Air-pollution requirements were based on a belief that ambient airborne particles should be treated as a single homogeneous mass representing potential harm to health and welfare. However, research showed that a distinction needs to be made between fine particles below 2 μm in size and coarse particles over 2 μm. Fine particles are acidic whereas the coarse particles are largely basic. Also the fine and coarse particles in general originate separately, are removed from the atmosphere by different mechanisms, and require different control techniques. Therefore regulations need to take this into account.

34.5.4   Health and Safety

Regulations on health and safety are wide ranging, from safety in the workplace to the operation of a clinical laboratory. Because of public concern that health applications should be safe and not produce worse effects than the illness being treated, regulations can become technology hindering. For example, regulations over new drugs or new medical techniques do not allow their use on humans until they have been demonstrated as safe for humans. While the regulation may give a route to demonstrate safety, often the alternative is to experiment with the drug or technique in a country which does not have strict regulations.

The number of regulations concerned with health and safety is multitudinous. Section 34.7.3, on Legal Liability, looks at a different approach to health and safety than that of having specific regulations. Many of the health regulations are medically orientated, but here are considered three representative issues which involve physical measurements; medical equipment, fire safety and visual display units (VDU).

Medical equipment in many countries has regulatory control governing safety, reliability, efficiency, and performance (Willingmyre, 1984). Developments in the US for the electrical safety of medical devices paralleled that of the IEC which produced an international standard in 1977. US requirements differed from the IEC consensus, thus creating a trade barrier between the US and countries which adopted the IEC standard for regulatory use. Under the GATT agreements (see Section 34.4.2) countries cannot afford to develop independent standards. On the other hand ISO/IEC need more effective and faster means to produce standards (Strawbridge, 1991).

Building codes normally require an assessment of fire risk and the means for prevention or protection, e.g. smoke alarms, sprinkler systems. To provide choice in building materials and methods some countries are using a performance-based code for fire ratings. Flammability ratings have been used for a variety of materials for some time. For a building it is important that the structures can contain a fully fledged fire for long enough for the building to be evacuated. Providing appropriate measurements are made on the building's structural components then an assessment of a new building can be made for its fire retardation potential, before it is built. Development of quantitative measures for fire retardation enables the development of more suitable materials and methods. Refinement of the measurement methods may be needed for wider acceptance amongst countries.

VDUs have become a common way to display textual material. Most VDUs do not meet basic human visual needs when compared to the printed page. Consequently operators using VDUs for extended periods often show symptoms of poor health. Regulations are needed to require that work stations are installed in a proper ergonomic manner, i.e. proper attention paid to the seating, desk heights, keyboard, and screen position. Because of the high voltages for the electron beam of the screen there have been concerns over the types of radiation emitted. While most are kept well under control there have been questions over the low-frequency radiation. The radiation is associated with the mains electrical wiring and if it proves to have harmful effects it will greatly affect how all electricity is reticulated and used.

34.5.5 Consumer Protection

One of the major areas of consumer protection was dealt with in Section 34.3 on Legal Metrology. Other quantities can be in short measure with the result that consumers do not obtain the product they thought they were purchasing. Usually the only recourse would be through contract law which covers unwritten contracts. However, this gives the consumer no real protection and the maxim 'let the buyer beware' applies. In order to give the consumer some rights nations may introduce consumer laws and regulations. Considered here are consumer concerns which should be addressed in regulations.

Mass production of most items, particularly foodstuffs, means products are bought in packages and not weighed out in front of the customer (Becker, 1975). Besides ensuring that the measuring equipment is correct it is also necessary to ensure that the quantities in the package are correct. Adequate sampling procedures are required with appropriately determined accuracy levels. Clearly small differences either way can affect the profit the packagers make as they work on small margins. Other packaging problems are concerned with settling, moisture content, and misleading packaging, e.g. a price increase may be hidden by keeping the package the same but reducing the contents.

Many products require adequate labelling to correctly identify them and their care. Foodstuffs need to give details of the contents, especially when additives are used. Clothes need to be correctly sized and the washing care given as labels on the garment.

Performance of the product should meet that claimed, either in specifications or in advertising. Policing such claims is a difficult and expensive business because of the wide variety of measurements involved, let alone finding measurement methods for some claims! It certainly is not something an individual consumer can do, but individuals can require their government to do so. The rise of QA systems may help to give the customer confidence that performance figures will be met, and any complaints dealt with promptly and fairly.

34.6   PATENTS

Scientists and engineers create intellectual property as a natural consequence of their work. If this property has a potential commercial value then the questions arise:

- how to protect it?
- who owns it?

Brief answers are given here. A lawyer specializing in intellectual property, i.e. a patent lawyer, should be consulted before legal actions are taken. The questions involve detailed points of law and do not necessarily follow any technical logic as they depend on the legal case histories of a country. Here the problem is generalized for the likely effect in various countries. This gives a starting point for issues to be raised with a lawyer. An employer who is involved with inventions should have procedures for staff to follow as well as a retained patent lawyer.

The ownership problem is fairly straightforward as the employer almost always owns the patent. If the work was done for a client then the issue is less clear and the ownership of intellectual property should have been covered in the terms of the contract. Intellectual property is real property in that it consists of paper and ink, or tapes, or photos or other devices. Information inside a person's head is not generally considered property. Any control of this internal information would be subject to special contracts involving a monetary consideration. An employee having concerns over ownership should consult his or her own lawyer.

There are four main forms of intellectual property, namely trade secrets, copyright, trademarks, and patents (Konold *et al.*, 1989).

(1) A *trade secret* is information which a company or individual does not wish others to know. Providing that a degree of secrecy is maintained by the company then any unauthorized use could be prevented by legal action. Trade secrets can be licensed and royalties collected. If 'reverse engineered' then the secret is lost and a patent could even be taken out on it. Otherwise a trade secret can be held indefinitely, with appropriate precautions.

(2) A *copyright* is a statutory right to exclude others from copying creations. Use of Copyright or © with the name of the owner and year of publication on the work is usually sufficient to ensure copyright. The owner of the work is the commissioner of the work, i.e. the employer or client. Copyrights can be assigned, e.g. in publishing scientific work the copyright is assigned to the journal publisher; also it can be inherited as it extends beyond the author's death. In some countries copyright can be registered.

With copyright, proof of the date of production may be needed to establish ownership. A simple method for this is to seal the work in an envelope and send it registered mail to yourself and leave it unopened until needed. The amount of protection the copyright gives and the duration of protection will depend on the medium used. Criminal or civil proceedings can be made against infringers. Details of the protection available will depend on the country. Note that copyright applies to the thing and not its principles, e.g. a circuit layout can be protected but the circuit could be copied in another layout.

(3) A *trademark* is a mark, word, symbol or device which is applied to goods moved in commerce. A trademark is really a two-party certification performed by the producer for the ultimate customer. The meaning or value of the trademark depends on what the producer wants it to be and what the customer comes to expect. Trademarks can become very valuable and are jealously guarded. Trademarks can be registered and in the US they have to be used to be valid. In other countries registration can be made without use and hence a conflict can arise in introducing a new trademark. The right to a trademark can be lost if the trademark becomes a generic term, e.g. aspirin was once a trademark.

(4) A *patent* is a grant by a government, through its patent office, to an inventor of the right to exclude others from making, using or selling his or her patented invention for a period of time (usually 14–20 years, depending on the country).

Obtaining a patent can be a very expensive and time consuming process. Due consideration is needed before embarking on the patent application. A patent is used to maximize the commercial return to the owner of the patent. If the owner (usually the inventor's employer) is not involved in developing and commercializing inventions, then there may not be much point in obtaining a patent. The inventor's career prospects may be better served by publishing the information in the open literature rather than spending long hours pursuing a venture with a low monetary gain. Commercially successful patents face a high probability of attempts being made to have the patent invalidated, engineered around, or infringed upon. Costly law suits would have to be brought to fight off these challenges. Therefore the costs and risks need to be well evaluated.

Patents can be given for almost all new inventions or discoveries. However, some things are not covered, such as perpetual motion machines. A patent can cover new measurement methods as well as instruments, providing they use new principles. If the subject matter was described in a printed publication anywhere in the world or was in use or on sale, a valid patent cannot be obtained. In the US there is an important exception which allows a public trial of the invention to be made up to one year before the patent application. Most other countries have an absolute novelty law which means that the invention must be kept secret before applying for a patent.

One extremely important criterion for determining patentability is that a patent will not be granted 'if the difference between the subject matter sought to be patented and the prior art such that the subject matter as a whole would have been obvious at the time the invention was made to a person having ordinary skills in the art to which said subject matter pertains' (Konold et al., 1989). The key word is *obvious*. Clearly a variety of interpretations can be made—and are made. Countries such as Germany and Japan have a tough interpretation, making it difficult to obtain patents unless the invention is a major advance in its field. An inventor needs to make the 'unobviousness' of his or her invention clear in the patent application.

Before making an application the inventor should have the patent records searched in his or her own country and if possible in other countries. A search of the general scientific and technical literature should also be made if it has not been done so at the start of the project. The information gathered will tell the inventor if the invention is new and if so how it differs from previous inventions.

The filing date of the application is important in that this usually establishes the priority of the claim against any similar inventions. Under international agreements the filing date of one country is recognized by other countries, providing a patent application is received within one year. In the US, however, priority is determined by the actual date of the conception of the invention and the diligence of the reduction to practice. Therefore, well-kept laboratory books are essential to record ideas, to have them witnessed and to keep a record of the diligent pursuit of the idea (Figure 34.5)

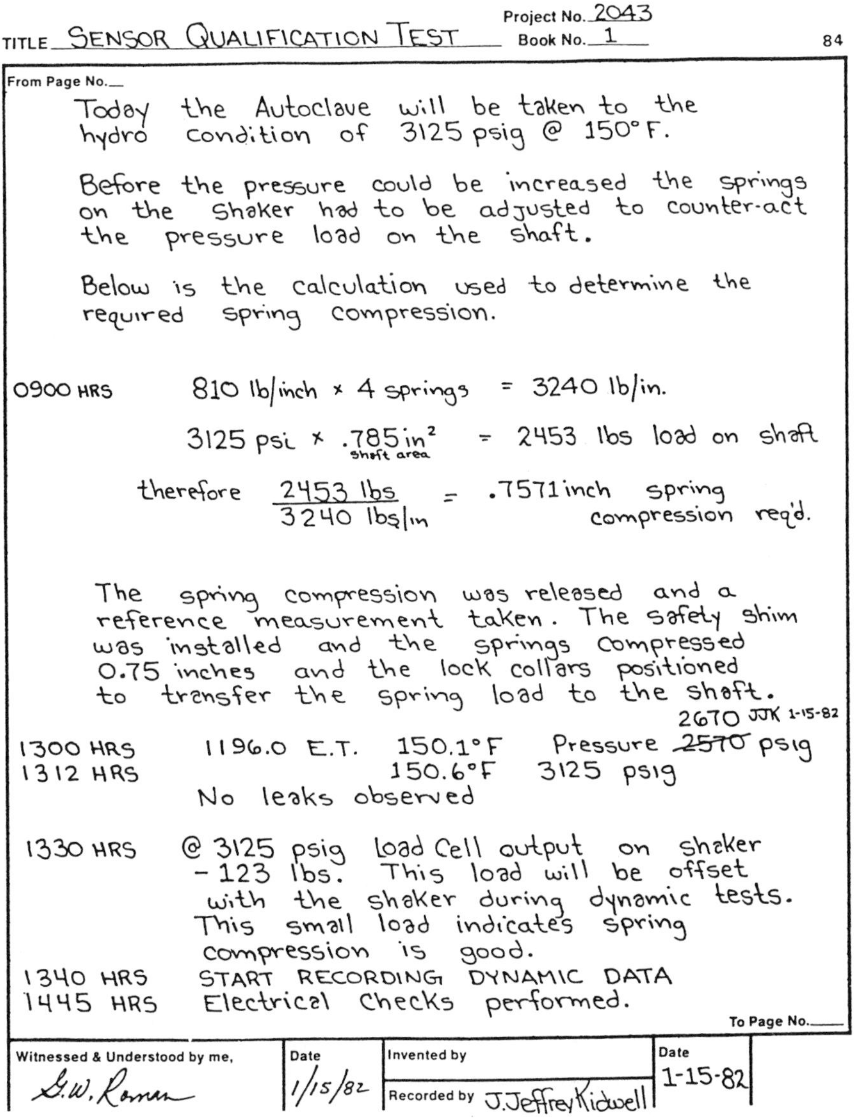

TITLE _Sensor Qualification Test_   Project No._2043_ Book No._1_   84

From Page No.__

Today the Autoclave will be taken to the hydro condition of 3125 psig @ 150°F.

Before the pressure could be increased the springs on the Shaker had to be adjusted to counter-act the pressure load on the shaft.

Below is the calculation used to determine the required spring compression.

0900 HRS   810 lb/inch × 4 springs = 3240 lb/in.

                  3125 psi × .785 in²   = 2453 lbs load on shaft
                           shaft area

            therefore $\dfrac{2453\ lbs}{3240\ lbs/in}$ = .7571 inch spring compression req'd.

The spring compression was released and a reference measurement taken. The safety shim was installed and the springs compressed 0.75 inches and the lock collars positioned to transfer the spring load to the shaft.
                                                    2670 JJK 1-15-82

1300 HRS   1196.0 E.T.   150.1°F   Pressure ~~2570~~ psig
1312 HRS   150.6°F   3125 psig
              No leaks observed

1330 HRS   @ 3125 psig load Cell output on shaker
            - 123 lbs. This load will be offset
            with the shaker during dynamic tests.
            This small load indicates spring
            compression is good.
1340 HRS   START RECORDING DYNAMIC DATA
1445 HRS   Electrical Checks performed.
                                                    To Page No.___

Witnessed & Understood by me,	Date	Invented by	Date
G.W. Roman	1/15/82	Recorded by J.Jeffrey Kidwell	1-15-82

Figure 34.5 Detailed laboratory notebooks are needed to assist with US patent applications (Reprinted from Roberts (1983), by courtesy of Marcel Dekker Inc.)

A patent application should be drawn up in conjunction with the patent lawyer, who normally does the writing. There are three main parts to an application:

- a specification,

- a drawing where appropriate, and

- an oath or declaration by the inventor.

Present US law (Konold *et al.*, 1989) describes the specification well.

> The specification shall contain a written description of the invention, and of the matter and process of making and using it, in such full, clear, concise, and exact terms as to enable any person skilled in the art to which it pertains, or with which it is most nearly connected, to make and use the same, and shall set forth the best made contemplated by the inventor of carrying out his invention. The specifications shall conclude with one or more claims particularly pointing out and distinctly claiming the subject matter which the applicant regards as his invention. . . .

Strong emphasis should be put on the advantages of the invention over the prior art. The claims should be as broad as possible without invalidating the patent. Great skill is needed to achieve this. Regard for the likely readers of the patent application should be taken. First and foremost is the patent examiner who, while having technical or scientific expertise, will not be a specialist in the inventor's field. Second in importance may be a judge (and jury) if the patent is ever contested. Clear specifications will help these people to understand the patent and perhaps come to a favourable judgement on it. Obscure specifications will only convince them that the inventor does not have a clear idea what his or her invention is and encourage an unfavourable decision. Courts dealing with patent cases often show great skill in understanding the issues (Black, 1988). Often they disregard the confusing expert witnesses and follow up on the references to the scientific and technical literature.

The drawings are for illustrative purposes rather than for design and they serve to aid the description.

The inventor (or inventors) have to make the declaration. The inventor is the person who conceived the idea and carried it through to realization. The requester of the invention is not an inventor, nor is, say, the skilled machinist who produced the parts for the invention. If other inventions are needed to realize the first invention then the situation is more complicated and the patent lawyer can advise on how to handle such a case. Basically inventions cannot be patented by other people even if they have a close association with the invention.

Ownership is a different matter. In the patent application the inventor will most likely have to assign the ownership of the patent to his or her employer. If the work from which the invention arose was done for a client then the ownership of the patent can be problematical. Hopefully this is covered by the contract with the client, otherwise the ownership depends on whether the client could have reasonably expected an invention when the work was commissioned. The patent lawyer should be consulted about the invention before the client is told. Co-ownership of a patent usually means either party can do what they want without consultation with the other, including holding on to the royalties. Other contractual agreements are needed if profit sharing is intended.

Finally if the patent is granted then the commercialization needs to be pursued. Most patent laws expect this and have mechanisms to encourage it. For example, regularly increasing fees may be required to keep the patent. In some countries there is compulsory licensing of patents when patents are not used. Note that the patent is a publication which gives a competitor information to get around the invention.

34.7 FORENSIC CONCERNS

34.7.1 Forensic Sciences

Forensic, as an adjective, is used to describe things belonging to or used in law courts. Thus forensic science refers to scientific knowledge used in legal matters and, in particular, that for police investigations. Here 'science' is used in its widest meaning, to include engineering, medicine, etc. Forensic engineering, though, would be more concerned with the investigation of liability under civil law for an engineering failure, or occasionally criminal law if there was loss of life or injury.

A forensic expert will be called as an expert witness in a court case because of the investigations he or she has made. Other experts can be called to assist in the understanding of any issue before the court. They are not usually directly involved in the investigation, e.g. a surveyor may need to testify that a map is correctly drawn. Section 34.7.2 gives a guide for the measurement expert called as an expert witness. Major issues have been raised in forensic sciences which involve measurement technology.

Science has been assuming an increasingly powerful role in the execution of justice since the discovery of fingerprinting. Scientific testimony is often the deciding factor for the judicial resolution of civil and criminal cases. Often the scientific evidence can seem more compelling to a jury than the testimony of eye witnesses. Unfortunately, many forensic tests, which have previously been accepted by courts of law, have been shown to be faulty. Also evidence given by forensic experts has been shown to be biased in order to obtain a conviction, where the evidence did not warrant it. As a result there is unease in the legal profession over the nature of scientific evidence, in particular the application of new techniques (Neufeld and Colman, 1990; Black, 1988; Freckleton, 1987).

Forensic experts have either developed their own measurement techniques, e.g. fingerprinting, or have adopted well-established techniques, e.g. gas chromatography. When a new technique is introduced the forensic expert is interested in whether or not it will be admissible in court. The new technique is a success if it results in a conviction. Therefore, the scientific procedures normally used to establish the validity and reliability of a method are sometimes short circuited. Normal scientific controls are absent in a judicial trial when applying science to a fact-finding process. Indeed the purpose of having expert witnesses is to avoid the detailed study of a subject by the court. The scientific community monitors research by subjecting new theories and findings to peer review and independent verification. Law courts rely on cross examination to show up potential flaws and, if lawyers are not aware of any, then they will fail to challenge faulty or inadequate scientific evidence. For example, in cases where new techniques were used for hair comparisons (Black, 1988) the forensic scientist had not tried to establish the statistical significance of the test by the random sampling of the population. Another example is blood tests based on well-established clinical tests using adequate samples of fresh blood. The validity of these tests had not been established for small dried blood samples which may have been contaminated, acted on by bacteria or degraded by other environmental factors.

Most jurisdictions have rules of evidence which are meant to prevent unsuitable evidence being admitted into court. None of these rules are based on scientific method (Black, 1988; Freckleton, 1987). Consideration of the scientific method (Black, 1988) leads to three criteria on which to assess scientific evidence,

• the underlying scientific theory must be considered valid by the scientific community,

• the technique itself must be known to be reliable, and

• the technique must be shown to have been properly applied in the particular case.

While these appear common-sense criteria the common sense can get lost in a law court due to the adversary system used. This results in (Black, 1988) '. . .two hostile camps, and prepared to attempt, under solemn oath, to uphold opinions diametrically opposed, yet supposedly derived from a single series of facts and observations'. Even amongst the scientific community what one group of peers considers an adequate measurement technique may be considered erroneous by another.

Solutions to these problems are needed. A part of the solution undoubtedly will have to come from the courts themselves. They have shown skill in dealing with complex scientific problems (Black, 1988). To introduce these skills into a normal trial would be to change the trial to a trial of the expert. A fundamental problem is that no overall principles are available to assist a lawyer in evaluating the myriad of techniques. Texts on scientific evidence are mainly catalogues of scientific techniques (Black, 1988). For example, there is no adequate theory of measurement that a lawyer could use.

The legislature in setting up legal metrology has had to deal with similar problems and their solutions have been covered in previous sections. There is no reason why the judiciary could not adopt a similar approach as a basis to a solution. Most forensic laboratories are run by the police and have no specification standards or accreditation schemes to assess their operation. Some nations, though, have independent forensic services and forensic reports are made available to the defence as well as the prosecution. Since many criminal defend-ants are poor, availability of forensic reports is an important feature to ensure justice.

It is vital that the immensely valuable role forensic science plays does not become discredited and the challenges need to be met.

34.7.2 Expert Witness

Rule 702 of the US Federal Rules of Evidence provides 'If scientific, technical or other specialized knowledge will assist the trier of fact to understand the evidence or to determine a fact in issue a witness qualified as an expert by knowledge, skill, experience, training, or education, may testify thereto in the form of an opinion or otherwise.' (Black, 1988). Most other countries provide for expert witnesses along similar lines even if expressed differently. It has been seen from the previous section that there is abuse of the expert witness role in both criminal and civil cases and so given here is an introductory guide to help a measurement expert avoid contributing to the abuse. Forensic scientists, of course, have their own procedures as do engineers regularly involved in civil forensic work (ASFE, 1987). While most of the comments here are directed to civil cases they will apply to tribunal hearings which are less formal and to criminal cases which are more formal.

Initial approaches to retain an expert witness may be indirect. For example, a lawyer may phone up for 'free' advice, or a request is received to evaluate a product, or a request to carry out a series of measurements. At this stage the expert or organization needs to have a clear policy on legal involvement, and it should be communicated to the client who should be required to disclose any possible legal involvement. Otherwise you may find, to your embarrassment, your tests or measurements being used out of context.

Three policies are:

- no involvement whatsoever;

- impartial involvement i.e. your testimony would be the same no matter who retains you;

- supportive involvement, i.e. the style but not the substance of your testimony may favour your client.

No involvement may be hard to achieve if the person contacted is the only expert around and especially if he or she has talked to the lawyer. However, it should be taken into account that if the most competent experts avoid involvement then less competent ones may be available. Before being involved make an honest appraisal of the skills to see if they match the requirements. Involvement may have personal costs, the most important of which is a reputation that may be attacked. For example, to discredit a university professional, the opposing lawyer may denigrate that person's practical experience.

Impartial involvement of an expert witness is preferred by many lawyers. Obtaining a favourable opinion from an obviously impartial expert witness will be a strong point for the lawyer's case. To remain impartial keep contact with the lawyer to a minimum and also knowledge of the case. The lawyer may require presence on the day only. Any tests or measurements made should be made to an agreed standard appropriate to the case, e.g. a surveyor besides using professional guidelines may need to know local bylaws to confirm a survey. On the day the lawyer may wish to establish a fact, e.g. the result of a test. Additionally the lawyer may also try to extract favourable opinions from an expert witness, who should avoid answering if they go beyond the expert's direct experience or field of expertise. For example in a criminal case a defence lawyer may seek opinions to merely cast doubt on some evidence because a criterion for conviction is 'beyond reasonable doubt'.

A supportive involvement can be a difficult task. An overriding consideration is that the substance of opinions conveyed by experts should be the same no matter who has retained them. Even with this consideration in mind, it is best to avoid the position taken by a few forensic scientists who believe their job is to obtain convictions because they are employed by the police. Withdrawal from the case may be necessary if the lawyer keeps trying to deviate the evidence into something that is unacceptable to the expert. Consider now a civil case to show how the expert can be supportive. At the beginning the client needs an impartial investigation from the expert in order to determine what position to adopt. Pre-trial meetings between both sides of the dispute try to establish what agreements or concessions can be made, and if possible make an out-of-court settlement. Thus if a trial does occur it will most likely be because of basic differences of opinion between opposing experts. The expert will therefore be supporting his or her own opinion. A positive presentation of the evidence can be made along the same lines as a research paper is presented. Support for the case can be made by working on the courtroom appearance and presentation so that the expert will be credible to the jury. Civil cases are decided on the 'preponderance of the evidence' and the quality of the expert's presentation can enhance this.

While there can be no hard and fast rules about court work, some pointers are:

Before the trial:

- Evaluate language differences. An expert's use of English will not be the same as the average person on the jury.

- In preparing evidence have answers to the basic legal questions: Who? What? Where? When? Why? Witnesses?

- Answers to questions should be based on well-researched principles.

- Opinions should be those held by the expert and not those of others.

- The fee for the service should be paid beforehand. (Never accept work for a contingency fee as this could discredit the evidence.)

For the court appearance:

- Arrive at the right place and time, well rested.

- Speak clearly and slowly for the court recorder.

- Speak to the judge or jury.

- Answer only the question asked, and say 'I don't know' if it applies.

- Do not worry how things are going, concentrate on the questions.

- Pause before answering to consider the question fully. In cross examination it gives the lawyer a chance to object.

- Be prepared for questions about qualifications and experience.

34.7.3 Legal Liability

A measurement expert's court appearance is most likely to be over a civil case involving liability. Occasionally in criminal cases of liability the expert may be called on by the defence. Legal liability action may be taken against the measurement expert who will then be the defendant. Liability, in the general sense, is a prime driving force for all the legal aspects of measurement covered in this chapter. What constitutes legal liability has tightened and avoidance of liability is difficult. Details on legal liability depend on the highly individual nature of the national legal systems, and important trends based on US and UK experience are now considered.

In the US the traditional rule for liability (Watson and Bernstein, 1988) is to measure the adequacy of a provider's delivery of care based on the custom of other providers in the defendant's locality. An example of the custom of other providers could be a specification standard which was voluntary law (see Section 34.4.2). The locality rule provides for very local practice such as an earthquake area requiring relevant building codes, but if appropriate, the locality can be extended to be international in scope.

A potential problem is that the rule allows an entire profession or industry to adopt too slowly newly available safety devices. This has led some courts to adopt the rule of 'reasonable prudence' as the measure of what custom should be. There are three prongs to the test for the imposition of liability:

- the technology must be available,

- the technology should not cause more harm, and

- the cost of the technology should be reasonable.

These features are illustrated with an example. A doctor is expected to have and use a stethoscope because it is readily available, causes no harm, and is low cost. However, the doctor would not be expected to own a body scanner, but if the local hospital has one then patients that benefit should be referred to it. In this case the patient decides if the cost is reasonable.

The effect of the criteria is to require a profession or industry not only to be aware of the latest technology but to realize its potential use. A similar situation exists in specification standards (Section 34.4.2) where the certifiers (e.g. test laboratories, accreditation bodies) need to be aware of changes and to push for them.

Product liability in the UK has undergone radical changes and the European Community will cause more (Tye and Bowes, 1978; Montague, 1987). Consider now some of the basic concepts behind the product liability and related laws, which have been tightened to be more effective in the prevention of death and injury. The traditional way to recover damages was with the law of Torts, which relies on proving negligence. The difficulty was in identifying who was negligent and proving it. Certainly no help could be expected from the retailer, the distributor, the manufacturer or other suppliers. For instance, a case involving injury from a shattered windscreen failed because it could not be proved if the installer or the manufacturer of the windscreen was at fault.

There are obvious faults with this system and specific laws were passed to cover workplace safety as well as safety from products which have associated dangers (e.g. automobiles). New product liability laws have aimed to be all encompassing. Producers of goods are held to be strictly liable for any defect in the product that causes injury or death. Strict liability reverses the onus of proof in that it assumes negligence until proven otherwise. How this principle is applied in practice is the point of interest.

If a defect is found then an investigation would be held to determine the cause and responsibility, starting with the seller to see if the seller was also the producer. The seller could be considered to be the producer even if the seller is engaged only in simple assembly of components. The seller would be required to disclose all records and safety procedures. If there are no records, or the safety procedures are inadequate, then this would be taken as negligence and the seller could be considered the producer and charged. Adequate records would lead on to the next person in the chain and so on, leading ultimately to a producer. The cause of the defect can be found from the producer's records. There are two options for the producer to transfer responsibility if the records are good enough. An employee may have been neglectful but if neglect was the usual custom rather than the official procedure, then the producer is still liable. That is, the producer has the responsibility to see that the safety procedures are actually carried out.

The other transfer of responsibility is to a supplier, and this may be affected to some extent by the contract between the supplier and purchaser. The supplier has the responsibility to ensure that the product supplied is used safely. At the retail level this means adequate instructions for use are included with the product. The principle applies also at other parts of the production chain. For example, a circuit board manufacturer should have in the contract with a TV assembler all the requirements for safe assembly and what final inspection and tests should be carried out. If the supplier becomes aware that the product is being used unsafely then the supplier may have to take preventative action to avoid liability.

Product defect laws cover anything that causes injury to any person during any part of the manufacturing chain, including the producer's employees, test personnel, visitors and neighbours to the factory. The concept of a producer is quite wide and would cover the importer of goods, persons reconditioning equipment and also assemblers of equipment for factory use. Quite detailed safety procedures are needed, such as those associated with QA systems. Indeed some QA systems are oriented to safety or, more generally, to loss prevention. Avoidance of product liability involves having the appropriate management systems rather than hiring a lawyer.

An aspect not covered above is liability due to unforseen defects. That is defects, faults or problems that were not known or not predictable at the time the product was designed or made. Drug laws usually require liability for unforseen effects. It is likely some responsibility for these will be included in product liability laws.

For the measurement expert there are three important principles arising from these liability concerns:

- *Learning*. Reasonable prudence cannot be exercised if there is no awareness of new or even old techniques.

- *Traceability*. QA systems can be seen as an expansion of the traceability care that a measurement expert should be exercising.

- *Education*. Responsibility to clients (in the general sense) may be achieved by providing training, publishing or preparation of standards.

The main thrust of the legal aspects covered in this Chapter centres around these three factors, even for cases where injury or death are not involved. Scientists and engineers should already be well aware of these principles, but what is new is the level at which they need to be implemented.

34.8 CONCLUDING REMARKS

The main conclusion of this chapter is that the measurement practitioner needs to be aware of the relevant laws and the likely future changes. The adage 'ignorance of the law is no excuse' is even more true today. Laws and their regulations are specific to each nation and may be changed and as a consequence this overview has not been specific since it is the practitioners' responsibility to find out the relevant legal information. Given is merely a guide to the types of information that may be required and details will come from tracking down the sources of regulations, be it at local, national, or international level.

Many of the legal issues raised here may not be familiar to lawyers, but a lawyer should be consulted over any legal action particularly in respect of patents, contracts and liability. An introductory text on national law could also be consulted for an overview.

The references given, while being a random sampling of the available literature, can also be used to locate sources of information. HBMS, Volume 1, Chapter 3 gives addresses of NSLs and national standards organizations and these can be contacted. GATT information is likely to be held by standards organizations as well as any standard cited in legislation. The components of the national measurement system need to be identified to know who to approach. Generally most organizations have regular publications which list current work, which is often driven by regulatory pressures.

Finding regulations can be difficult as there are few electronic data bases containing legislation. Most legislative bodies publish their laws and regulations but finding the origin of them is more difficult. For research areas, the topics can be found by computer literature searches.

34.9 SOURCES OF FURTHER ADVICE

The reader is referred to the following textbooks and journals for further information on legal issues in measurement.

Books

Anon (1989) *International Directory of Laboratory Accreditation Systems and Other Schemes for Assessment of Testing Laboratories* National Association of Testing Authorities, Sydney.

Dunham, C. W., Young, R. D. and Bockrath, J. T. (1979) *Contracts, Specifications and Law for Engineers* McGraw-Hill, New York.

Knight, P. and Fitzslmons, J. (1990) *The Legal Environment of Computing* Addison-Wesley, Sydney.

Taylor, J. K. (1987) *Quality Assurance of Chemical Measurements* Lewis Publishers Inc, Chelsea.

Berman, G. A. (ed.) (1980) *Testing Laboratory Performance: Evaluation and Accreditation* National Institute for Standards and Technology, Gaithersburg.

Journals

ISA Transactions ISA, USA.

Measurement and Control Institute of Measurement and Control, UK.

Measurement Science and Technology (formerly *Journal of Physics E: Scientific Instruments*), Institute of Physics, UK.

Measurement IMEKO, UK.

REFERENCES

Andrus W.E. Jr (1975). 'International Organization of Legal Metrology—the emerging US role', *NBS Special Publication 442*, 138–56.

ASFE (1987). *A Guide to Forensic Engineering and Service as an Expert Witness*, ASFE, USA.

Becker, M.H. (1975). 'Weights and measures and the consumer', *NBS Special Publication 442*, 67–73.

Bell, M.R. (1988). 'NZ Code of Laboratory Management Practice', *Proc. International Seminar-Laboratory Accreditation and Trade*, Telarc, 41–49

Berrens, R.P. DeWitt, J.S. Baumann, D.D. Nelson, M.E. (1988). *Examination of Noise Management Approaches in the United States*, Report No. IWR-88 R-8, Institute of Water Resources (Army).

BIPM (1987). *Le BIPM et la Convention du Mètre*, BIPM, Sèvres.

Black, B (1988). 'A unified theory of scientific evidence', *Fordham Law Review* 56, 595–695.

Cammell, J.E. (1990). 'The European community's global approach to conformity assessment: The quality implications', *Quality Forum*, 16, 122–4.

Christmas, P. (1984). 'Traceability in radionuclide metrology', *Nuclear Inst. and Methods in Phys. Res.* 223, 427–34.

Freckleton, I.R. (1987). *The Trial of the Expert*, Oxford, Melbourne.

Gallagher, W.M. (1988). 'Electrical testing and approval. International cross recognition—personal views and experiences', *Proc. International Seminar—Laboratory Accreditation and Trade*, Telarc, 51–53.

Gilmour, A. (1988). 'Mutual recognition of accreditation systems and acceptance of test data', *Proc. International Seminar—Laboratory Accreditation and Trade*, Telarc, 17–26.

Glazebrook, R.T. (1931). 'Standards of measurement, their history and development', *Phys. Soc. Proc.* 43, 412–57.

ISO (1980). *Certification—Principles and Practice*, ISO, Geneva.

ISO (1982). *Units of Measurement*, Handbook 2, ISO, Geneva.

ISO (1984). *International Vocabulary of Basic and General Terms in Metrology*, ISO, Geneva.

ITU (1990). *Radio Regulations*, ITU, Geneva.

Konold, W.G., Tittel, B., Frei, D.F., Stallard, D.S. (1989). *What Every Engineer Should Know about Patents*, Marcel Dekker Inc., New York.

Montague, J.E. (1987). *Business Law*, Chambers, Edinburgh.

Moss, C. (1978). 'Quality of precision measurement equipment', *NCSL Newsletter*, 18, 23–24.

Neufeld, P.J. and Colman, N. (1990). 'When science takes the witness stand', *Sci. American*, 262, 18–25.

Roberts, G.W. (1983). *Quality Assurance in Research and Development*, Marcel Dekker Inc., New York.

Smith, R.M (1990). 'The importance of certification and accreditation to the single market', *Quality Forum*, 16, 133–35.

Stein, A.T. (1985). 'Legal aspects of the proposed revised TSP secondary ambient standards', *Proc. 78th APCA Meeting*, paper 85–8.4.

Strawbridge, G. (1991). 'Look beyond Europe for standards', *New Scientist*, 23 March, 6.

Telarc (1991). 'Inside quality', *Telarc Talk*, March, 3–5.

Tye, J. and Bowes, E. (1978). *The Management Guide to Product Liability*, Brehan Press, London.

Watson, B.L. and Bernstein, J.M. (1988). 'Liability for not using computers in medicine', *Proc. Annual Symposium on Computer Applications in Medical Care*, 898–901.

Willingmyre, G.T. (1984). 'Role of standards: international commerce for medical devices', *IEEE Engineering in Medicine and Biology Magazine*, March, 26–30.

Chapter

35 A. E. BALDIN

Condition Monitoring

Editorial introduction

To squeeze more out of capital plant it has become necessary to find ways to obtain more effective use of that plant. To wait until a machine fails to operate before any maintenance is done is a wasteful method as it brings with it long down-times and unnecessarily high repair bills. For this reason modern plant is now being provided with large numbers of sensors that monitor machine variables in order to assess what will soon need repair and when it is best to make the repairs. The availability of low-cost electronic processing has allowed a revolution in this concept as plant moves progressively toward improved quality control with in-process measurement.

This chapter provides an insight into industrial plant health monitoring but as the concept is widely applicable material in Chapters 36, 42, 44, and 45 should also be read to gain an appreciation of how multi-sensor systems are needed in many areas of modern life.

35.1 DEVELOPMENT OF CONDITION MONITORING IN PROCESS AND MANUFACTURING SYSTEMS

35.1.1 Towards 21st Century Maintenance

The society of the 1990s can be considered as the maintenance and preservation society. Examples of this attitude towards maintenance and preservation are:

● preservation of historical city centres;

● protection of rare animal and vegetal species;

Handbook of Measurement Science, Volume 3
Edited by P. H. Sydenham and R. Thorn
© 1992 John Wiley & Sons Ltd

- support for continuance of minorities and old dialects.

In future, maintenance activities can be expected to increase, even at the domestic level, due to the wider use of electrical appliances and private transport. In the industrial sector, maintenance costs are equal to approximately 5–6% of fixed assets as an average of all sectors, and can reach 10–12% in the heavy industries. To this direct maintenance cost must be added the value of lost or poor quality production.

Maintenance practices are related to safety and environmental issues, to the inaccurate operation of a machine or production of a product, and to accidents and injuries (this being particularly true in the case of public transport).

The type, content and depth of professional knowledge needed by maintenance personnel is increasing with plant complexity. As a result, a new type of professional maintenance person who is able to cope with the technological advance of installations is emerging.

35.1.2 Enterprise as 'System' and Maintenance as 'Support'

Generally speaking a system is defined as the sum of a whole (that is a group of people, machines and investments) and its relevant purpose:

$$SYSTEM = WHOLE + PURPOSE$$

In this way the enterprise represents a system that, in its turn, is a component of a bigger system (the national and international community) inside which are recognizable subsystems. Therefore, maintenance is a subsystem of the enterprise system which, in its turn, is a part of a macroeconomic system. Owing to the definition given to the word 'system' we have now to determine the purpose of the enterprise and as a consequence those involved in maintenance must be in harmony with the enterprise goals.

The aim of an industrial organization is to produce goods according to a plan, at costs foreseen by the budgetting process. As production processes have been improved to create up-to-date, longer lasting, and less expensive goods, many problems have developed, the solution of which has become more and more difficult. At the same time rationalization of production processes and the introduction of the 'productivity principle' (required for an enterprise's survival), has led to greater importance of the complementary activities of production which has, in turn, generated new costs.

Over the years, maintenance has acquired a still greater influence among the different complementary production activities. It generates considerable interest due to the continued expansion of machines into every production sector at all possible stages of the operating cycle. Technical and technological levels are of ever-growing complexity.

The following points are pertinent:

- the trend is for greater investment of capital instead of manpower;

- initial plant costs seek to ensure full and rational exploitation at conditions of maximum profitability;

- cessation of operation of fully integrated systems has an adverse effect on production and considerable influence on production cycle costs;

- high levels attained by mechanization and increasing machinery complexity in turn call for greater time required to carry out a diagnosis (Table 35.1);

- labour costs are rising;

- rising prices of some raw materials is increasing the purchase costs of spare parts which can represent from 30 to 50% of the total maintenance costs, depending on the industry sector and local situations.

Table 35.1   Diagnosis and repair time versus kind of system

Kind of system	Diagnosis time (% of total down time)	Repair time (% of total down time)
Electronic	90	10
Electric	60	40
Hydraulic	20	80
Mechanical	10	90

From the above it can be stated that the primary purpose of an enterprise's management is the pursuit of optimized economic results from the following points of view:

- *Products*. Quantity and quality maximization.

- *Costs*. Total cost minimization.

- *Societal issues*. Safety and environment, and personnel maximization.

Thus maintenance purposes, and the means to be used to attain it, can now be outlined.
    Maintenance must contribute to:

- *total costs minimization*, by means of:
    —minimum maintenance costs;
    —reduced number of failures;
    —reduced number of preventive maintenance shut-downs;
    —reduced shut-down time.

- *Personnel gratification maximization* by:
    —improving work safety.

In order to achieve these aims maintenance must:

- carry out programs for preventive (time-based and condition-based) and modification maintenance work to raise plant reliability and availability;

- provide correct management of the available resources, namely personnel, materials, equipment and operating systems.

35.1.3 Monitoring and Technical Diagnostics

Monitoring and technical diagnostics, as applied in industry, includes all technical aspects covering any phase of the life cycle of a plant from the project initiation stage through construction to operation. Aspects can include the following:

- critical examination of project solutions, to prevent faults during operation and to ensure inspection capability and maintainability;

- materials choice and qualification;

- materials testing;

- choice and qualification of manufacturing process (welding, heat treatments, surface treatments);

- materials and components, destructive and non-destructive tests at pre-service, in-service and on-line levels;

- on-line monitoring of components (structural, early detection of malfunctions);

- on-line monitoring of process performance (optimization and control of process variables);

- failures analysis.

Research and technological innovation have greatly improved industrial diagnostics in terms of:

- sensor development;

- development and dissemination of digital electronics, microprocessors and computers and applied industrial instrumentation;

- theoretical search and development for explanatory models of basic phenomena of different diagnostic technicalities.

In choosing the diagnostic technicalities to be implemented it is also important to note they must be:

- applicable in industry and give the expected diagnostic response under actual field conditions and constraints (execution times, response time, costs);

- reliable;

- transferable to staff with a non-scientific background;

- within satisfactory cost/benefit requirements.

35.2 RELATIONSHIP BETWEEN MAINTENANCE POLICY AND RELIABILITY

35.2.1 Maintenance and the break-even point

A maintenance program which allows lowering of the repairs frequency and shut-down duration will greatly reduce plant costs. As shown in Figure 35.1, a decrease in the variable costs causes a positive fluctuation of the break-even point (BEP) to the advantage of plant productivity.

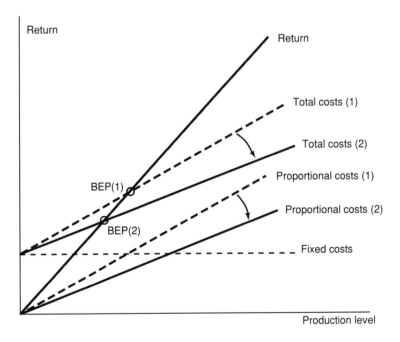

Figure 35.1   Influence of maintenance costs on break-even point (BEP). For a constant return value, a decrease of costs shifts the BEP towards a lower production level

Maintenance can contribute to the reduction of the production costs that are variable with output (less frequent stops = lower consumption of raw materials, lower losses of finished products, lower energy consumption, less production labour requirements in order to face emergency causes) and of the variable part of its own costs (according to experience the share of maintenance costs variable with production output is about 50% of the total).

Maintenance also affects fixed costs and in this way further advantages in cost reduction and profitability increases will be obtained through improved maintenance.

35.2.2 Plant availability

The influence of maintenance on general plant economy may also be considered from the point of view of availability defined by the following ratios:

(1) *Intrinsic availability*, A_i, is given by

$$MTBF/(MTBF + MTTR) \tag{35.1}$$

where, MTBF—mean time between failure;
MTTR—mean time necessary to repair.

(2) *Technical availability*, A_t is given by

$$MTBM/(MTBM + MTTR) \tag{35.2}$$

where, MTBM—mean time between maintenance.

(3) *Archieved availability*, A_a is given by

$$MTBM/(MTBM + MDT) \tag{35.3}$$

where, MDT—mean down time which includes time to organize the intervention, obtain the materials and the MTTR.

Maintenance can have an influence on the plant availability in the following ways:

- *by reducing the MTBF* through improvement of machinery performance; optimization of inspection frequencies; research for more suitable or better-designed materials; improvement of working conditions; research for more suitable lubricants; improvement of the repair operation quality, etc;

- *by reducing the MTTR* through improvement of the maintainability; improvement of personnel training;

- *by reducing the MTBM* through introduction of diagnostics monitoring and condition-based maintenance;

- *by reducing the MDT* through improvement of organization of teams; avoidance of idle times; improvement of the information flow between maintenance and production and between maintenance and store or purchasing.

35.2.3   Maintenance policies

According to British Standard recommendations the following maintenance policy is applicable to industrial and utilities undertakings:

- *preventive maintenance*, carried out at predeterminated intervals or to other prescribed criteria intended to reduce the likelihood of an item not meeting an acceptable condition;

- *condition-based maintenance*, preventive maintenance initiated as a result of knowledge of the condition of an item ascertained from routine or continuous checking (monitoring).

- *condition monitoring*, the continuous or regular measurement and interpretation of data carried out to indicate the state of an item in the interests of safe and economic operation. Condition monitoring is normally carried out with the item in operation, in an operable state, or removed but not subject to a major stripdown.

Another possible policy is *contingency maintenance* consisting of removal and substitution of components upon occasion of any shut-down. The decision for this type of intervention is taken after statistical analyses of data, such as calculation of the MTBF.

An *item*, in this context, denotes any engineering part, component, subsystem or system that can be individually considered.

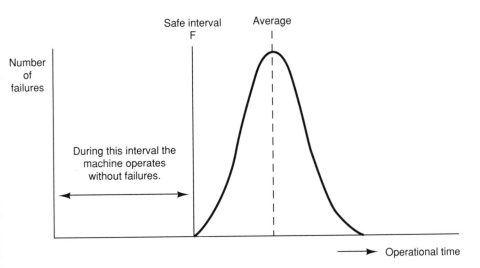

Figure 35.2   Normal distribution of failures and selection of a safe preventive maintenance interval

All of the above policies, each having a different degree of importance, are usually implemented in a plant in order to optimize the economic and technical performance.

Machines can be run until they fail and then be repaired. This method can be very costly since it increases production losses, risks major damage to machines and injury to personnel. In order to avoid this (mainly in the case of costly, complex and large installation), it is more economic to perform preventive maintenance activities consisting of interventions at regular predeterminated intervals.

The application of preventive maintenance is based on the assumption that failures are distributed according to a Gaussian curve (Figure 35.2).

The basic difficulty with this type of maintenance is to select an appropriate interval between intervention and the subsequent failure. In actual operation the elapsed time needed before a preventive intervention varies greatly from one situation to another one because of different working conditions and different component lifetimes. Figure 35.2 shows how selection of a safe interval can give unacceptably short times of operation.

If a preventive maintenance interval is chosen that is too far on the safe side the machine will be overhauled unnecessarily, with consequent loss of production and possible introduction of costly human errors during reassembly. Only seldom does a machine deteriorate in a very short time with a sudden failure. More often breakages are the apex of slow deterioration taking place over months or years.

Preventive maintenance, therefore, consisting of the thorough dismantling of a machine into its items with substitution of all rotating and related parts, at pre-determined intervals, is not profitable in most industrial situations. Such 100% preventive maintenance is, however, practised in fields where human safety requirements demand it—but with subsequently higher costs of supply of services.

35.2.4   Bath tub curve and monitoring

The prevailing approach to satisfactory maintenance is now described. From machine history data the failure rate λ (where $\lambda = 1/\text{MTBF}$) can be calculated and plotted against time (Figure 35.3). From this it can be concluded that:

- during the initial 'teething trouble period' λ is decreasing;

- after this period, most machinery failures are constant versus time within certain limits, that is, failures are random (λ is constant);

- in several cases failures increase after an overhaul, that is, a transient situation of 'teething trouble' is recreated;

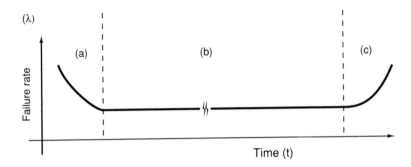

Figure 35.3   Probability distribution of failure, 'bath tab' curve: (a) running in; (b) useful life (c) wearout

• during the decay phase, the number of failures is increasing owing to wear and corrosion (λ increases).

The *bath tub* pattern can be interpreted with respect to maintenance in this way:

• preventive maintenance intervention carried out at predetermined intervals is not useful during the λ constant period because the general machine condition cannot be upgraded;

• conditions may actually be worsened after preventive intervention as a result of human error;

• preventive intervention carried out at intervals linked to real conditions of machine components is feasible.

An acceptable compromise for a maintenance policy is, therefore, to perform maintenance at intervals determined by the real condition of machine components. To obtain this, it is necessary to know the actual condition of the machine and its trends. This is found through monitoring the various aspects of the system.

If predeterminated parameters are measured at regular intervals while the machine is working, it will be known if the measurement value is within the maximum allowable limit. By gathering these data the trend can be checked (Figure 35.4) and acted upon accordingly. Prediction of the lead-time within which to carry out the maintenance intervention is then possible with reasonable certainty. Checks of machine condition while working are usually performed as a daily practice in industry, but with monitoring something different is implied.

Monitoring is a diagnostic process originating from the application of fundamental engineering laws and specific empiric relationships enabling the maintenance staff to know (monitor) beforehand, the condition of every machine component.

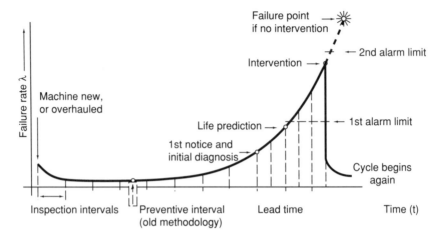

Figure 35.4   Condition monitoring representation curve showing warning, alarm, failure limits and lead time

Knowledge of a machine's *mechanical health* helps operators to plan and schedule repair and overhaul work based on real physical conditions, rather than on the time that has elapsed since the last intervention. This approach considerably increases the up-time of factories and avoids unnecessary substitution of components.

35.3   ADVANTAGES AND LIMITATIONS OF CONDITION MONITORING

35.3.1   Advantages

The main advantages of implementing a monitoring programme are that it lowers maintenance costs and reduces the likelihood of failures and shut-down.
Other benefits are:

● A lead-time between alarm and failure can be calculated allowing machines to be stopped before they reach critical conditions. Machinery handling dangerous materials and machines transporting people can be operated at peak efficiency and safety.

● Machine up-time can be increased, overhaul time can be reduced, resources prepared beforehand and induced damages eliminated.

● During commissioning of new installations, or installations of a new design, weak components are highlighted. Better specifications can be written for future plants, stating maintenance jobs and expected operational performance of existing machinery.

● Efficient plant operation and better quality is achieved by the production of goods in the required quantity and at the prescribed quality without waste and energy loss, thus assuring operation of plants at full capacity.

● Remote installations can be operated more satisfactorily as data can be transmitted to other, more convenient centres. Maintenance staff are able to be used more effectively and correct spares taken on visits.

● Lead-times made possible by condition monitoring largely avoid the inconvenience of unexpected breakdowns. Advanced planning assures production schedules will be adhered to thus better satisfying customer's expectations.

● Measurements, at the end of the guarantee period, or after overhaul, can be compared with specified values to help assure effectiveness.

35.3.2 Limitations

Certain limitations exist with use of the monitoring method. These are:

● When plants are not working at full capacity not all monitoring methods can be carried out because the plant is not in a fully operating condition.

● If the number of similar components is too small it may not be feasible to build sufficient experience for measurement interpretation. An alternative is to contract out monitoring services to agencies specializing in particular areas.

● In the case of mobile machines working at distant locations from central services it may not be possible to implement a worthwhile program.

● It is not always possible to obtain later access to a parameter not foreseen at the design time.

● Machines often have inaccessible parts.

● Components suddenly collapse through unpredictable fatigue.

35.3.3 Condition Monitoring and JiT (Just in Time)

Knowledge of machinery condition through condition monitoring has a positive effect on planning and scheduling of maintenance interventions, and thus on production programmes.

Planning means making a commitment to complete a certain task by a certain time, at a given cost. It has been determined that use of planned maintenance allows tasks to be carried out in approximately 20% less time, sometimes better.

Lack of maintenance planning in a company may lead to a series of short-, middle- and long-term consequences.

In the short term, that is during work execution, absence of planning leads to:

- excessively high number of urgent works;

- excessive interruption of work;

- lack of precise and detailed instruction being issued to the personnel with potential for more error;

- shortage of personnel having adequate skills;

- stoppage of work due to shortage of materials or spares;

- stoppage of work due to the shortage of special tools necessary to carry out the work needed;

- discontent and frustration among personnel being compelled to work under confused and uncertain conditions;

- impossibility of prearranging safety measures to prevent labour accidents during the work.

In the medium term, the lack of organized maintenance works leads to:

- higher number of shut-downs because works are carried out less accurately;

- lengthening of the duration of the intervention.

- higher production losses.

In the long term, lack of organized maintenance will negatively affect company management because of:

- negative influence on clients;

- costs exceeding profitable operations;

- control becoming impossible;

- services progressively degrading.

This is even truer for a firm organized according to Just in Time (JiT) principles where Work in Process (WiP) is reduced to a minimum or even stopped. In this system of management every machine stoppage, even for short times, causes stoppage of the production line and delay in the production programme; this negates the beneficial effects of JiT organization.

It should be considered that the JiT factory operation has a production line that processes parts in very small lots, where the processing time has been reduced from weeks, or days, to hours or even minutes. The total work time is now the sum of the work time of every single machine. Transportation and queue times do not exist, providing no catchup periods when failures occur. Whereas increased process stability has been gained, obviously any machine failure can rapidly lead to stoppage of the whole line, thus stopping production.

The solution is to invest in preventive maintenance and condition monitoring. This is regarded as an acceptably small price to pay.

35.3.4 Condition monitoring and spare parts management

Consider now the effect of condition monitoring on lock-up of warehouse materials. The main aim of JiT is to reduce WiP. The same concept should be extended to materials stocked, and to maintenance. When dealing with spare parts management, the maintenance manager is confronted with two options:

- maintain a spare parts stock sufficiently high to overcome any emergency;

- delay a maintenance job until the spare part reaches the factory.

The first solution is often applied. In both instances the uncertainty of the real machine condition negatively influences costs on account of:

- too much stock held in hand;

- delayed restoration of machinery with subsequent production losses;

- inability to arrange shut-downs to advantage due to lack of spares.

In one actual case, for example, provisioning problems accounted for 25% of all down-times.

From the spare parts cost point of view, condition monitoring and condition-based maintenance is more economically feasible than preventive maintenance because under the latter a component is often discarded only for the reason that the predetermined period has elapsed even though there is no evidence of a deterioration justifying removal of the component.

As an example, the maintenance manager of a large petrol refinery reported a spare parts stock reduction in the range of 16% in value after introducing a condition monitoring program applied to rotating machines at the refinery.

35.3.5 Condition Monitoring and Saving of Maintenance Costs

According to the experience of the Department of Industry, UK (DOI 1979) the savings which could be made by the application of condition monitoring arose from two sources:

- two thirds of the savings arose from reduced production losses;

- one third of the savings arose from reduced maintenance costs.

These proportions are only an average indication and any particular industrial company may find considerable variations from these mean values. The actual amount of savings varies between different companies. In general, however, annual savings are likely to result from:

(1) reduced production losses corresponding to 75% of days lost per year when condition monitoring is not used multiplied by the AVO. day^{-1} where AVO is added value output, calculated from the total sales revenue minus the cost of raw materials and energy bought in;

(2) reduced maintenance costs corresponding to about 5% of the total annual maintenance costs.

Only 75% reduction is expected for the production losses because incipient failures will still require the plant to be shut down when it might otherwise be in production. Furthermore, the monitoring process cannot be 100% efficient.

The 5% value for the maintenance cost saving is derived from the fact that most of the savings is in breakdown maintenance labour cost. Since breakdown costs are usually about one third of total maintenance costs and since about one third of these are labour costs, breakdown maintenance labour costs amount to 10% of the total. A saving of half this gives the 5% figure.

Reasoning in terms of added value, the economic advantage of condition monitoring has been estimated to be in the range:

● from 3% for gas, electricity and water costs

● to 0.7% for printing and publishing costs.

In general the following industrial sectors are more responsive to use of condition monitoring:

● capital intensive industries (with high assets per employee);

● installations where machines are arranged in series, without stand-by and intermediate storage of in-process goods and where they are working round the clock (with higher added value per installation). Examples here are mines, manufacturing, public services and utilities.

35.4 ORGANIZATION AND IMPLEMENTATION OF A CONDITION MONITORING PROGRAMME

35.4.1 Initial Trial Period

Implementation should be gradually applied after top management approval has been gained pointing out to staff the possibility of lowering existing maintenance costs and production losses and, at the same time, raising quality and safety (Baldin, 1986).

At this first stage the anomalous conditions to be detected should be listed (misalignments, vibrations, corrosion, etc.) and instruments selected to suit these parameters.

Experiments should be defined for each parameter to enable:

- the *baseline* to be drawn representing the machine operating properly;

- the *severity limits* (alarm and emergency) to be set for each parameter;

- *trend* of deterioration with time to be calculated;

- *influence of deterioration* on the whole plant performance to be evaluated;

- *lead time* elapsing before failure to be estimated;

- *measurement consistency* to be determined.

During the experimental phase maintenance crew, designers, machine and instrument manufacturers and plant operators should all be involved for verification at global goals to be achieved, namely reduction of:

- checking time spent by instrument operators;

- amount of overhauls;

- down-time;

- maintenance costs;

- maintenance intervention.

35.4.2 Programme Implementation

Information gained during the trial phase will allow preparation of a monitoring programme that will include the following factors:

(1) Listing and numbering of all machines, in order to record their identification and location.

(2) Selecting critical machines—those where a shut-down could cause production loss or danger to personnel. This distinction does not depend on quantity alone but also on product quality. The degree to which a machine is critical is one of the factors influencing the frequency of examination.

(3) Establishing a programme specifying the parts to be examined, and methods of examining them.

(4) Selecting a suitable monitoring instrument (Table 35.2) taking into consideration the following points. The equipment must be strong enough to stand misuse; must be reliable and have a facility for checking its calibration; must be accurate and consistent to enable it to be used for checking trends and must be intrinsically safe when used in hazardous areas. Moreover, it should be remembered that the reliability of condition-based maintenance depends not only on the instrument but largely on the skill and sense of responsibility of the examiner who is entrusted with a group of machines.

(5) Establishing, for each part of the machine, the allowable limits of the parameters (such as vibration, sound, temperature), to be measured. Sources of such information are obtained from:

Table 35.2   Monitoring instruments and applications

Parameter	Instrument	Application
Visual observation	Endoscope	Internal parts of machines possible to reach through small inspection parts
	Fibre optics	As above, but a more flexible instrument
	Stroboscope	Belts, coupling, gears and any related moving parts can be 'frozen'.
	Penetrating liquids	Surface flaws
	Thermographic paints	Surface and steam trap temperatures
	Fluorescent liquids	Surface flaws (mainly for light alloys)
	Scintillograph	Cracking of acid-proof linings
Temperature	Instantaneous thermometer	Ball bearing, steam traps, all cases when fast measurement is required
	Infra-red thermometer	Surface temperature, flame
	Thermography with Infra-red rays	Condition of insulating refractory linings, thyristor heat sinking, gas pipelines, heat losses during heat exchange cleaning
Sound	Stethoscope	Amplification of audible frequencies, noises of mechanic parts
	Noise meter	Measuring noise in rooms, emitted by machines
	Ultrasound	Identification of fluid loss under pressure, vacuum, compressor valves, piston rings, chain drive spreading of cracks and other structural faults

Table 35.2 (continued)

Parameter	Instrument	Application
Vibrations	Vibrometer	Measuring of vibrations
	Analyser	Frequency analysis for fault diagnosis
	Phase indicator	Accessory for balancing purposes
	Real-time analyser	Frequency analysis of transient and complicated cases
	Monitor	Measuring and monitoring alarm system for continuous control
	Acoustic emission	Condition of ball bearings
	Data collection and software system	Field collection of measurements, inspection programming, trend and waterfall
	Expert system	Where heuristics define variables
	Shaft orbital analysis	Misalignment, oil whirl, bent shaft, rubbing
	Signal averaging	Defects in gear trains and rolling elements
	Shock pluse	Defects in ball bearings and lubrication
	Kurtosis	Bearing damage
NDT (non-destructive) testing	X-rays	Thicknesses, corrosion, cracks, slags, inclusions, blowholes, errosion
	Ultrasound	Thicknesses, corrosion, cracks, slags, inclusions, blowholes, errosion
	Eddy currents	Surface and underneath faults, condition of exchanger tubes
	Dye and fluorescent penetrants	Detection of cracks
Wear debris	Magnetic plugs	Identification of magnetic parts and their shapes
	Automatic counter	Counting of wear debris as well as determining their sizes
	Emission and absorption spectrophometry	Total quantity of metallic particles according to the metal quantity present in a lubricating oil
	Ferrography	Total quantity of particles according to their size
	X-ray fluorescence	Identification of the presence of a given element
Lubricating oil	Lubrisensor (Tradename)	Rough measuring of condition of lubricant

(continued overleaf)

Table 35.2 (continued)

Parameter	Instrument	Application
	TBN analyser	Measuring of total basic number to define the influence of sulphur in fuel
	Setaflash (Tradename)	Control of presence of polluting elements (fuel, water) in the lubricant
Other instruments	Spinterometer	Dielectric constant of oils of transformers
	Gas chromatography	Titration of gases present in oil of transformers in order to know the condition of the insulating materials
	Mobil electronic compression tester (Tradename)	Diagnosis of condition of Otto and Diesel engines
	Leak detector	Gases and vacuum leakage
	Proximity detector	Relative motion, piston ring wear
	Flammable/toxic/explosives gas detector	Measurement and detection of such

(a) machine manufacturers;

(b) instrument manufacturers;

(c) national or international standards (such as ISO, BSI, DIN, API);

(d) experience of maintenance technicians with similar machines.

(6) Selecting suitable examination frequencies. In determining the best frequency of examination, it is necessary to consider, for each machine the:

(a) importance in the process flow chart;

(b) availability of stand-by machines;

(c) standardization of the item;

(d) operating conditions;

(e) failure statistics (MRBF, MTBM, MTTR);

(f) cost of examinations;

(g) overall costs of failures;

(h) costs of maintenance.

(7) Setting up an information system and recording the data gained. The examiner may encounter two different situations:

(a) the condition of the machine is approaching the lead-time. In this case normal procedures of the planning office will be carried out—(work requisition; work order with specification; programming; scheduling; carrying out of work; reporting work carried out);

(b) the condition of the machine is already within the lead-time (and thus near to shut-down). Here the information will need to be passed directly to the foreman for emergency maintenance.

In order to achieve correct operation of a maintenance programme based on monitored conditions, it is necessary that the following information is given to the maintenance foreman by the examiner:

(a) condition of each machine;

(b) parts of machine that are probably defective;

(c) probable defect;

(d) time before which the work has to be done to avoid failure.

Feedback is very important to assure success of the new philosophy.

(8) Training examiners. Examiners are preferably taken from the workshop, with a high standard of experience, a sound knowledge of the machinery and a fair degree of analytical skill. It is necessary to train the personnel for this new job by:

(a) making clear the aims of monitoring and condition-based maintenance;

(b) illustrating the principles and the operations of the instruments;

(c) revising or teaching the mathematical or physical principles essential for correct performance of the measurement task;

(d) running refresher courses on related technical subjects such as lubrication and ball-bearing assembling.

(9) Establishing a schedule of maintenance jobs to be carried out. Under monitored conditions based on a maintenance regime, the maintenance engineer makes decisions and coordinates the maintenance teams on the basis of hard data collected by the monitoring instruments.

(10) Updating the maintenance organization. To carry out monitoring it is necessary to introduce new functions to existing maintenance services. Table 35.3 shows tasks and monitoring actions. Two alternatives exist.

Table 35.3   Maintenance technician tasks and monitoring

Phase	Task	Maintenance engineering	Maintenance support office	Maintenance area manager	Checking operators	Workshop	Maintenance manager	Production technicians
Study	Condition-based maintenance procedures	X						
	Condition-based maintenance planning	X	X					
	Critical machinery selection						X	X
	Checking frequency selection	X	X	X			X	
Operation	Parameters measurement (e.g. vibration)				X			
	Measurements analysis			X	X			
	Decision making			X	X		X	X
Correction	Anomalous component correction (workshop)					X		
	Anomalous component correction (area)			X	X			
	Collection of substituted components (for further study)					X		
Information	Severity limit fixing			X	X			
	Trend diagram preparation (cards)				X			
	Maintenance job record					X		
	Entering job records on trend cards		X	X				
	Recording job costs on trend cards		X					
Development	Probability studies (MTBF, MTTR, expected life)	X		X			X	
Supervision	Information procedure supervision	X	X					
	General supervision						X	

First, the examination team is located as a centralized function in order to optimize the use of instruments and to optimize the employment of examiners by allowing them to be transferred from one area of the factory to another.

The alternative is to allocate to departments their own technicians to carry out the checking tasks. The advantages of this are personnel motivation, better information flow and higher personnel engagement.

(11)   Establishment of a budget for maintenance costs, inclusive of the cost of monitoring services. Costs include:

   (a)   crew and supervisor salaries;

   (b)   instrument's appropriation;

   (c)   training of personnel;

   (d)   information system updating;

   (e)   instrument maintenance;

   (f)   consulting services, if needed.

35.5   TECHNIQUES AND APPLICATIONS

35.5.1   Vibration monitoring and spectrum analysis

All machines vibrate and measurement of the vibration supplies information of the condition of rotating parts.

In the process of channelling energy into the job to be performed by the machine, forces are generated which will excite the individual parts of the machine.

Whilst the process is constant the nature of vibrations measured will be practically constant. As faults develop in the machine some of the dynamic processes in the machine change resulting in changing forces that influence the vibration spectrum. A basis for using vibration measurements and analysis in machine health monitoring is the principle that an increase of vibration is the sign of an incipient failure if the unit is continued under the same operating conditions.

Due to wear the tolerances change, with the consequence that rotating eccentricity and dynamic imbalance are generated. In its turn these cause vibrations that increase the wear, the result being a cumulative effect of the phenomena.

A machine vibrates when the mass of its casing is subject to periodic forces generated by components attached at the casing, reaction forces, and forces transmitted from rotating parts to the casing through the supports.

Forces can be centrifugal due to imbalance or arise as impulses, due to gears meshing or to fluid flowing through rotor blades.

Generally, vibrations are monitored on the bearing supports of the machine because force is mostly transmitted through bearings. Other points of monitoring are casings, baseplates and flanges.

Periodic vibration may be looked upon as an oscillating motion of a body, about a reference position, the motion repeating itself exactly after certain periods of time. A periodic vibration, called *harmonic motion*, when plotted as a function of time, is represented in its most basic form by a sinusoidal curve (Figure 35.5). The period of vibration T is the time elapsed between two successive, exactly equal, conditions of motion.

The frequency f (Hz) of the vibration is given by

$$f = 1/T \qquad (35.4)$$

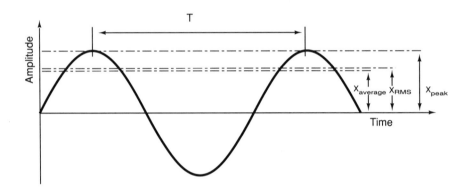

Figure 35.5   Example of a harmonic vibration signal with indication of the peak, r.m.s. and average absolute values

The magnitude of the vibration along the time axis is the instantaneous displacement of the body from the reference position. The respective values of this amplitude are given in Figure 35.5. The vibrating motion can also be described in terms of velocity and acceleration as these are derivatives of motion.

The major reason for the importance of the r.m.s. value as a descriptive quantity is its relationship to the *power* content of the vibrations.

Most vibrations encountered in industry are not pure harmonics because different generating causes are superimposed on each other to form complex wave shapes. Other methods of description must be used. Most are based on use of *frequency analysis*.

The Fourier theorem states that any periodic curve can be built up as a combination of a number of pure sinusoidal curves with related frequencies, amplitudes and phases (Figure 35.6). The frequency elements of the Fourier transformed signal constitute the vibration frequency spectrum.

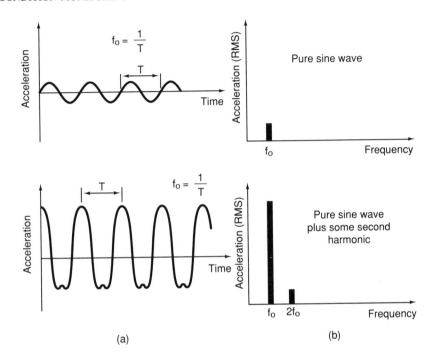

Figure 35.6 Examples of periodic signals and their frequency spectra: (a) time domain; (b) frequency domain

It is found that the vibration frequency spectrum has a characteristic shape. This is established when the machine is in good condition (Figure 35.7).

If the machine kinematic details are known (such as, shaft rotation speed, number of teeth of gears), monitoring the vibration caused by each component allows the source of the increasing vibrations to be traced. In this way a diagnosis of the condition is possible. Utilizing the fast Fourier transform (FFT) processing method an analysis can be carried out relating the actual vibration frequencies to the fundamental generated by the rotating speed of the machine.

Vibrations are generated by free forces, with different directions, so it is necessary to measure their magnitude according to which of the six degrees of freedom are able to be excited.

Most constituents of vibrations are generated by mechanical defects or by mechanical characteristics of the machine. Vibration can, however, be caused by hydraulic, aerodynamic or electromagnetic phenomena.

Sources of malfunction that can be monitored through vibration analysis are listed in Tables 35.4–35.7.

The basic measuring element is the *accelerometer*, a transducer which produces at its output terminals, a voltage proportional to the acceleration to which it is subjected. Piezo-electric accelerometers exhibit the most suitable charac-

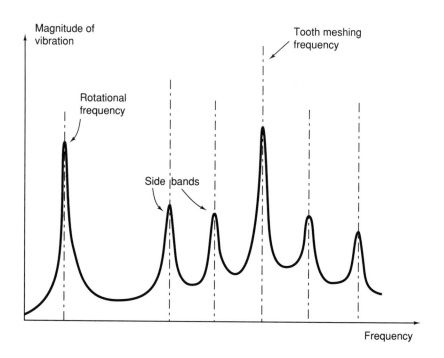

Figure 35.7   Simplified gear train spectrum and side bands

teristics for this kind of application and are preferred for measurements covering a wide frequency range.

In theory it is irrelevant which of the three parameters, acceleration, velocity or displacement are chosen to measure vibration for they are strictly related to each other. Plotting a narrow band frequency analysis of a vibration signal in terms of the three parameters shows they have different frequency response

Table 35.4   Machinery malfunctioning causes expressed in terms of frequency of vibration and magnitude

Cause	Vibration level	Frequency of vibration	Remarks
Unbalance	Proportional to unbalance. Largest in radial direction	$1.n$ where n is shifts normal rotational frequency.	Most common cause of vibrations
Bent shaft	Large in axial direction	$1n$	—

Table 35.4 (*continued*)

Cause	Vibration level	Frequency of vibration	Remarks
Misalignment (couplings or bearings)	Large in axial direction (50% or more of radial vibrations)	$1n$, $2n$ most common, $3n$ sometimes	Best found by appearance of large axial vibrations. Positive diagnosis can be obtained using dial indicators. If sleeve-bearing machine and no coupling misalignment, problem is probably unbalanced or bent rotor
Eccentric journals	Usually not large	$1n$	If on gears, largest vibrations in line with gear centres
Rubs	No particular characteristic if continuous	Mainly $1n$ plus $2n$	Can excite high natural frequencies of machine. Amplitude at same speed may vary between different runs
Mechanical looseness	Variable	$2n$	Usually accompanied by misalignment and/or unbalance. Not a loose bearing assembly
Loose bearing assembly	Variable	$0.5n$	Sometimes manifest at rotor critical speed. Quite a common problem
Oil whirl	Severe, radial motion	$0.43n$–$0.47n$	Only in high speed or vertical rotor machines with pressure lubricated sleeve bearings
Friction-induced whirl	Can be severe radial vibrations	Usually less than $0.4n$ and equal to first critical speed of rotor	Rare. Can be caused by loose rotor components
Rolling-element bearing distortion	Depends upon amount of distortion	$1n$	Large component in either horizontal or vertical planes. Taper roller bearings will also have axial components of vibration
Rolling element bearing damage	Unsteady	High frequencies	

(*continued overleaf*)

Table 35.4 (*continued*)

Cause	Vibration level	Frequency of vibration	Remarks
Bad gears or gear noise	Low	Very high T.n (T-number of teeth)	
Faults in belt drives	Erratic or pulsing	1, 2, 3 and 4 times rotational frequency of belt	Stroboscope can be used to diagnose belt defects
Electrical	Low. Disappears when power is turned off	$1n$ or 1 or 2 times synchronous mains frequency	
Aerodynamic or hydraulic	Variable	$1n$ or bn (where b is number of blades or lobes)	Rare as a cause of trouble, except in cases of resonances
Reciprocating forces	—	$1n$, $2n$ and higher orders	Inherent in reciprocating machinery
Faulty combustion in diesels	High	$0.5n$	Faults with injectors, fuel pumps, calibration or timing show unbalance. resiliently mounted unit rocks at 0.5n which is close to natural frequency of mounted unit, and causes large amplitudes
Foundation faults	Random and can be high	Low and erratic	Check foundation bolts

Table 35.5 Vibration frequencies associated with ball and roller bearings defects

Race unbalance or distortion	n	Not a common problem
Cage unbalance	$\dfrac{n}{2}\left(1 - \dfrac{r}{R}\cos\Theta\right)$	Not a common problem
Defect in inner race	$\dfrac{en}{2}\left(1 + \dfrac{r}{R}\cos\Theta\right)$	Equal to frequency of element impacts upon race defect

(*continued overleaf*)

Table 35.5 (*continued*)

Defect in outer race or stiffness variations around bearing-housing	$\dfrac{en}{2}\left(1 - \dfrac{r}{R}\cos\Theta\right)$	Equal to frequency of element impacts upon race defect
Defect in element	$\dfrac{Rn}{r}\left\{1 - \left(\dfrac{r\cos\Theta}{R}\right)^2\right\}$	Equal to frequency with which element defect comes into contact with the inner and outer races consecutives

n = shaft rotational frequency (Hz)
r = radius of element
R = pitch circle radius of element train
e = number of elements
Θ = contact angle (ball bearings only)
$\quad\Theta$ = 0 for roller bearings

Table 35.6   Vibration frequencies associated with electrical machine defects

Type of machine	Cause	Characteristic frequency
D.C. machines	Armature slots and commutator segments	Number of slots or segments × rotational frequency
Synchronous machines	Magnetic field	Twice supply frequency
Induction motors	Magnetic field. Rotor slots	Twice supply frequency. (i) Number of slots × rotational frequency. (ii) Number of slots × rotational frequency. ± twice supply frequency.
All	Unbalanced magnetic pull	Rotational frequency and twice rotation frequency

slopes (Figure 35.8). It can be seen that displacement measurements give low-frequency components most weight and acceleration measurements better weight than the high-frequency components. This leads to a practical consideration that can influence the choice of parameters.

Displacements with practically measurable amplitudes only occur at low frequencies so they are of limited value in the general study of mechanical vibrations. Displacement is used as an indicator of imbalance in rotating machine parts because it usually occurs at the shaft rotation frequency which rarely generates high frequencies.

Velocity measurements are used to monitor vibration severity. Vibratory velocity is related to vibratory energy and is, therefore, a measure of the destructive effect of the vibration of similar constructions vibrating in the same mode.

Table 35.7   Sources of wear materials in combustion/compressor machines detectable by
spectroscopy

Aluminium	Pistons, bearings, turbo charger casing
Copper	Bearings
Chromium	Piston rings, cylinder liners
Iron	Cylinder liners, crank shaft, auxiliary drive, piston rings and camshaft
Lead	Bearings
Molybdenum	Piston rings
Nickel	Valves
Silver	Bearings
Tin	Bearings
Silicon	Ingested dirt, core sand

Acceleration is preferred where the frequency range of interest involves high frequencies.

Moreover it is found that the higher frequencies contain information on faults that are developing well before they influence the actual condition of the machine. Lower frequencies are involved after they have occurred.

For practical reasons machine vibration is usually measured at accessible points. The best vibration measurement conditions are those taken when the machine is in operation at its rated capacity. Direction and magnitude of the vibration can vary if the machine is controlled at different points, these depending on its configuration (such as, shape, stiffness and anchor points). For this reason it is a good rule to mark (by a conic-shaped hole or a blind hole of

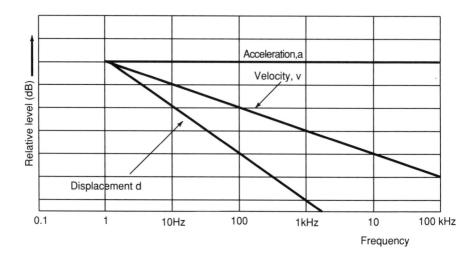

Figure 35.8   The relationship between vibration acceleration, velocity and displacement

diameter 5 mm, depth 5 mm) the points where the accelerometer should be positioned to assure measurement reproducibility. Sometime measurements can be made directly on the rotating shafts. In this case vibration magnitudes can be from 2 to 10 times higher than when measured on the body.

It is not enough to monitor the highest peaks of the spectrum because important machine health indicators may have low levels (Figure 35.9). Trend monitoring is better suited because periodic measurements can be extrapolated with reasonable accuracy to indicate vibration limits.

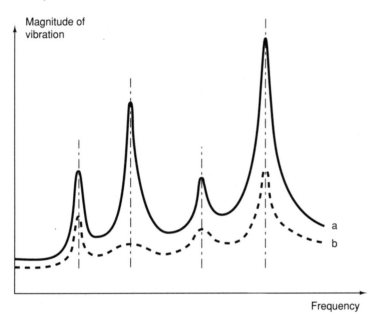

Figure 35.9   Measured spectra for a machine: (a) actual; (b) signature.

The vibration level measured on the bearing housing of a rotating machine is compared with vibration standards in order to check the acceptability of the vibrations and of the machine condition. Several vibration standards are in common use. The two shown here are:

● IRD Mechanalysis recommendations (Figure 35.10), and

● ISO standards 2372, 3945 (Figure 35.11).

Such standards can be used only as rough guides and cannot be expected to be accurate for every type of machine and its operating conditions of monitoring. The actual level is, therefore, determined by experience and is expressed as a ratio of the vibration level when the machine is first running satisfactorily and the measured value at times thereafter. For this reason it is recommended that

GENERAL MACHINERY
VIBRATION SEVERITY CHART

For use as a GUIDE in judging vibration as a warning of impending trouble.

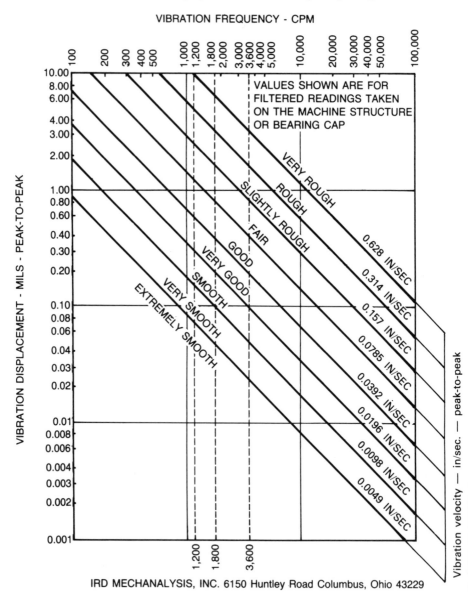

Figure 35.10   IRD Mechanalysis recommendation for vibration severity limits (Rathbone chart) (IRD Mechanalysis, Inc, 6150 Huntley Rd, Columbus, Ohio 43229)

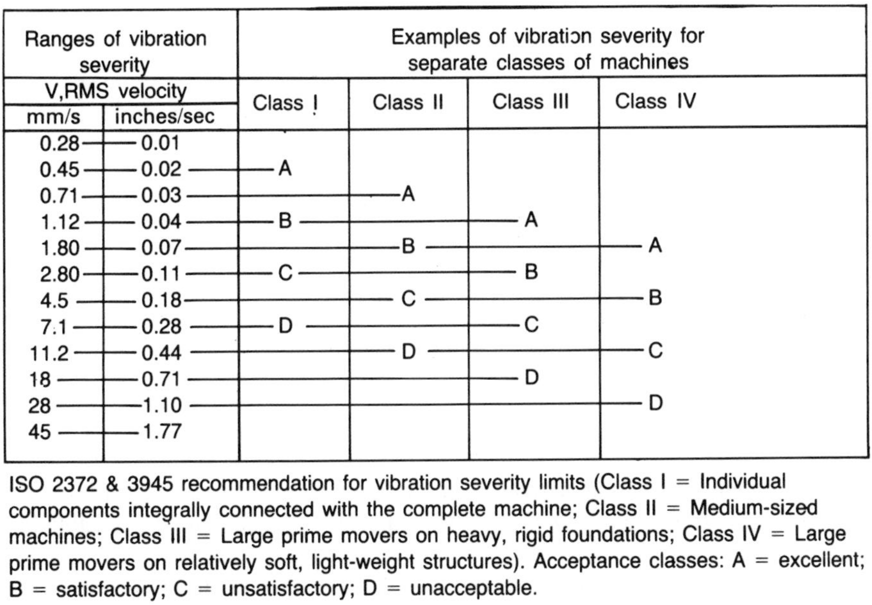

Ranges of vibration severity		Examples of vibration severity for separate classes of machines			
V,RMS velocity		Class I	Class II	Class III	Class IV
mm/s	inches/sec				
0.28	0.01				
0.45	0.02	A			
0.71	0.03		A		
1.12	0.04	B		A	
1.80	0.07		B		A
2.80	0.11	C		B	
4.5	0.18		C		B
7.1	0.28	D		C	
11.2	0.44		D		C
18	0.71			D	
28	1.10				D
45	1.77				

ISO 2372 & 3945 recommendation for vibration severity limits (Class I = Individual components integrally connected with the complete machine; Class II = Medium-sized machines; Class III = Large prime movers on heavy, rigid foundations; Class IV = Large prime movers on relatively soft, light-weight structures). Acceptance classes: A = excellent; B = satisfactory; C = unsatisfactory; D = unacceptable.

Figure 35.11   ISO 2372 and 3945 recommendation for vibration severity limits (Class I = individual components integrally connected with the complete machine; Class II = medium-sized machines; Class III = large prime movers on heavy, rigid foundations; Class IV = large prime movers on relatively soft, light-weight structures). Acceptance classes: A = excellent; B = satisfactory; C = unsatisfactory; D = unacceptable

when a machine is first put into service and running satisfactorily, a *signature* spectrum of the machine is taken under normal running conditions. This signature provides a basis for later comparison in order to locate those frequencies in which significant increases in vibration level have occurred.

35.5.2 Other Vibration Measurement Techniques

Signal averaging

This technique can detect defects in gear trains and can diagnose which particular gear is defective, and the nature of the defect.

The output signal from the accelerometer is sampled over a period equal to the time for one revolution of the gear being examined. This sample is stored. Another sample of equal period and timing is then taken. The *average* of this and the first sample is then stored.

When this procedure is repeated many times those frequency components of the signal related to the gear being examined are enhanced because much of the signal energy is random and this averages toward zero leaving the systematic signal. The result is a plot which exhibits the same number of complete cycles as the number of teeth on the gear. A tachometer is required to generate synchronizing pulses at a frequency equal to the rotation frequency of a reference shaft in the system.

Kurtosis method

Kurtosis is a statistical concept relating to the *peakiness* of the distribution of vibration amplitude. Condition of a bearing is related directly to the kurtosis value displayed on a meter, independent of bearing speed and geometry.

Cepstrum analysis

Signal *sidebands* can be generated in the frequency spectrum of a gearbox or a machine. The frequency *spacing* of the sidebands is equal to the modulating frequency.

When a spectrum displays a number of sidebands the spectrum becomes very difficult to interpret. The cepstrum technique provides a method of determining the sideband spacing. It is defined as the *power spectrum of the logarithm of the power spectrum.*

Spike energy (SE) circuits

In this proprietary method (of IRD Mechanalysis Inc.,) energy is generated when repetitive transient mechanical impacts occur as a result of surface flaws in rolling-element bearings or gear teeth.

The accelerometer signal generated by the vibration is processed by filtering and detection circuitry to produce a single *figure of merit* related to the intensity of the original impacts. This figure of merit is expressed in 'gSE' units. Spike energy gSE readings can be correlated with the severity of the surface flaws.

Proximity transducers

In case of slowly rotating, large diameter, shafts accelerometer-based vibration analysis is not always effective. Here an assembly of two proximity pick-ups working in the following way can be used.

A small coil facing the moving shaft is excited at a radio frequency. Power dissipated in the shaft's conducting material varies the transducer energy balance allowing a signal proportional to the distance between pick-ups and shaft to be generated. Such transducers are usually mounted in the bearing housing to monitor the relative motion of the journal to its sleeve bearing by reading the peak-to-peak displacement.

X - Y Display	Malfunction
	Unbalanced or bent shaft
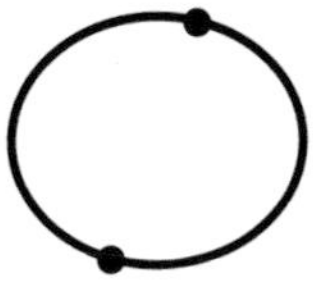	Oil whirl
	Misalignment
	Rubs

Phase marker ●

Figure 35.12 Oscilloscope-displayed malfunctions of the vibrations of a sleeve bearing obtained by proximity transducers

When proximity meters are fitted with signal output connections the vibration waveforms can be analysed for defect diagnosis and observed with an oscilloscope. The motion of the journal can be displayed in the *time base* and *X–Y* mode.

Faults monitored by this method include bent shaft, oil whirl, misalignments and rubbing (Figure 35.12).

Gear sidebands

Beside the existence of gear rotation and tooth meshing frequency, gear systems tend to generate *sidebands* which result as sums, differences and products of the fundamental frequencies. For example, if one gear of a pair is mounted eccentrically on its shaft the characteristic frequency of this defect can *modulate* the tooth meshing, causing families of sidebands. These appear in the frequency domain as spikes equally spaced on each side of the tooth meshing frequency by an amount equal to the gear rotation frequency and multiples—see Figure 35.7 given earlier.

35.5.3 Thermography—Thermovision

The operation of this instrument is based on the principle that every warm object emits invisible infrared radiation (IR). In this particular case the radiation is sensed by an IR television camera and converted into an image by a special television system. Thermography shows the surface temperature of objects as a television picture.

Cold and hot areas, relevant to the observed object, are seen respectively as dark and light spots according to the selected discrimination, which can be very small (0.2°C).

Remote measurement of temperature and the thermal state of an object, carried out with the help of this instrument, can have great importance for industry either for research purposes or for monitoring if correct plant settings are in place.

The performance of thermography has been improved by use of colour monitor screens which show the *isotherm* values of the surface under observation. The display colour varies according to the temperature range to be measured. For example, from a maximum range of $-30°$ to $+2000°C$ to a minimum range of $425°$ to $450°C$.

In maintenance, thermography can be applied to the following cases:

- refractory condition inspection;
- presence of ash deposit in hot fluid, distribution pipes;
- plant and building heat loss measurement;

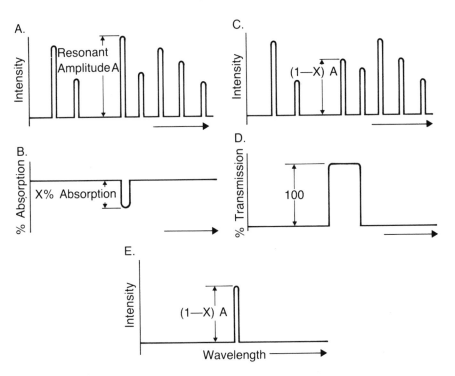

Figure 35.13 Atomic Absorption Spectrophotometer (AAS) instrument principle used for oil analysis: (a) hollow cathode lamp emits line spectrum of element to be determined; (b) sample absorbs energy at the resonance line; (c) resultant spectrum after absorption by sample; (d) Monochromator isolates resonance wavelength, and rejects all others; (e) Photo-detector sees only the resonance line, diminished by sample absorption

- electronic integrated circuit inspection;

- electrical component inspection;

- hydraulic component inspection;

- process variable examination;

- high tension (HT) distribution lines aerial inspection.

This technique allows monitoring of rotating and stationary objects without contact probes and without inducing a surface temperature change. The instrument supplies photographs of invisible thermal radiation emitted by an object. An isotherm may be superimposed on the thermogram so that any object whose temperature will correspond to those of isotherm will appear of brilliant white colour.

35.5.4   Spectrophotometric Oil Analysis (SOAP)

SOAP analysis must not be confused with the analysis of chemical and physical properties of lubricants in laboratories. SOAP is not monitoring the quality of lubricants but is, instead, concerned with the wear conditions of the machine using the lubricant as a carrier of material. It is based on the measurement of the amount of wear particles which go through oil filters and remain suspended in lubricating oils. Measurement is mainly carried out with an atomic absorption spectrophotometer (AAS). The AAS is able to determine wear particle percentages in the order of a fraction of one thousandth of a gram. The spectrophotometer works on the principle that atoms of every element absorb light of a given but different specific wavelength (Figure 35.13).

When looking for a wear element, the spectrophotometer is tuned to measure only the light with the characteristic wavelength of the examined element. The quantity of light absorbed in the flame is proportional to the quantity of such element in the sample of oil.

For example to carry out a Diesel engine oil analysis the spectrophotometer was arranged to measure five elements: aluminium, iron, copper, chromium and silicon. The five elements chosen detect condition of parts as follows (Figure 35.14).

- *iron*: shows wear of oil pumps, driving shafts, piston barrels;

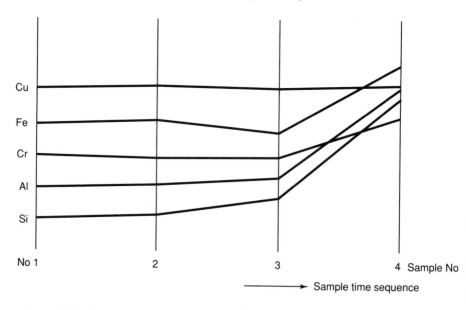

Figure 35.14   Example of spectrophotometer oil analysis of a Diesel engine. The increase of aluminium and of silicon indicates the presence of dust or earth from the air suction. The increase of chromium and of iron shows that the dust started wearing out cylinders and piston rings.

- *chromium*: shows wear of piston rings, bearings, valve stems;
- *copper*: indicates wear of thrust bearings, infiltration of water from the radiator and wear of the power shaft and/or wheel clutch plates;
- *aluminium*: shows wear of pistons or bushes;
- *silicon*: measures the amount of dirt which is found in the machinery.

Lubricant samples can also be examined to find out whether there are traces of water or, in case of the internal combustion engine, anti-freeze and petrol in the oil.

Sampling takes place at every oil replacement for the motor, transmission and gear-box hydraulic system. It is carried out at regular intervals of 500 working hours. The sampling process must ensure that the oil sample is not contaminated by other elements extraneous to the item under examination. SOAP provides knowledge only of the quantity of metal in the oil and no information on the size and shape of the metallic particles.

35.5.5 Ferrography

Ferrography, which separates particles by magnetic force, captures for examination those particles which are ferromagnetic, including the slightly ferromagnetic materials. Aluminium, bronze and some polymers which have worn against ferrous material can also be slightly ferromagnetic.

Comparison of particulates detectable by spectrographic and ferrographic techniques is summarized in Tables 35.7 and 35.8.

Table 35.8   Sources of metalic particles in the circulating oil of a Diesel engine

Cylinder liner	Cast iron, chromium
Piston rings	Cast iron, Chromium, molymbenum, copper
	Aluminium, silicon alloy, malleable cast
	Iron, tin or lead coated
Crankshaft	Low-carbon alloy steel.
Main bearings	Pb–Sn, Cu–Pb–Sn, Al–Si, Al–Sn, Cd, In
Big end bearings	Pb–Sn, Cu–Pb–Sn, Al–Si, Al–Sn, Cd, In
Thrust bearings	Pb–Sn, Cu–Pb–Sn, Al–Si, Al–Sn, Cd, In
Camshaft	Cast iron
Valve train	High alloy steel, nickel
Auxiliary drive	Phosphor bronze, low carbon alloy steel

A high proportion of particles generated from surfaces under stress would be expected to be larger than 5µm. These particles carry a history of the wear process but are ignored by spectroscopy owing to its failure to atomize this size of particle.

The ferrography technique separates wear debris and contaminant particles from lubricant and arranges them on a transparent substrate for examination. The precipitate particles, deposited according to size, may be individually examined and the characteristics of all sizes of particles can be established.

Significant numerical data can be derived from the direct reading section of the ferrograph, such as the fractional area coverage A_S of particles in the small 1–2 μm (S) size range and the area coverage A_L of large (L) particles (> 5 μm). If normal rubbing wear is the predominant wear mode and most of the particles are small, the L readings will be comparable with the S readings. If the operative wear modes are severe, the L reading will be large compared with the S reaching. Use of the formula

$$I_S = (A_L + A_S)(A_L - A_S) = (A_L^2 - A_S^2) \tag{35.5}$$

gives the figure I_S which is designated the *severity wear index*. I_S increases with increase in the severity of wear.

As an example, Table 35.9 gives summarized results from use of a DR Ferrograph readings of lubricant samples taken at various times during the life test of a large gearbox (Scott and Westcott, 1977).

Table 35.9   The severity of wear index is from DR Ferrograph reading of a gearbox from Scott and Westcott, 1977

Sample	Test duration life of gearbox (%)	Area coverage (%)		Severity of wear index I_S	Comments
		Large particles A_L	Small particles A_S		
1	7	7.1	2.3	45.1	Miscellaneous debris
3	13	0.8	0.2	0.6	small normal rubbing
5	27	1.2	0.2	1.4	wear platelets
7	40	1.2	0.5	1.2	
10	53	1.0	0.2	1.0	
12	60	0.9	0.2	0.8	
14	73	3.2	1.0	9.2	Cutting-type
15	77	3.7	1.3	12.0	wear particles and
16	80	5.0	1.6	22.4	spherical debris

Information on the morphology of the deposited particles is obtained with the aid of a bichromatic microscope which uses simultaneously reflected red light and transmitted green light. Metal particles reflect red light and block green light and thus appear red. Particles composed of compounds allow much of the green light to pass and appear green or, if they are relatively thick, yellow or pink.

Particles generated by different wear mechanisms have characteristics which can be identified with the specific wear mechanisms. Rubbing wear particles found in the lubricants of most machines are platelets and indicate normal permissible wear. Cutting or abrasive wear particles take the form of miniature spirals, loops and bent wires similar to swarf from a machining operation. A concentration of such particles is indicative of a severe abrasive wear process; a sudden increase in the concentration of such particles in successive lubricant samples signals imminent machine failure. Particles consisting of compounds can result from an oxidizing or corrosive environment. Steel spherical particles are a characteristic feature associated with fatigue crack propagation in rolling contacts. The concentration level of spherical particles indicates the extent of crack propagation.

Exposure of certain metals to heat produces temper colours which can be discriminated (Barwell et al., 1977). Clearly this method can be selected for further investigation of separated particles (Table 35.10).

Table 35.10   Heating test of metal plaques

Temperature	Surface colour changes				
	AISI 1090	AISI 52100	Cast iron	Nickel	304 stainless steel
204 °C (400 °F)	Blue	Part blue	Bronze	No change	No change
232 °C (450 °F)	Blue	Blue	Bronze	No change	No change
260 °C (500 °F)	Blue	Blue	Blue	No change	No change
287 °C (550 °F)	Blue-grey	Blue-grey	Blue	No change	No change
315 °C (600 °F)	Grey	Grey	Grey	No change	No change
398 °C (750 °F)	Grey	Grey	Grey	Bronze	No change
420 °C (800 °F)	Grey	Grey	Grey	Blue	Bronze
471 °C (880 °F)	Grey	Grey	Grey	Blue	Blue (mottled)
510 °C (950 °F)	Grey	Grey	Grey	Blue	Blue (Mottled)

35.5.6   Ultrasonic Testing

The principle of non-destructive ultrasonic testing is based on the propagation of high-frequency mechanical vibrations and their reflection when meeting a discontinuity in the material. Frequencies used are between 0.5 and 15 MHz. For more detail refer to Chapter 38.

The advantages of non-destructive tests (NDT) made using ultrasound are as follows:

● very high sensitivity;

● possibility of checking metallic thickness in the order of metres;

● immediate knowledge of test results;

- absence of danger for the operators;

- easy location of fault position in the item;

- possibility of testing in hardly accessible positions.

On the other hand they present the following restrictions:

- good propagation in many materials, but not in all;

- sometimes hard to locate superficial faults;

- shape of the item limits the use of ultrasonic tests;

- require high specialization from the operators;

- do not leave proof of test results except for photograph of the image (oscillogram) which appears on the oscilloscope screen.

The phenomena used to generate ultrasonic energy for use in NDT are *piezoelectricity* and *ferroelectricity*.

As ultrasonic radiation is attenuated by gases, it is usually necessary to interpose a *coupling fluid*, such as tap or emulsified water, mineral oils or silicones, between the probe and the part to be checked.

According to the surface conditions of the application, it may be used in one of the following ways:

In the *reflection method (or pulse–echo method)*, the ultrasonic beam is reflected when meeting a discontinuity in the material. It displays the fault position in terms of distance and can also give information about fault dimensions (Figure 35.15).

Evaluation of sub-surface faults is difficult as these lie close to the reflection from the first surface. Here probes are employed, in which two piezoelectric crystals, a transmitting one and a receiving one, are set at a slight inclination.

- The *transparency method* or *through-transmission method* uses a probe which transmits the ultrasonic beam and another one which receives it. The presence of a fault is displayed by decrease of the received energy. Here the probe to item coupling becomes important because an imperfect coupling reduces the transmitted energy as if there were a fault. Another restriction is the impossibility of locating the fault position or depth. It does, however, offer the advantage to be applied to very thin bodies. A way to overcome an imperfect coupling is to use the *immersion test* wherein the ultrasonic beam is transmitted to the part after first going through a fluid.

Choice of the most suitable method is assisted by the following pointers:

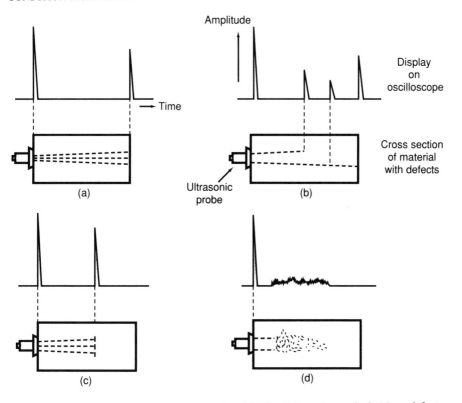

Figure 35.15   Ultrasonic non-destructive testing (NDT)—Pulse-echo method: (a) no defects; (b) defects; (c) large defects with no back echo response; (d) internal structure discontinuities

- with non-corroded parallel surfaces most methods are suitable;

- in case of corroded or non-parallel surfaces the reflection method is recommended;

- with pitting corrosion present, double probes are recommended;

- for testing refractory or plastic materials, wood or other material with a high acoustic absorption coefficient, for testing coupled items (such as shrink fitted parts and welding of two different materials) and for testing structural variations in materials, the transparency method is recommended;

- for checking pipes use angled probes and the reflection method;

- for checking welding, use angled probes and the reflection method with a frequency between 2 and 5 MHz.

The smaller the probe the better will be the matching capability to the discontinuities of the surface resulting in improved fault position display. However, probes with a greater diameter allow quicker inspection as they cover more area. The probe angle must be chosen according to the thickness of the part—guides are given in manufacturers' user handbooks. To display cracks or inclusions the probe must be slightly rotated. Evaluation of the fault dimensions generally must be made by referring to the result obtained by testing sample pieces in which have been found faults of known dimensions. More information is available in the International Institute of Welding specifications on these tests.

35.5.7   X-ray Testing

In radiology the quantity of radiation or *dose*, which is linked to the voltage (kilovolts) applied to the instrument, must be considered.

- Low kilovolt energy levels produce soft X-rays which yield high contrast and noticeable tonalities of black and white but show little evidence of object discontinuities.

- High-energy radiation gives low contrast, but discontinuities in the object are displayed.

Defects displayed by radiography are either cavities containing gas, air or materials with different densities (cracks, blow-holes, lack of welding penetration, slag inclusions) whose radiation absorption and, therefore, their contrast, will be different. The higher the absorption the softer the radiation needed, that is, the lower is the voltage applied to the X-ray tube.

Use of the X-ray technique is governed by many factors:

- distance inverse square law;

- radiation intensity or output;

- radiation quantity or dose;

- formation of the radiological image—enlargement;

- geometrical lack of sharpness;

- X-ray exponential absorption law;

- film lack of sharpness;

- screen lack of sharpness and intensification factors;

- absorption contrast (radiological contrast);

- contrast and definition (sharpness);

- density (or blackening) of a film area;

- sensitivity or speed;
- photographic contrast;
- inherent fog.

Ionizing radiations are harmful to the human body. It is necessary to set 'doses' (calculated as radiation intensity for time of exposure) and methods of measurement according to declared safety standards.

Measurements of intensity are carried out by Geiger counters. Operators must take precautions (stay as far as possible from the radiation source) and adopt passive defences (gloves, lead aprons—which are bulky) or, where possible, use an appropriate room with walls built for this purpose.

35.5.8 Dye Penetrant Inspection

Special dyes penetrate, by capillary action, into very small cracks on the surfaces of items. By means of special powders added to the liquid cracking can be observed. Successful tests require the surface to be clean and degreased. Red coloured liquids and fluorescent dyes are used.

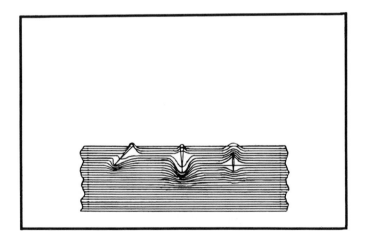

Figure 35.16   Magnetic particles testing showing magnetic field distortion by cracks in a magnetized steel piece

35.5.9 Magnetic Particles

This method can only be applied to magnetic materials, such as iron and steel, and is based on the feasibility of magnetizing the examined part. At cracks and surface, or sub-surface, faults there will be a deflection of the free magnetic lines of force, (Figure 35.16).

Particles are attracted to lie on the surface according to the distribution of the magnetic leakage flux, thereby displaying the discontinuances. Due to the skin effect, use of alternate current generated fields restricts the measurements to 2–3 mm depths. Using direct current generated fields there is no skin effect and up to about 10 mm thick pieces can be inspected. After the test the part retains residual magnetization, making it necessary to eliminate it either by heating the part to temperatures higher than its Curie point or by submitting it to an alternating magnetic field that is progressively reduced to zero.

Powders of iron Fe304 (which is magnetic) are used for the fault detection mixed with a pigment usually of red colour, magnetic liquids that are visible in daylight, or fluorescent liquids.

35.5.10   Endoscope

The *endoscope* is an instrument made by a bundle of coherent image optical fibres. These use a set of transparent fibres to transmit light and images along guided ways. By this means, an image can be transmitted along complex paths. A characteristic of a coherent optical fibre bundle is that within the resolving power of the bundle there are no aberrations of the image, which is consequently transmitted with high precision. Fibre dimensions are not critical provided the diameters are large compared to the wavelength of the radiation used. Optical fibre instruments, allowing inspection inside inaccessible equipment, are of great interest to the maintenance crew. An upper temperature limit of 80°C exists for these instruments.

35.5.11   Stroboscope

This is an instrument which allows visual inspection of a periodic phenomenon at a required time of its cycle. Using a rapidly pulsed light source, synchronized to the motion, the motion is apparently frozen. The instrument is called a *stroboscope*.

Electronic stroboscopes can give up to 400 flashes per second of 0.3 s duration. When the pulsed frequency corresponds to that of the event, or to multiples or divisors of it, the object will appear stationary. Slight pulse frequency changes allow gradual movement through the full cycle of the event.

Stroboscope applications for monitoring are very numerous—vibration phenomena, gears, springs, ventilators, belts, turbines and rotating shafts being examples. Stroboscopic observation can also be made using cine film, synchronizing the frame frequency with the stroboscope frequency.

35.6 DEVELOPMENTS IN MONITORING TECHNIQUES AND APPLICATIONS

35.6.1 Introduction

Interesting new monitoring applications include the extension of traditional technical methods to industrial sectors where until now there was little application, application of traditional technical methods to specific tests and new applications, new methods and technologies.

35.6.2 Extension to New Sectors

Hydroelectric and other electric machinery

During operation, hydroelectric machinery is subject to various dynamic phenomena combined with disturbances of a mechanical and hydraulic nature. Traditional practice monitors dynamic variation of the main operating parameters, pressure, motor torque and speed. Pressure is measured at the significant places of hydraulic passages to observe pulsations with which vibration phenomena are associated. Vibrations tend to be localized in certain areas or are induced into the machine, particularly along the axle line. Induced vibrations can be used to determine material stress, machinery degradation and eventual secondary effects (such as erratic control systems behaviour and effects on building). It is very important to realize that all vibration phenomena are related to natural frequencies of the system parts as well as to the excitation source. The high-quality factor (Q) of such systems can severely distort signals if not left long enough to develop peak amplitudes.

Recently continuous monitoring of hydroelectric machinery has been implemented because they are increasingly operated from remote control rooms. Important factors being monitored are, imbalances on the axle line due to the magnetic field or mechanical loosening, clearance of supports going out of tolerance and hydraulic imbalances.

Developments in this sector currently call for an expert system to correlate vibration values to operational conditions.

Machine tools

The trend of transforming mechanical industry machinery from single-station machines into numerically controlled machines, working centres and flexible production lines imposes a need for the critical analysis of maintenance strategies in the same way as has been done by the process industry. Machine tool monitoring applications are:

Driving heads:

● vibration analysis

● torque measurement.

Speed reducers:

● vibration analysis

● response curve.

Bearings:

● vibration analysis

● spike energy.

Hydraulic valves:

● response curve.

Hydraulic motors and pumps:

● vibration analysis

● oil analysis

● debris count.

Pumps and fans:

● vibration analysis.

Another monitoring strategy is concerned with tools wear, breakages and collisions responsible for rejected machined parts, scrap, rework and low productivity. A system developed by IRD Mechanalysis Inc. to monitor tool conditions works on the following principles (IRD, 1987).

All machine actions have distinct and repeatable patterns. The monitoring systems detect, measure and record these patterns. Changes in these patterns provide early warnings of tool wear or breakage. Once the optimum tool actions are known, setpoint levels, called the *tool numbers*, are programmed into the system. As tool wear increases the tool number changes. Should the number exceed the setpoint, the system will warn the operator or shutdown the system.

Buildings

In modern building complexes such as hospitals, hotels, schools, offices, stores and residences, it is necessary to ensure a series of services are provided—heating, air conditioning, lighting, water supply, firefighting, effluent disposal and more. Monitoring methods, of industrial derivation, have been applied to these plants using:

● thermography for electrical circuit control;

● endoscope inspections of inaccessible parts (for example air-conditioning plant);

● vibration analysis of pumps and fans;

● refrigerator temperature monitoring and fan operation. The trend to installation of data networks in buildings makes monitoring easier as the necessary information network then exists at most locations.

35.6.3   Specific Monitoring by Traditional Technical Methods

Inspection of ferromagnetic heat exchanger bundles

By means of special probes, working at relatively low frequencies, waves are emitted that move along the walls of tubes. With these it is possible to inspect the whole length of the tube by placing the probe at the edge of the tube on the tube plate. This technique enables detection of sudden variations of the tube's cross section, caused by corrosion and scouring.

Automatic monitoring of joints in tube plates

By means of double probes with tandem crystals at 70°C, used either in the transmission/reception or transparency mode, it is possible to test joints in considerably less time by a completely automatic monitoring system set up to assure the repeatability of the examination and to graphically present the examination results.

35.6.4   New Applications

Monitoring of fluid losses by acoustic or ultrasound methods

It is possible to detect acoustic energy travelling through the metal structure by means of a non-invasive sensor placed on the outside part of components. This can be used to automatically monitor pressurized piping, high- and low-pressure heaters, overheating steam piping, valve blow-by and high-pressure pumps blow-by.

Radiographic technicalities

Innovations regarding radiographic methods include:

● films with emulsion and grading controlled enabling improved definition and contrast;

● radiography using over-exposure techniques to obtain radiographs of pieces with variable thickness or of heterogeneous manufactured goods made of materials with different absorption coefficients;

● digital conversion and post-processing of radiograph images to assist interpretation, exposure time reduction and obtain higher reliability;

● computerized multidimensional *tomography* where the image is built from signals from several sensors.

(a)

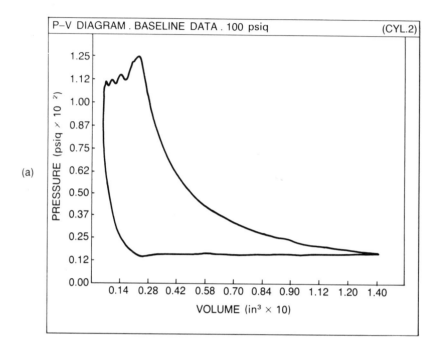

(b)

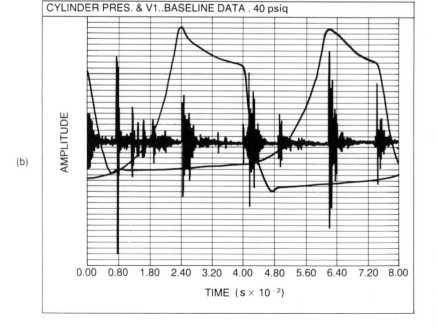

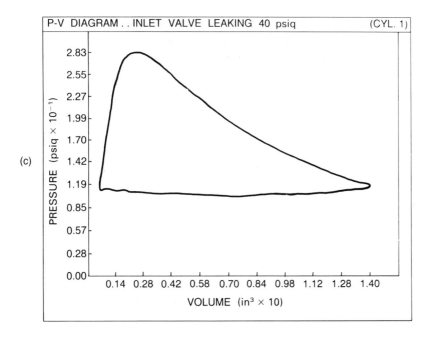

P-V DIAGRAM . . INLET VALVE LEAKING 40 psiq (CYL. 1)

(c)

PRESSURE (psiq × 10⁻¹)

2.83
2.55
2.27
1.99
1.70
1.42
1.19
0.85
0.57
0.28
0.00

0.14 0.28 0.42 0.58 0.70 0.84 0.98 1.12 1.28 1.40

VOLUME (in³ × 10)

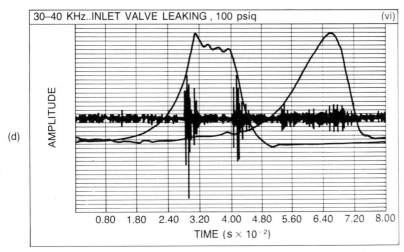

30–40 KHz..INLET VALVE LEAKING , 100 psiq (vi)

(d)

AMPLITUDE

0.80 1.80 2.40 3.20 4.00 4.80 5.60 6.40 7.20 8.00

TIME (s × 10⁻²)

Figure 35.17 Monitoring compressor valve leakage: (a) *P–V* diagram baseline; (b) vibration amplitude–time baseline; (c) *P–V* diagram, inlet valve leaking; (d) vibration diagram, inlet valve leaking. (Source Johnston and Stronach, 1986)

Monitoring of reciprocating compressor valve conditions

During the operation of reciprocating compressors a noisy band of signal is created for each cycle that is continuously repeated. Sources are mechanical rubbing, gas flow and structural resonance. Transients (due to such events as valve blow-by and stuck valves) can be detected as these can be seen as 'rare' events compared with normal operations, see Figure 35.17 (Johnston and Stronach, 1986).

Motorized valve actuator monitoring

Regular inspection of actuators cannot always determine its real condition. By obtaining the *profile* of the actuator operating conditions in terms of motor torque to engine speed it is possible to detect, see Figure 35.18 (Derry, 1986):

- the linear part of movement;
- the starting delay;
- the motion roughness;
- the over-run;
- the greatest deviation in respect to movement linearity.

It is therefore possible to diagnose the condition at any time by comparing actual profiles to that of the actuator when new.

35.7 INFORMATION PROCESSING

35.7.1 Introduction

Traditional information systems cope mainly with problems related to the efficiency of maintenance services. Originally information-processing systems related to management aspects with the objective of supplying managers with information that enables them to take actions to reduce or control maintenance costs. Later systems also aimed to systematically gather technical data relative to maintenance interventions, processing them mathematically with statistic techniques.

A third aspect of maintenance information-processing systems, however, concerns problems such as has how accurately are faults diagnosed, how to calculate their malfunctional trends, and how these functions are integrated into the global information system. The system related to monitoring will be set up to correlate the gathered data about plant, decide the technological families to which they belong, suggest the faults to which they can be subjected and enable cross-relationships between the above classifications to be determined.

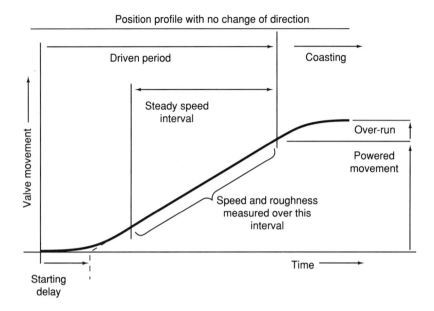

Figure 35.18   Valve actuator monitoring response curve (Source Derry, 1986)

To this end the maintenance manager must define, in an unequivocal and correct way, the description of plant layout, their technological classification, fault classification and cross-relationships between any two classifications.

The task of defining and rationalizing the data base is a major task which can be gradually undertaken by starting from initial synthetic classifications to which are gradually added more analytical detail.

35.7.2   Functions Supporting Monitoring and Technical Diagnostics

Monitoring and diagnostic problems concern maintenance information system modules dealing with historical analysis and standards and work request/orders.

The first step in setting up a system is to create a data-base, the second step is to improve the gathering of monitoring and diagnostic data gathering.

As shown in Figure 35.19 the two above modules are correlated both to the plant layout and its performance and to the Production Information System (PIS).

Collection and processing of data must be timely, accurate and comprehensive. This calls for greater distribution of measuring instruments with interactive terminals situated in the production departments near the plant. These must provide interfaces to analog or digital instrumentation existing in the Production Departments or within Maintenance and Quality Control Services.

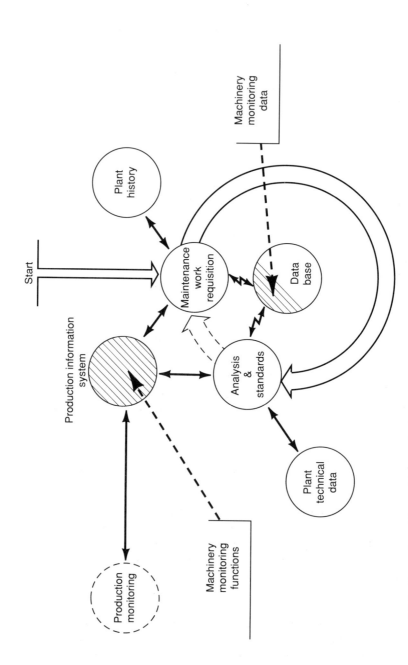

Production and machinery status monitoring are integral parts of maintenance work order processing.

Figure 35.19   Maintenance information system and monitoring process

It must be easy to set up reliable data inputs within production departments and this system must resist the stresses and harsh environments that arise in manufacturing plants. Three methodologies are now explained.

Monitoring through an advanced production control information system

This system is based on the use of plant terminals and is able to detect data through keyboards that are simple to use, through connection with analog or digital sensors to various instruments and through reading of precoded and pre-printed records set according to the goals of planned production. In this way it is possible to track a product through its various phases of work or testing, to track a piece of equipment or special tool in its movements, to signal plant operational requirements and to report signals in order to know, at any time, the state of each plant or machine being referred to by a plant, product, tool, employee, crew or maintenance intervention code. It can also indicate the machine stoppage cause. Such a system can also signal, for each plant or machine, maintenance intervention requests referred to precoded faults or malfunction data. It can also be used to measure weight, count pieces and measure physical dimensions such as thickness and temperature.

With this level of information capability it is possible to compare quantity and quality trends of production with planning schedules and budgets in time for management to take action.

Data about production performance trends and their comparison with plans, which is of basic importance for control of production phenomena, is also useful for maintenance services purposes. Implementation of this level of information system allows a significant jump toward more efficient management of the whole maintenance phenomenon.

Monitoring through automatic control of machines and processes

Through production control and plant condition information systems it is possible to monitor continuously, and in real time, the condition of plant, machines and components along with the trend of physical quantities related to wear, malfunction and faults.

Data inputs to the maintenance system allow generation of:

● alarms of serious conditions in a timely and already diagnosed form;

● historical trend data of physical quantities for different plants;

● evaluation reports about trends of physical quantities, comparing them with prescribed allowance limits;

● intervention plans, in a preventive and already diagnosed form, to reduce future faults or malfunctions.

Table 35.11 Examples of vibration reports obtained by an acquisition and spectrum analysis system: (a) Alarm report for selected route; (b) Machine report; (c) Spectral points on a machine.

IRD Mechanalysis Data Analysis System
Dec-03-88

Alarm Report for Selected Route
Route: DEMO1   Location: 2nd Floor compressor area

Machine	Pos	Dir	Date	Time	Ampl	Units	% change	% of Alarm	Alarm
01 BLOWER	2	H	Dec–17–87	13:24					
Band 1					.324	IN/S		216%	.150
Band 2					.143	IN/S		143%	.100
Band 4					.054	IN/S		108%	.050
Band 5					.099	IN/S		283%	.035
Band 6					.044	IN/S		200%	.022
01 BLOWER	3	H	Dec–17–87	16:54					
Band 1					.199	IN/S		133%	.150
Band 3					.087	IN/S		145%	.060
Band 6					.123	IN/S		559%	.022
01 BLOWER	4	H	Dec–17–87	07:34					
Band					.422	IN/S		141%	.300

Table 35.11 (b) *(continued)*

IRD Mechanalysis Date Analysis System

Dec-03-88

Machine Report (Latest Data) for: PUMP 101

Date	Time	Pos	Dir	Ampl	Units	Change	Alarm	Inspection
Jul–1–10–86	07:56	1	H	.110	IN/S	5%	.300	Machine Normal
Jul–1–10–86	06:35	1	H	.135	g/SE	4%	.500	Machine Normal
Jul–1–10–86	11:23	2	H	.160	IN/S	–3%	.300	Abnormal Normal
Jul–1–10–86	07:46	2	H	01.45	g/SE	–3%	.500	Machine Normal
Jul–1–10–86	10:04	3	H	.120	IN/S	–9%	.300	Machine Normal
Jul–1–10–86	08:46	3	H	01.66	g/SE	6%	.500	Loose or Broken Parts
Jul–1–10–86	14:54	4	H	.122	IN/S	–21%	.300	Machine Normal
Jul–1–10–86	07:25	4	H	.322	g/SE	–10%	.500	Machine Normal

8 lines this report

(continued overleaf)

Table 35.11(c) (continued)

IRD Mechanalysis Data Analysis System

Dec-03-88

Spectral Points on a Machine for: 3A VERT PUMP 101

Pos	Dir	Units	1×RPM	FMax	% of Lines	% of Avge	Spectral Capacity	Lower Limit	Upper Limit	Band Alarm
2	H	IN/S	1755	90K	400	2	6			
							Band 1	900	2700	.300
							Band 2	2700	5400	.200
							Band 3	5400	10800	.200
							Band 4	10800	21600	.100
							Band 5	21600	54000	.100
3	H	IN/S	1755	90K	400	2	6			
							Band 1	900	2700	.300
							Band 2	2700	5400	.200
							Band 3	5400	10800	.200
							Band 4	10800	21600	.100
							Band 5	21600	54000	.100

Implementation of this level of function requires a deep technical knowledge of plant.

Monitoring through inspection procedures

Wherever the above approaches are not economically feasible, at the least, *inspection procedures* should be adopted that enable periodic measurement of particular conditions and physical quantities. These must be carried out by suitably trained staff.

Utilizing the maintenance system, a calendar of inspections is worked out. Inspections are planned, documented, released and monitored during execution, treating them as forms of maintenance intervention activities. At fixed frequencies, for example at the end of each shift, data are input to the system. This enables management to obtain, although in less timely form and for greater time intervals, similar results to those mentioned in the previous functions. Obviously the costs of losses may be higher due to less data being available. Clearly, the more automated the measurements and means to accept the data the more reliable the information is likely to be.

35.8 PERSONAL COMPUTERS AND EXPERT SYSTEMS

35.8.1 Data Acquisition and Spectrum Analysis

Efficient monitoring programmes tend to need expansion because they lead to longer machinery life, increase in production, reduced maintenance costs and less down-time. A substantial increase arises in time spent in recording data and in managing the programme. As soon as this situation develops, it is time to use a dedicated computer system having a data-acquisition interface (manual or automatic) and software. A personal computer often suffices.

The operator follows a set programme inspection routine, making the prescribed measurements and entering them into the computer. Beside recording measurements of standard parameters for analysis purposes (such as vibration, velocity, acceleration, displacement and spike energy), results of visual inspections (hot bearings, unusual noises) the operating parameters (such as pressures and temperatures) can also be recorded.

Software can process the data to display the FFT analysis spectrum in frequency or harmonic terms. It is also possible to set alarm levels, thus detecting specific problems such as disalignment, cavitation and ball or roller bearing problems.

Once data have been gathered, they are transferred and filed on a larger host computer. Software keeps the data-base files, supplies trend paths to date and prints results concerning machine conditions (Table 35.11).

It is important that data is processed in terms useful to the maintenance team and is not left as raw data. The instantaneous data of machine conditions can be compared to the initial parameters to detect significant spectrum variations. *Waterfall* graphics are available for examination of past spectrum tendencies as well as giving several print-outs which enable the operator to determine the necessary actions, causes of actions and how to program a shut-down or maintenance intervention. The programs can also suggest the probable machine fault related to measurements taken.

A comprehensive menu of a software system dedicated to vibration monitoring would include the following functions (taken from an existing system by IRD Mechanalysis Inc). Configuration options are also available.

(1) *Vibration alarm report.* This report lists all measurements points that exceeded the alarm set-point at the last check date for the route specified.

(2) *Inspection report.* This lists all measurement points where an abnormal inspection code was entered at the last check date for the route specified. Abnormal inspection codes are user-defined and are selected in the *modify inspection codes* option in the *configuration* selection.

(3) *Machine report.* This lists the current measurement data on a selected machine. Amplitudes exceeding alarm levels are given and are highlighted in the display.

(4) *Vibration trend display.* Entry of measurement point details (machines, position, direction, unit) will call up amplitude versus time trends. A machine can be quickly selected by its name yielding a list of the measured points. In one example a maximum of 26 data points can be retained: when the 27th is entered the oldest point will be eliminated. Parameters other than vibration can also be displayed as trend data. After identifying the desired measurement point a graphic display is presented of three measurements. The amplitude scale, measurement, dates and other information are automatically scaled for the best presentation.

(5) *Multiple point trend display.* Up to four measurement points can be displayed on the same graph. The graph for each will be normalized to their individual alarm set points.

(6) *Action report.* This report is selected on a route basis and lists all points that exceed the alarm level or had an abnormal inspection code entered the last time the data was taken.

(7) *Estimated time to alarm.* This uses trend data to estimate, by extrapolation, when a measurement point will reach the alarm level.

(8) *Latest measurement route.* This report generates a machine report for each machine on the selected route. Current measurements on all machines on a route can then be quickly reviewed.

(9) *File modification.* A *route file* establishes route numbers, locations, check intervals, machines on the route and the sequence numbers. A *machine measurement point file* establishes all measurement point details such as machine number, position, direction, units and alarm level.

(10) *Route file modification.* A file enables the user to establish a new route and location to modify an existing route's machine, to remove a route, to make global route changes, to renumber or resequence machines on a route and to modify a route location or cycle.

(11) *Machine file modification.* With this it is possible to add a new machine or a new measurement point, modify existing machine data and remove a machine measurement point.

(12) *Spectral reports.* This option provides a baseline spectral comparison. This selection generates graphic displays showing a comparison between two spectra. It also gives a selected spectral comparison, waterfall display, which is the history of spectra stored over the past. The multiple spectra display function allows any number of current spectral records to be displayed on the graph. Other menu options are to modify the spectral parameters, delete a spectrum, display a single spectrum display and set the baseline definitions.

(13) *Scheduling* generates another menu for selecting routes due on a specific week number, on a specific date, this and next week and overdue.

35.8.2 Computer Based Guide to Assist Balancing

Balancing a machine using traditional instruments requires a series of time-consuming operations.

- filter retuning at each launch of a balance test;

- phase reading through stroboscopic flash light or other method;

- vector diagram plotting or data input in a programmed calculator;

- weight and angle calculation;

- repetition of launch owing to problems due to cross linkages in balancing;

● repetition of several launches to solve such problems as static coupling in the case of overhang or where multistage rotors and rollers are involved.

On-site balancing of rotating parts is made easier and shorter by use of computer programs which guide operators, by means of messages, through each stage of balancing procedure.

35.8.3 Expert System for Heavy Machinery Monitoring

Vibrations, temperatures, shaft positions, oil and phase analysis are monitored during the deceleration phase of machinery (Scheibel *et al.*, 1986). From this experience and a set of measurements a data base, containing a hundred rules and diagnostic strategies, is derived. These rules, in the form 'if/then' or 'condition/action', are used to verify if a condition is true or false (Table 35.12).

Table 35.12   Expert system misalignment analysis

Fa-cts	Misalignment analyses: symbolic facts			
		True range	True	False
A	Are there any abnormal d.c., positions?	> 9 mils	T	
B	Are there any abnormal bearing metal temperatures?	> 15 F		F
C	Is the 1/rev phase steady?	< 10 deg/hr	T	
D	Is the 2/rev phase changing?	> 20 deg/hr	T	
E	Is there A significant difference between adjacent bearings' metal temps?	> 30 F		F
F	Is there a significant difference between adjacent bearings' orbits?	manual	T	
G	Is there a significant difference between adjacent bearings' d.c. positions?	> 9 mils	T	
H	Are any coupling d.c. positions abnormal?	> 9 mils	T	
I	Are any axial metal temperatures abnormal?	> 15 F		F
J	Are any axial d.c. positions abnormal?	> 9 mils		F
K	Is the 2/rev vibration component abnormal?	> 0.8 mils	T	
L	Did the relative d.c. positions of adjacent coupling probes change?	> 9 mils	T	
M	Did the relative phase of adjacent coupling probes change?	> 10 deg	T	
N	Is the 1/rev vibration component abnormal?	> 1.8 mils		F
O	Is the sub-synchronous vibration component abnormal?	> 0.5 mils		F

'If true' significance is weighed as a fault syndrome. The total of the weights for each fault syndrome is used as a parameter of probability assessment of the suspected fault.

35.8.4 Expert System to Diagnose Rotor Conditions of Asynchronous Motors

When a broken rotor bar exists, harmonic fluxes are produced in the airgap which induces current components in the stator windings at frequency $f_i = (1 \pm 2s)f$ where f is the supply frequency and s the slip of the motor. Due to the change in the flux spectrum, the stator vibration will be modulated at $\pm 2sf$. The amplitude of the sideband components, relative to the main frequency, are used to estimate the degree of the fault.

An expert system can be used to isolate these peaks by analysing the feeding current. (Entec, 1988). The method can detect:

● presence of rotor broken bars;

● aircore eccentricity;

● cracked rotor and rings;

● high-resistance joints in copper-cage windings;

● casting porosities or blowholes in aluminium die-case rotors;

● poor brazed joints in copper or aluminium rotors;

● imbalanced magnetic pull;

● mechanical imbalance;

● bent shaft;

● out of round bearings;

● air-gap irregularities.

35.8.5 Expert System for Rotating Parts Vibration Analysis

Software has been designed to use the stored vibration data and machine details to automatically diagnose machinery problems. Three separate programs are used. The main program oversees and executes the others, which are expert system shells.

The *interface shell* obtains data from the vibration monitoring data base, examines the spectra, extracts specific frequencies and makes it available for the second shell to run the rules.

The *inference engine shell* uses rules and data to solve the problem. This shell was also used to build the expert-system knowledge bases.

The system requires a data base of machine details that consists of machine measurement point identification, revolution speeds, bearing details, machine type, number of gear teeth, belt length, pulley diameter, paper roll diameters and other machine facts. This data base was created when the expert system was first installed and was used to calculate vibration frequencies, symptomatic of the different faults of the machine under examination.

When a diagnostic report is requested the system checks the vibration data base to see which machines have excessive vibration. It then gets the overall vibration readings, vibration spectra and the corresponding machine details and by applying the diagnostic rules, identifies the problems, outputting the results in a report format. For each fault found an explanation is provided. This describes the logic used in making the diagnosis and lists additional tests that may be used to verify the problem (Table 35.13).

Table 35.13   Example of a vibration expert system. (a) Diagnostic report; (b) Explanations

	Alarm: IN/S
Route PUMPS	Total number of faults: 3
Machine P-04015 DEPR RE	
Pos 2—Dir H—IN/S	
Mach Type motor	Unbalance is partly the problem
	Looseness
Rotating speed RPM 3405	Eccentricity possible
Other Shaft RPM 0	
Overall Ampl 0.164 IN/S	
Alarm Limit 0.157 IN/S	
bpfo 15234.52	
bpfi 22220.48	
GMF1 0.00	
GMF2 0.00	
Blade pass 0.00	
Belt pass 0.00	

(b)

UNBALANCE IS PARTLY THE PROBLEM
   Vibration amplitude at 1× RPM is high.
   Amplitudes greater than 5% of the 1× RPM exist else where in the
      spectrum.
   Therefore, unbalance is a likely problem, but there are others.
   AMETHYST has listed all the problems for you.
   Verify and correct all other problems before balancing.

Table 35 13 *(continued)*

LOOSENESS

Amplitude at 1 × RPM multiples is greater than 50% of 1 × RPM amplitude.

2 × RPM amplitude is high. Axial vibration was NOT high.

Looseness is the likely problem but additional checks are recommended.

Check at the machine for visual signs of looseness.

Tighten loose parts and recheck vibration.

Looseness is directional. Vertical vibration is largest for looseness at the base or at the bearing split.

Study the machine with pickup locations as described in IRD textbooks.

35.9 CASE STUDY

35.9.1 The Machine Park

To bring the above material into perspective this section describes an extensive vibration monitoring installation at an oil refinery. Monitoring of rotating machines was experimentally introduced at the refinery in 1973 and was soon extended to be used on 30% of all machines (Baldin and Di Alessio, 1989). In 1976, after doubling the size of the plant, the monitoring programme was extended to cover 50% of the whole machine park. In 1990 some 1300 machines, of a total of 2700, are monitored. Machines installed and monitored in the refinery are listed in Table 35.14.

Table 35.14   List of rotating machines
installed in refinery premises

Type of machine	Number installed
Electrical motors:	1143
Centrifugal pumps:	892
Compressors:	46
Fans:	163
Steam turbines:	103
Ljngsytrom:	9
Electrical generators:	28
Reducers:	123
Gas turbines:	2
Other rotating machinery:	252
Total machines monitored: out of which subjected to:	1300
— weekly control:	192
— monthly control:	429

35.9.2 Organization of the Team

The plant machines have been classified as super-critical, critical and non-critical machines. Super-critical machines are monitored weekly with critical ones being monitored monthly. Non-critical machines are subjected to corrective intervention only.

Experience and continuous feedback enable determination, for each machine, of allowable values for the vibration parameter under control—displacement, velocity, acceleration and spike energy.

The control program is carried out by three operators situated in the mechanical workshop (condition based maintenance—CBM—group). Experience proved that, from the technical point of view, it was most convenient to entrust the same group with the coordination and control of machine lubrication plans. The CBM group was also responsible for corrective maintenance interventions on rotating machines, carried out by three other operators under their supervision. This method obtained important information via plant operators, who made correct interpretations of measurements given by the instruments. It also entrusted responsibility to each CBM inspector to enhance the man–machine relationship in order to improve the operator understanding of machine performance variations and anomalies.

35.9.3 Procedure

Each machine has allocated its own allowable reference values for parameters being monitored. A maximum of 10 measurements on each machine are taken at predeterminated points recorded on a card where a sketch of the machine is simply outlined representing the group. Data are recorded on a computer for quick reference.

In case of an anomalous measurement arising a CBM operator sends a report to the maintenance planner and those responsible for the plant where the machine is installed. On the basis of analysis the report indicates the severity of the condition and the suggested intervention to be carried out. The duty of the CBM operator is also to point out increases in measured values, intensify controls when necessary and request further administration by the emergency or lubrication teams. Feedback on diagnosis is recorded on a computer terminal.

35.9.4 Operator Training

Good operator training is the basic element for success of a monitoring program. Operators were selected from existing employees who had the following characteristics:

● good mechanical/fitter with knowledge of electric motors;

- have a mature knowledge of machinery;
- have an understanding of the process where the machines operate;
- have top-level knowledge of instruments;
- be able to interpret measurements in order to suggest possible faults;
- be able to work in harmony, have a sense of responsibility and be aware of the consequences of decisions made.

Professionalism has been developed through attendance at courses on technical subjects and instrumentation and development of diagnostic capability.

35.9.5 Information Support

Rotating machine management uses computerized programs based on lapsed working time of machines. For continuously operating machines time is incremented automatically. The computer automatically updates the programs of vibration measurement, lubrication, greasing and machine turnover.

35.9.6 Advantages Achieved

After implementing monitoring a study was carried out to evaluate monitoring policy advantages compared to the time-based preventive maintenance previously implemented. According to these studies, a 12% gain in working hours and 16% reduction on spare parts costs has been achieved.

35.10 SOURCES OF FURTHER ADVICE

The reader is referred to the following textbooks, journals and professional groups for further information on condition monitoring.

Books

Baldin, A. E., (1989) *Condition Based Maintenance*, European Federation of National Maintenance Societies, Project No. 11.

Baldin, A. E., (1988) *La manutenzione secondo condizione*, Angeli, F., Milano, 3rd reprint.

Baldin, A. E., (1989) *Gli impianti industriali, la manutenzione e il rinnovo*, Angeli, F., Milan, 3rd reprint.

Broch, J. T. (1980) *Mechanical Vibration and Shock Measurements*, Brüel & Kjaer.

Collacott, R. A., (1977) *Mechanical Fault Diagnosis and Condition Monitoring*, Chapman and Hall, London.

Collacott, R. A., (1979) *Vibration Monitoring and Diagnosis*, UKM Publishing Ltd., Leicestershire.

Gilardoni, A., Orsini, A., and Tacconi, M. and (1981) *Handbook of Nondestructive Testing NDT*, Gilardoni, Mondello Lario, Como, Italy.

Journals

Diagnostic Engineering, Institution of Diagnostic Engineers, Leicester.

Journal of Condition Monitoring, BHRA, Bedford.

Maintenance, Conference Communication, Monks Hill, Tilford, Farnham, Surrey.

Measurement, IMEKO, Budapest.

Mantenimiento, Puntex Calle More de Deu del Coll, Barcelona.

Maintenance et Enterprise, Achats et Entretien, 16 rue Guillaume Tel, Paris.

La Manutenzione—Dimensione Editoriale snc, Milan.

Plant Engineering, Cahners Publ., 44 Cook Street, Denver, 10.

Professional groups

TC10 Committee (Technical Diagnostics) IMEKO
C/o Josef Kozak
Aeronautical Research and Test Institute
Praga 9
Latmany
Czechoslovakia

Institution of Diagnostic Engineers
3 Wycliffe Street
Leicester
UK

Mechanical Failures Prevention Group
National Bureau of Standards
Department of Commerce
Washington, DC 20234
USA

REFERENCES

Baldin, A.E., (1986) 'Technical diagnostic and condition based maintenance for better plant availability', *Measurement*, **4**(1), 7–22.

Baldin, A.E. and Di Alessio, C. (1989) 'La diagnostica degli impianti elettrici utilizzatori industriali', *Giornata di Studio AEI—ANIE—Milano*.

Barwell, F.T., Bowen, E.R., Bowen, J.P. and Westcott, V.C., (1977) 'The use of temper colours in ferrography', *Wear*, **44** 163–71.

Derry, J.J.S., (1986) 'Condition monitoring of motorized valve actuators and similar devices', *International Conference on Condition Monitoring*, Brighton, (UK), organized by BHRA.

DOI, Department of Industry (1979) *A Guide to the Condition Monitoring of Machinery*, HMSO, London.

Entec Scientific Co. (1988) *Motormonitor*, Entec Scientific Co., Cincinnati, Ohio, USA.

IRD Mechanalysis Inc (1987) *ATAM System*, IRD Mechanalysis Inc, Columbus, Ohio, USA.

Johnston, A.B. and Stronach, A.F., (1986) 'Valve fault detection in reciprocating compressors', *International Conference on Condition Monitoring*, Brighton (UK), organized by BHRA.

Scheibel, J.R., Carlson, G.J., Imam, I. and Azzaro, S.H., (1986) 'An expert system based on-line rotor crack monitor for utility steam turbines', *International Conference on Condition Monitoring*, Brighton (UK), organized by BHRA.

Scott, D. and Westcott, V.C., (1977) 'Predictive maintenance by ferrography', *Wear*, **44**, 173–81.

Chapter

36 M. J. WILLISON

Environmental Monitoring

Editorial introduction

As already mentioned in the previous chapter there is a much heightened interest in having a clean environment free of pollution. As many pollutants cannot be sensed with our natural sensors and as the problem is of a distributed nature multi-sensor systems are needed that meet both technical and legal requirements. This chapter introduces an aspect of sensing that will find increasing importance. Allied to this need are the shortcomings of many current sensors—see Chapters 37 and 40.

36.1 INTRODUCTION

This chapter provides an outline of current issues and highlights the important techniques available for environmental monitoring.

Section 36.2 examines the reasons for environmental monitoring. Then follows, in Section 36.3, discussion of the design of monitoring programmes. This aspect is covered in some detail as it is perhaps the most important aspect. If the objectives of a monitoring programme are not well defined, then expensive and scarce resources will probably be wasted. Considerations in monitoring land contaminants are also discussed there, as it is an increasingly important area that should be included in any discussion of environmental monitoring.

Section 36.4 addresses the analytical techniques which are commonly used to detect pollutants. It focuses on the sampling and sensing of pollutants of air and water. Some examples from the literature are given as case studies. Section 36.5 considers standardization and calibration, while Section 36.6 gives a brief discussion of future trends. The last section is devoted to sources of advice.

Handbook of Measurement Science, Volume 3
Edited by P. H. Sydenham and R. Thorn
© 1992 John Wiley & Sons Ltd

Radiation monitoring is not covered. For this topic the reader is referred to the chapter by Mitchell (1982) for water and the chapter by Schulte (1976) for air.

36.2 REASONS FOR ENVIRONMENTAL MONITORING

36.2.1 Introductary Remarks

The 'environment' in environmental monitoring refers to the three media of air, water and land. Natural and man-made emissions in the environment create the conditions in which monitoring becomes necessary. Monitoring of the environment is, of course, generally undertaken to establish information about the levels of pollutants in emissions, both in initial discharges and in the wider environment once these discharges have been dispersed. The information is used to understand the effects of pollutants on living creatures, on building materials and on climate. These general statements will now be clarified by some examples of detailed reasons for environmental monitoring.

36.2.2 Compliance with Legislation

Monitoring to comply with legislative controls can take three basic forms which can be illustrated by examples from European Economic Community (EEC) directives. Firstly there is legislation relating to emission levels from specific processes. For example Table 36.1 lists the limit values for different industries and processes of the concentrations of cadmium (Cd) (CEC, 1983) and mercury (Hg) (CEC, 1982a, 1984) in liquid effluent discharges. Secondly, legislation is often introduced to meet environmental quality objectives (EQOs) and it sets environmental quality standards (EQSs). Table 36.2 lists the concentrations of nitrogen dioxide NO_2 (CEC, 1985a), sulphur dioxide SO_2, particulate matter (CEC, 1980a) and lead (Pb) (CEC, 1982b) which must not be exceeded in the ambient atmosphere of EEC countries. Another example is given in Table 36.3 which lists some of the parameters and their concentrations which must not be exceeded in water intended for human consumption. Table 36.4 shows the maximum permissible concentrations of potentially toxic elements in soil after the application of sewage sludge to agricultural land in the UK. Thirdly, there is legislation which has neither specific emission levels nor EQSs. An example here is the EEC Directive on Environmental Impact Assessment (EIA) (CEC, 1985b), in which there is a requirement for an EIA to be produced for proposed new developments. In producing an EIA there is often a need to carry out monitoring of all three media (air, water, land) to establish baseline data upon which can be superimposed the projected emissions from the new development. The projections are used to assess whether the development would have any adverse impact on the local environment.

Table 36.1   EEC limit values for the discharge of cadmium and mercury from various industries and processes

Cadmium (Cd)	Concentration $(\mu g\ l^{-1})$
Zinc mining, lead and zinc refining, Cd metal and non-ferrous metal industries	0.2
Manufacture of	
● Cd compounds	0.2
● pigments	0.2
● stabilizers	0.2
● batteries	0.2
Electroplating	0.2

Mercury (Hg)	Concentration $(\mu g\ l^{-1})$	Quantity
Chloralkali plants (all Hg containing water)	50	
● Recycled brine		
– from Cl₂ production unit		0.5 [a]
– total quantity of Hg in all Hg containing waters discharged from site		1.0 [a]
● Lost brine		
– total quantity of Hg in all Hg containing waters discharged from site		5.0 [a]
Industries using Hg catalysts		
● vinyl chloride production	50	0.1 [b]
● others	50	5.0 [c]
Manufacture of		
● Hg catalysts	50	0.7 [c]
● other Hg compounds	50	0.05 [c]
● batteries	50	0.03 [c]
Non-ferrous metal industry and waste treatment plants	50	

[a] $g\ t^{-1}$ of installed chlorine production capacity.
[b] $g\ t^{-1}$ of vinyl chloride production capacity.
[c] $g\ kg^{-1}$ of mercury processed.

Table 36.2   EEC air-quality standards $(\mu g\ m^{-3})$ for smoke, SO₂, NO₂ and Pb

Smoke and SO₂	Smoke	SO₂
Median of daily values during year	80	120 (if smoke < 40) 80 (if smoke > 40)
Median of daily values during northern winter (October–March)	130	180 (if smoke < 60

(continued overleaf)

Table 36.2 *continued*

Smoke and SO$_2$	Smoke	SO$_2$
98th percentile of daily values during year	250	130 (if smoke > 60) 350 (if smoke < 150) 250 (if smoke > 150)
NO$_2$		
98th percentile of mean hourly values measured over the year		200
Pb		
Annual average concentration		2

Table 36.3 Concentrations of some parameters in drinking water required to meet the EEC directive (80/778/EEC) (CEC, 1980b)

Parameter	Unit	Maximum admissible concentration
Arsenic	mg As l^{-1}	0.05
Chloride	mg Cl l^{-1}	400
Cyanide	mg CN l^{-1}	0.05
Nitrate	mg N l^{-1}	11.3
Phosphorus	mg P l^{-1}	2.2
Aluminium	μg Al l^{-1}	200
Cadmium	μg Cd l^{-1}	5
Chromium	μg Cr l^{-1}	50
Iron	μg Fe l^{-1}	200
Lead	μg Pb l^{-1}	50
Magnesium	μg Mg l^{-1}	50 000
Mercury	μg Hg l^{-1}	1
Nickel	μg Ni l^{-1}	50
Zinc	μg Zn l^{-1}	5000
Hydrocarbons	μg l^{-1}	10
Tetrachloroethylene	μg l^{-1}	10
Trihalomethanes	μg l^{-1}	100
Phenols	μg l^{-1} as C$_6$H$_5$OH	0.5

This is not exhaustive, refer to the Directive for complete list of parameters and their concentrations.

36.2.3 Pollution Damage

Environmental monitoring may be required to assess health effects on humans and other biological systems, or effects on building materials or on climate. Here environmental monitoring is used for the establishment of dose–response relationships between pollutant concentrations and observed effects. Table 36.5 summarizes the main features of the ecological and health effects of environmental pollutants. In the UK the Department of the Environment (DoE) has set up a number of review groups to examine the effects of air pollution on the environment. These include the Acid Rain Review Group (DoE, 1987a, 1990a), the Terrestrial Effects Review Group (DoE, 1988), the Photochemical Oxidants Review Group (DoE, 1987b), the Building Effects Review Group (BERG) (DoE, 1989, 1990b). All of these groups have made initial reports which have recommended monitoring programmes. The latter BERG report, for example, recommended the implementation of a National Materials Exposure Programme whereby different materials (stones, bare metals and painted metals) are exposed at several sites of differing pollution and meteorological character.

Table 36.4 Maximum permissible concentrations of potentially toxic elements in soil after application of sewage sludge and maximum annual rates of addition (From Department of the Environment (1989) Code of Practice for Agricultural Use of Sewage Sludge, HMSO, London.)

PTE	Maximum permissible concentration of PTE in soil (mg/kg dry solids)				Maximum permissible average annual rate of PTE addition over a 10 year period (kg/ha) (2)
	pH(1) 5.0 < 5.5	pH(1) 5.5 < 6.0	pH 6.0–7.0	pH(3) > 7.0	
Zinc	200	250	300	450	15
Copper	80	100	135	200	7.5
Nickel	50	60	75	110	3
	For pH 5.0 and above				
Cadmium	3				0.15
Lead	300				15
Mercury	1				0.1
Chromium	400 (Provisional)				15 (Provisional)
Molybdenum(4)	4				0.2
Selenium	3				0.15
Arsenic	50				0.7
Fluoride	500				20

* These parameters are not subject to the provisions of Directive 86/278/EEC

(1) For soils of pH in the ranges of 5.0 < 5.5 and 5.5 < 6.0 the permitted concentrations of zinc, copper, nickel and cadmium are provisional and will be reviewed when current research into their effects on certain crops and livestock is completed.

(2) The annual rate of application of PTE to any site shall be determined by averaging over the 10-year period ending with the year of calculation.

(3) The increased permissible PTE concentrations in soils of pH greater than 7.0 apply only to soils containing more than 5% calcium carbonate.

(4) The accepted safe level of molybdenum in agricultural soils is 4 mg/kg. However there are some areas in UK where, for geological reasons, the natural concentration of this element in the soil exceeds this level. In such cases there may be no additional problems as a result of applying sludge, but this should not be done except in accordance with expert advice. This advice will take account of existing soil molybdenum levels and current arrangements to provide copper supplements to livestock.

36.2.4   Requirement for Legislation

There may arise a need for environmental monitoring to assess the need for the introduction of legislative controls. For example the demonstration of a cause and effect relationship between emissions to air and soil acidification or building damage, the depletion of stratospheric ozone by chlorofluorocarbons, or the persistence of organochlorine compounds in the environment and their concentration in the food chain would be a reason for the introduction of controls. However, before such controls can be introduced the extent of the damage, the value placed by society on the damaged caused and the costs of implementing controls must be assessed in a cost–benefit analysis.

Table 36.5   Main features of ecological and health effects of environmental pollutants

1. Pollutants adversely altering the physical and chemical environment.
    (a)   Radiation balance of the Earth
          (i)   depletion of stratospheric ozone, e.g. by chlorofluorocarbons
          (ii)   increases in absorption of outgoing infrared radiation by carbon dioxide
    (b)   Energy and nutrient flow in aquatic systems
          (i)   organic pollutants
          (ii)   role of nutrient pollution in eutrophication e.g. nitrates and phosphates
2. Pollutants directly toxic to living organisms:
    (a)   Indigenous substances
          (i)   gaseous air pollutants (SO_2, and NO_2)
          (ii)   acid deposition
          (iii)   metals
          (iv)   hydrocarbons
    (b)   Synthetic chemicals
          (i)   organochlorine compounds

36.2.5   Environmental Awareness

Increased awareness of pollution and its effects on the environment have underlined the need for monitoring, be it on a continuous or case-by-case basis. For example, there is a need for water quality management to maintain natural waters in a suitable state for various purposes (e.g. use as drinking water, recreation). Such water quality management requires that the concentrations and effects of pollutants can be accurately assessed. Land contamination as a result of industrial activity is increasingly being identified. For example, land contamination can be a result of the abandonment of certain processes (e.g. town gas from coal carbonization) and the decline in other sectors of industry (e.g. iron and steel making) leaving areas of derelict and contaminated land.

Soil contaminants can be harmful to humans, animals, plants, water quality and building materials. Such land, therefore, requires investigation and chemical analysis before it can be reclaimed and brought back into some beneficial use.

Waste disposal sites which have taken municipal and/or industrial waste are also possible sources of contamination through leachate and landfill gas migration.

Leachate is potentially highly polluting foul drainage with a very high oxygen demand which is typically 50–100 times that of domestic sewage. Landfill gas is a mixture of methane and carbon dioxide, both of which are asphyxiating gases. In addition methane forms explosive mixtures in air in the concentration range 5–15% v/v.

36.2.6 Pollution Episodes

The London smog covering a period of four days in December 1952 resulted in about 4000 premature deaths of chronically ill individuals probably suffering from lung or heart complaints. It was not until some weeks later when the death statistics became available that the link was made with the high concentrations of smoke and sulphur dioxide which were measured during that period. This episode resulted in the Clean Air Acts to combat general urban pollution in the UK, especially from coal combustion.

Another type of episode is exemplified by the interaction of various gaseous pollutants such as oxides of nitrogen and unburnt hydrocarbons, both of which are of traffic origin, under hot sunny conditions. These *photochemical* smog events, common in such cities as Los Angeles and Athens (nephos-cloud), not only include products such as ozone and peroxyacetyl nitrate (PAN) which are harmful to the health of humans and crops, but also inorganic aerosols like sulphuric acid produced by the oxidation of sulphur dioxide. These aerosols react with ammonia generating ammonium sulphate which, because of its sub-micron particle size, readily scatters light resulting in visibility impairment.

Knowledge and extent of such episodes can only be defined by environmental monitoring. The levels of pollutants involved are important in terms of health implications. Authorities can impose restrictions on fuel use (e.g. change to low sulphur fuels), limit the use of motor vehicles, and advise the elderly and chronic sick to stay indoors. Monitoring of air pollution episodes also adds to knowledge and understanding of the chemical processes taking place in the atmosphere.

In the field of water quality the appearance in warm weather of blooms of blue-green algae, which result from the build up of the nutrient pollutants nitrate and phosphate, have human health implications. Monitoring the toxins produced by these algae enables authorities to issue appropriate warnings and advice.

36.2.7 Accidental Releases and Spillages

Such events taking place in, or near, water courses have potentially adverse effects on potable water treatment. Continuous monitoring at the intake to a water treatment works can alert plant personnel to either modify the water treatment process or to close off the intake to the works. In the UK subsequent analysis undertaken by the National Rivers Authority (NRA) would enable the source of the spillage to be identified and a prosecution brought by that body under the Water Resources Act, 1991.

36.2.8 Source Monitoring

This type of monitoring may be undertaken to determine the mass emission rates of pollutants from a production plant and to assess how these emissions are altered by changes in the process. Furthermore, source monitoring may be used to evaluate pollution abatement equipment, and also to determine the impact on local air or water quality.

36.2.9 Model Verification

Computer models describing the behaviour of pollutants in the environment are becoming increasingly important because of the developing trend towards environmental quality objectives and standards. The need for environmental quality management systems is evident in order to be able to advise on pollution control policies. For example, air quality monitoring programmes can be designed specifically for model validation and calibration. Noll and Miller (1977) summarize the conditions under which this type of programme is most appropriate:

• When local topographical features violate the basic assumptions of a simulation model, field measurements can be used to amend the model for different types of terrain.

• If no applicable model exists, field measurements can be used to develop an empirical model.

It is important to note that only by measuring all the input variables of source strength and meteorological parameters, and accounting for all errors in the field measurements, can any remaining error due to the inaccuracy of the model prediction be quantified.

36.3 THE MONITORING PROGRAMME

36.3.1 Objectives

This section examines the factors involved in the design of a monitoring programme. It takes a general approach which can be applied to the three environments of air, water and soil. Specific reference will be made to each area where

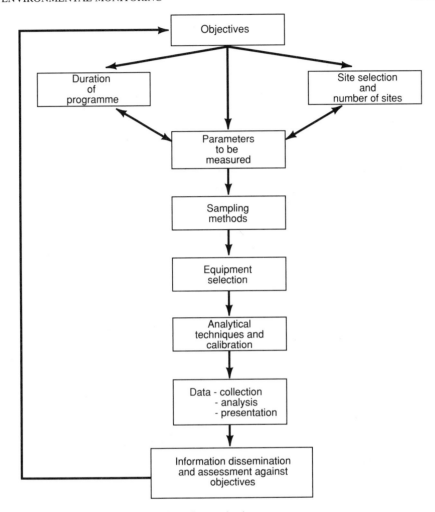

Figure 36.1   The stages in the design of a monitoring programme

necessary. The steps which have to be considered in a systematic way are identified in Figure 36.1. Let us consider the main steps in some detail.

Section 36.2 described some examples of the reasons for environmental monitoring and these form the basis for the objectives of a monitoring programme.

Objectives are perhaps the most important part of the monitoring strategy, and
it is essential that the objectives should be clearly and precisely formulated,
otherwise too many or too few results will be obtained. Furthermore it may prove
difficult to assess whether the objective has been achieved. A clearly defined set
of objectives is important to avoid wasting scarce financial and personnel resources. It is also important to establish what the results are to be used for and who
is going to make use of the results.

36.3.2   Site Selection

Water

The objectives of a monitoring programme will sometimes define the sampling location, for example when it is desired to know the impact of the discharge of a factory effluent on a receiving river. In this case sampling locations will be needed both upstream and downstream of the works, and sampling of the effluent itself may also be required. However, the objective may be very general such as 'to measure water quality in a river basin', and so give no indication as to which of the infinite number of possible sampling locations are of interest. Hunt and Wilson (1986) have stressed the need to explicitly define sampling locations in the objectives of any programme.

Once sampling locations have been identified the next step is to select the best sampling points at these locations in order to satisfy the programme objectives.

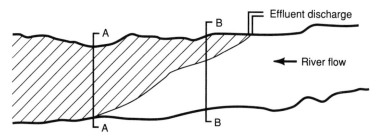

Sampling location should be downstream of A-A.

(a) Lateral dispersion of effluent

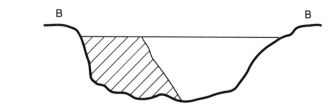

(b) Vertical and lateral dispersion of effluent

Region of mixing of effluent and river

Figure 36.2   Schematic representation of an effluent with a river: (Source Wilson, A. L., 1982)

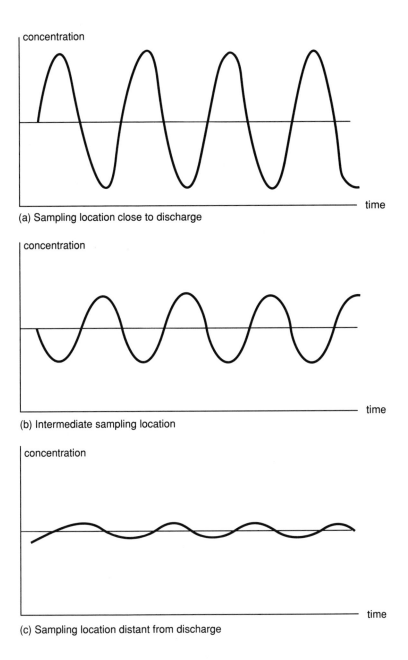

(a) Sampling location close to discharge

(b) Intermediate sampling location

(c) Sampling location distant from discharge

Figure 36.3 Schematic diagram of the dependence of variations in water quality on the distance downstream from a cyclically varying waste discharge: (Source Wilson, A. L., 1982)

Water bodies tend to be heterogeneous and, therefore, sampling should be at points which are representative of the bulk of the water body.

In general, sampling points should be away from river or lake banks and away from channel or pipe walls. Wilson (1982) has pointed out that if sampling takes place where the water body is not homogenous then any result might reflect differing rates of mixing rather than true variations in overall water quality. Figure 36.2 shows a schematic representation of the mixing of an effluent discharge with a river. When an effluent discharge varies temporally in volume and/or quality, and the interest is in the short-term changes in river quality, then the distance of the sampling point downstream should be as short as possible as adequate mixing allows. Conversely, monitoring should be carried out further downstream where mixing and longitudinal dispersion will have attenuated short-term variations if long-term average water quality is required (see Figure 36.3). In the case of source monitoring where there is a need to monitor process waste streams, sampling from horizontal pipes can be affected by deposition of solids, and allowance of approximately 25 pipe diameters should provide adequate mixing (Wilson, 1982).

Air

Site selection for the monitoring of stationary sources for gaseous emissions has been summarized by Pio (1986) and must include:

● safety of sampling personnel;

● availability and of a platform for personnel and equipment;

● access via a suitable port to the stack interior;

● power supply;

● avoidance of flow disturbance;

● relationship to points of interest (eg. pollution control equipment, process efficiency testing).

Safety of personnel is of paramount importance, and a site that satisfies all the above criteria except safety should be abandoned until suitable modifications are made to allow safe working.

Criteria for avoiding disturbances to flow have been specified by a number of organizations such as ASTM (1981), EPA (1977), BSI (1961, 1971) and ASME (1957). In general a sampling location in a vertical flue situated eight flue equivalent diameters downstream and two flue equivalent diameters upstream from a flow disturbance such as an inlet, outlet or bend is considered to be good. At this location ten or twelve points are sufficient to overcome any deficiencies resulting from any flow disturbances. For a circular stack or flue, sampling points should be at the centroids of equal area annular segments: in rectangular flues sampling points should be located at the centroids of smaller, equal area rectangles.

Ambient air monitoring sites may be summarized by the following categories (Moore, 1986a):

(1) Discrete source surveys for the verification of dispersion models. Here the minimum requirements for the sampling pattern are:

- three instruments arranged in an arc with an angular spacing of samplers of about $7\frac{1}{2}°$ and at three distances from the source;

- or, as above but at two distances from the source supported by single instruments nearer to and further from the source than the arcs, and if possible with an instrument located between them.

(2) Discrete source surveys to determine the frequency of occurrence of pollution levels have samplers set out in a ring or concentric rings centred on the source. The angular separation of the sites is such that when one site is in the plume the adjacent sites are unaffected. On the basis that an unaffected site experiences a concentration <10% of the average concentration seen by an 'in-plume' sampler, Moore (1986a) has calculated that the angular separation of sites should be at least 13°.

(3) For area surveys such as the UK National Survey of Air Pollution for smoke and SO_2 (Warren Spring Laboratory, 1972), the number and spacing of sites depends upon the exact objective and the resources available, but they may be located in areas where the pollution concentrations are expected to be greatest or where there is a high population density. In the latter case Moore (1986a) states that the number of sites should be $N^{1/2}$, where N is the population in tens of thousands.

(4) Distant source/regional scale surveys should avoid using sheltered locations, such as orchards, near hedges, woods, high walls, tall buildings, etc., especially if the pollutant is expected to be taken up by these surfaces. The effects of local and low-level sources should be avoided. Valley sites where there may be sheltering from reduced atmospheric dispersion should be recognized and considered in relation to the overall objectives of the survey.

(5) Global scale monitoring is generally concerned with long-term trends in background levels of pollutants likely to have a long residence time in the atmosphere. Here the sheltering effects of sampler locations will not be important. However, the sampling sites should be remote from centres of population, major highways and air routes.

In general inlets to samplers should be at least 2 m above ground level, the maximum height being determined by the objectives of the survey, and should be consistent from site to site within a given network. The site must be readily accessible and yet secure from vandalism; it must also provide an adequate and reliable power supply.

Land

The investigation of contaminated land is usually carried out in order to fulfil the objective of reclaiming that land to return it to some beneficial use so that users of the reclaimed land are not exposed to any health risks by virtue of its previous history. Lord (1987) describes the sequence of steps required in a site investigation. Once a site has been identified as requiring further investigation a sampling and analytical programme is designed and implemented.

The following factors, in relation to sample location, need to be addressed at the design stage:

- the sampling pattern(s) that should be used;
- the depths at which samples should be taken.

Sampling classes can be categorized as follows.

(1) *Composite sampling.* A number of spot samples are taken along the outline of a letter 'W' or 'X', bulked together and mixed giving one sample for analysis. This method of sampling is unlikely to identify the locations of hot spots, and any high concentrations encountered are likely to be 'diluted' in the mixing of the spot samples. Thus the ease of sampling by this method is far outweighed by the disadvantage of not revealing hot spots, and as such is a method which is only of limited use in contaminated land investigations.

(2) *Random sampling.* Although random sampling can make use of statistics to determine the probability of locating contaminated areas of various sizes, this technique is likely to be very expensive for two reasons. Firstly, the distribution of sampling points will be uneven unless the number of sampling points is large in relation to the number of places left unsampled. Secondly, random selection of subsurface samples will commonly select points below the surface where the samples above them have not been selected. Again this problem is reduced as the number of samples increases.

(3) *Grid sampling.* This technique provides an easy means of locating and mapping sampling points once the size of the grid has been determined. Statistical analysis can define the grid size once the boundary conditions have been set. For example, to be 95% confident of finding a contaminated area 5 m × 5 m requires a grid spacing of about 7 m between sample points

(Wilson and Stevens 1981). Such a grid size will produce 200 sample points per hectare, and is therefore likely to be expensive. An alternative approach is to use a larger grid size, but this will reduce the confidence limits of finding the 5 m × 5 m patch of contamination. This dichotomy of cost versus confidence can be addressed as follows:

- some after-uses are less demanding in terms of residual contamination, e.g. carparks;

- heavily contaminated sites may not need extensive quantitative analysis; semi-quantitative or qualitative visual examination may be sufficient to establish the necessary remedial action;

- it may be possible to re-orientate the development so that the amount of contaminated material left exposed is minimized.

(4) *Sampling depth*: The sampling depth is often dictated by the development, e.g. for housing, schools or recreational use a depth of 2.5 m is appropriate. However, on some sites an undisturbed clay layer may lie close to the surface and act as a barrier to downward movement of contamination and therefore will describe the maximum sampling depth.

36.3.3 Parameters to be Measured

Table 36.6 lists some examples of the different types of parameters which are commonly of interest in environmental monitoring. The selection of parameters

Table 36.6 Examples of parameters determined in environmental monitoring

Air

1. Gases — SO_2, NO_2, NO, O_3, CO, CO_2, NH_3, hydrocarbons
2. Particles — total particulate matter, composition of particles e.g. Cl, SO_4, NO_3, NH_4, metals, dioxins, polynuclear aromatic hydrocarbons (PAHs)

Water

1. Non-specific parameters — colour, turbidity, suspended solids, biological oxygen demand (BOD), chemical oxygen demand (COD)
2. Specific parameters — pH, dissolved oxygen, PO_4, NO_3, NO_2, NH_3, SO_4, S, Cl, Cd, Hg, Cu, Ni, Zn, Pb, Cr, Fe, hydrocarbons, chlorinated organic compounds

Land

Sulphates, sulphides, cyanides (free and complex), phenols, PAHs, dioxins, polychlorinated biphenyls (PCBs), solvents (e.g. benzene, toluene, xylene), asbestos, landfill gas (CH_4, CO_2)

which are to be determined should be based on satisfying the objectives of the monitoring programme. The selection should indicate the type of parameter (e.g sulphur dioxide in air, cadmium in water, polychlorinated biphenyls in soil) and the required concentration range. Although the parameter(s) may be readily identified, the concentration range likely to be encountered may not be known There may be no previously reported data for the particular parameter at the particular sampling location(s). In this case some preliminary screening tests may be required to ascertain the likely concentration ranges. In the case of ambient air monitoring meteorological data should also be obtained. The effect of meteorology on air pollutants is outlined by Moore (1986b), and for a more detailed discussion the reader is referred to Pasquill and Smith (1983).

36.4   ANALYTICAL TECHNIQUES

36.4.1   General Introduction

Analytical techniques employed in environmental monitoring are the same techniques which are employed in quantitative analytical chemistry, and are reliant upon the detection of physical or chemical properties of the parameters of interest. It is important that the analytical technique adopted must:

- produce results of the required accuracy;

- measure the parameter of interest;

- have detection limits appropriate to the concentration range of interest;

- be precise and accurate;

- be specific to the parameter of interest;

- be compatible with data obtained from other laboratories;

- be easy to operate;

- be cost effective and reliable;

- be easy to calibrate.

Table 36.7 lists the analytical techniques which are employed in environmental monitoring. Some of these techniques will be dealt with in a little more detail in the following sections. However, before then mention should be made of how the sample to be analysed is taken and presented to the sensor. As shown schematically in Figure 36.4 there are three basic ways of collecting the sample and presenting it to the sensor. Firstly there is sample collection which may be either manual or automatic with transport of the samples to the laboratory for analysis. There may be a requirement for sample pre-treatment for stabilization purposes or for conversion of the pollutant species to one which is more easily determined, and this may be either manual and take place either on site or in the laboratory, or else be automatic and linked to the sampling. Secondly, there is automatic sampling, which may be either continuous or discrete, sample pre-treatment (if required) and analysis, all of which take place on site. Lastly there is *in situ* monitoring where the sensor is placed directly in the environment to be measured.

Table 36.7   Examples of analytical techniques used in environmental Monitoring

1. Physical

  Gravimetry
  Spectroscopy

Absorption	— atomic	
	— ultraviolet (UV)/visible	
	— infrared (IR)	
Emission	— flame	
	— electron	
	— X-ray	
	— mass	
Fluorescence	— atomic	
	— spectrofluorimetry	
	— chemiluminescence	
	— X-ray	

  Chromatography

Gas	*Detectors*
	thermal conductivity
	flame ionization
	electron capture
	flame photometric
	mass spectrometric
Liquid	refractive index
	UV fluorescence
	conductivity

2. Chemical

  Titrimetry
  Electrochemical

    Amperometry
    Coulometry
    Voltammetry
    Anodic stripping voltammetry
    Polarography

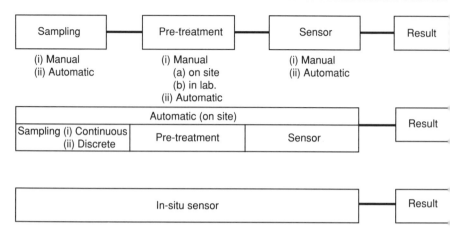

Figure 36.4   Schematic representation of the three basic ways of presenting a sample to the sensor

36.4.2 Sampling and Sensing of Water Pollutants

Water sampling—manual

In general there is no special water-sampling system required for the manual collection of water samples. Collection can essentially be achieved by the immersion of a container in the water of interest. Care must be taken to ensure that the water sample once taken is not altered before the analysis is undertaken.

The two important factors to consider in ensuring sample integrity are sample preservation and the choice of sample container. Wilson(1982) has summarized generally useful preservation techniques for different types of parameter and the reader is referred there for further details. The two types of container most commonly used for collecting water samples are glass and plastic. The advantages and disadvantages of the two types of material are given in Table 36.8. They show that due consideration must be given to the compatibility of the parameter(s) of interest with the container material. For example, if it is required to know the concentration of trace organic compounds in river water then a plastic container should not be used for sampling.

Generally these factors become more and more important as the concentration of the parameters of interest becomes smaller. Manual collection of samples, transport of the samples to a laboratory and their subsequent manual analysis has been common in environmental monitoring. Hunt and Wilson(1986) summarize the limitations of this approach to water quality monitoring as follows:

● The delay between sample collection and the reporting of the result may be too long, especially when either knowledge or control of water quality is required on a continuous basis, e.g. the automatic control of a potable water-treatment plant.

● There may be difficulties when samples have to be collected and/or analysed outside normal working hours.

● Laboratory facilities (e.g. space) and personnel may not be sufficient to permit collection and/or analysis of all the required samples.

● Analytical accuracy can be markedly dependent on the skill of the analyst and the conscientiousness and accuracy with which analytical protocols are followed in routine manual analysis.

● The instability of samples with respect to some parameters especially when samples have to be collected at sites remote from the laboratory.

Table 36.8 Advantages and disadvantages of different containers for water samples

Advantages	Disadvantages
Glass	
(i) Condition of internal surfaces readily apparent	(i) Absorption of trace metals into surface
(ii) Can be cleaned more vigorously	(ii) Heavy and easily broken
(iii) Can be sterilized so used for microbiological samples	(iii) Leaching of materials e.g. sodium silica
	(vi) Fluoride may react
Plastic	
(i) Light and robust	(i) Absorption of organic materials
	(ii) Leaching of trace metals

Water sampling—automatic

The advent of automatic samplers for collection of water samples has arisen out of the need to address some of the points above. There are basically two types of automatic sampler. First there is the sampler which takes samples at fixed time intervals. The sample volume, the sampling time, the number of samples collected within the sampling period and the length of the period itself can all be varied. This type of sampler is used when concentration variations rather than changes in mass flow of a particular pollutant are required. Secondly, there is the sampler which can either take samples of equal volume at a frequency proportional to the flow rate, or the sample volume collected at equal time intervals is directly proportional to the flow. Such sampling regimes would be adopted where the mass flow of a particular pollutant is required and where there is a large variation in flow. Automatic samplers:

● are easily portable;

● can be used for any type of water;

● can collect composite samples;

● can be used to study variations in water quality;

● allow unattended sampling from remote locations.

Sensing techniques

Hunt and Wilson (1986) Friedman (1990), and Mancey and Allen (1982 a,b) deal extensively with various sensing techniques. HBMS, Chapter 27, Volume 2 also gives an overview of chemical sensing in general. The important methods used in environmental measurements are now summarized.

(1) *UV/visible spectrophotometry.* In this technique light absorption by components of a solution in the 190–900 nm wavelength range is used to determine their concentrations. Some components absorb light sufficiently well but others require the addition of reagents to produce a strongly absorbing reaction product. The basis of the method is the Beer–Lambert–Bouguer relationship as follows:

$$A = \log_{10}(I_0/I_T) = \varepsilon c l$$

where A = absorbance or optical density = $\log_{10}(100/T)$
        T = % of light of wavelength λ transmitted by the solution = $100\,I_T/I_0$
        I_0 = intensity of incident radiation
        I_T = intensity of transmitted radiation
        ε = molar extinction coefficient of absorbing species at wavelength
            λ (1 mol^{-1} cm^{-1})
        c = concentration of absorbing species (mol l^{-1})
        l = length of light path through solution (cm).

A simpler version of the technique is colorimetry which uses filters to select a wide range of wavelengths as opposed to a monochromator that selects a narrow range of wavelengths. A colorimeter is cheaper and provides more robust equipment, which can easily be used in the field. The technique is well established with relatively cheap and reliable instrumentation. However, it is not applicable to the determination of several parameters at once in the same portion of sample. The technique, applied to the analysis of waters and associated materials, has been reviewed by Thompson (1980).

(2) *Atomic Absorption Spectrophotometry (AAS).* In this technique the sample is converted to its constituent atoms and the ground state atoms of the particular element of interest absorb light of a characteristic wavelength. The amount of light absorbed is proportional to the concentration of the element being determined. The conventional method of atomization is an air/acetylene flame, although some elements require the higher temperature of a nitrous oxide/acetylene flame. Increased sensitivity can be achieved by using electrothermal atomization techniques, conversion of some elements to their hydrides before atomization (e.g. arsenic, selenium), and conversion of compounds containing mercury to its elemental form. This is sufficiently volatile to be purged from the sample without the need for thermal methods, and as such this method is referred to as the cold vapour technique.

(3) *Flame Emission Spectrometry*. This technique relies on the detection of light emitted from the electronic transition of thermally excited atoms to their ground state. The equipment used for flame atomic absorption spectrophotometry, minus the light source, can be used. In this case the light source is the excited atoms in the flame and the detector is tuned by means of a monochromator to the characteristic wavelength of the element of interest.

(4) *Chromatographic techniques*. Chromatographic techniques are particularly important for the detection of organic compounds in waters, since other measurement techniques are not appropriate for the determination of specific organic compounds in the presence of others. This is because of the similarities in the physical and chemical properties of the large number of related compounds which may be present. Separation techniques are therefore required to enable individual components to be measured.

The principle behind chromatographic separation is the distribution of components in a mixture between two immiscible phases as one phase continuously passes through the other. A sample consisting of a mixture of compounds is injected at one end of the stationary phase and the individual components eluted by the mobile phase appear at the other end spread out in time.

There are two basic types of chromatographic separation—one in which the mobile phase is a gas (gas chromatography GC) and one in which the mobile phase is a liquid (liquid chromatography LC). GC is essentially used for volatile organic compounds whereas LC can be applied to the separation and determination of non-volatile organic species which are not amenable to GC. Common types of detectors used in environmental monitoring in each of these systems are now described.

In gas chromatography three main detectors are used.

- *Flame ionization detector (FID)*. The ion current generated in a hydrogen–air flame, when two electrodes placed near that flame are held at a constant potential, increases when an organic compound enters the flame. These detectors respond in varying degrees to nearly all organic compounds, but they show their greatest sensitivity for hydrocarbons and those materials not containing halogens.

  Flame photometric detector (FPD). Compounds containing sulphur and phosphorus, when burnt in a reducing hydrogen-rich flame, produce species of these elements which emit characteristic radiation in the UV region. The radiation level is determined by a photomultiplier after passage through a suitable optical filter.

  Electron capture detector (ECD). Compounds such as organochlorine pesticides, polychlorinated biphenyls (PCBs) and other halogenated hydrocarbons which have an affinity for electrons, produce ions by capturing electrons from a β-emitter, giving rise to an ion current which is measured by an electrode system.

In Liquid chromatography the three mainly used detectors are as follows.

- *UV/visible detector.* This can take the form of a fixed wavelength detector using the 254 nm line from a mercury lamp, a detector offering a range of fixed wavelengths, many forms of variable wavelength detectors or rapidly scanning wavelength detectors with diode arrays which produce a spectrum of each component eluted from the system. Most organic molecules can be identified using these types of detector.

- *Fluorimetric detector.* Some organic molecules absorb UV radiation and re-emit some of the energy at a longer wavelength. The fluorescent radiation is measured at right angles to the beam of exciting radiation after passing through a suitable filter or monochromator.

- *Conductivity detector.* This type of detector is commonly employed in ion-exchange chromatography which, as its name suggests, is appropriate for the determination of ionic species. In particular it is widely used to determine common inorganic parameters such as sulphate, nitrate, nitrite, chloride, phosphate and is accepted as a recommended method for water analysis (Standing Committee of Analysts, 1984).

(5) *Electrochemical techniques. Ion-selective electrodes (ISE)* are potentiometric membrance electrodes consisting of electrochemical half cells which respond to changes in the concentration of ions in solution when compared to a reference electrode according to the Nernst equation:

$$E = E^0 + (2.303RT/zF) \log a$$

where E^0 = constant for a given electrode pair (selective and reference)
and temperature
R = gas constant ($8.31432 \, J \, K^{-1} \, mol^{-1}$)
T = absolute temperature (°K)
F = Faraday ($9.64845 \times 10^4 \, c \, mol^{-1}$)
z = charge on the ion to which the electrode responds
a = activity of the ion in the sample.

A decade change in the activity of a monovalent ion at 25°C will produce a change of 59.16 mV in the e.m.f. of the cell. The commonest of these electrodes is the glass pH electrode. There are two other electrode types:

- solid-state electrodes which contain an inorganic salt membrane;

- ion-exchange electrodes with an organic membrane capable of selective ion-transport.

Gas-sensitive membrane probes are complete electrochemical cells and use a gas permeable plastic membrane (e.g. polyethylene, teflon) to separate an electrolyte solution from that under test. The sensing device, usually a glass and reference electrode pair, is in direct contact with the electrolyte. A soluble gas diffuses through the membrane, distributing itself between the test solution and the electrolyte, the change in pH of which is measured by the glass electrode and is proportional to the concentration of gas in the test solution. The most common use of ISEs and gas sensing probes are for the determination of pH, fluoride, nitrate and ammonia. Chapter 37 explains micro electronic forms of ISE.

(6) *Voltammetry and other techniques.* In *DC polarography* there are two electrodes, one of which is provided by mercury flowing through a capillary, the tip of which is immersed in the solution to be analysed. Successive drops of mercury grow and detach themselves from the capillary which is known as a dropping mercury electrode (DME). A voltage applied across the two electrodes is increased slowly at constant rate while monitoring the current passing between the electrodes. At a certain electrode potential a reaction takes place at the surface of the mercury drop involving the transfer of electrons to specific metal ions. The resulting metal atoms form an amalgam with the mercury. This transfer results in an increased current, the magnitude of which is proportional to the concentration of the metal ions. Increasing the voltage further produces no further increase in current until the potential of the DME reaches a value at which another species reacts.

Anodic stripping voltammetry is a two-stage process involving plating out of the metal as in the DC polarography above followed by a reversal of the applied electrode potential while at the same time monitoring the current as the metal is oxidized back into the ionic form.

These techniques are not as robust as AAS and solution forms of spectrophotometry but find their uses in the analysis of trace metals in sea water, a matrix not suited to AAS.

Case studies in sampling and testing of water pollutants

Case Study 1. Dispersion of Liquid Industrial Waste Discharged at Sea

Byrne *et al.* (1988) carried out a study to determine whether the rate of discharge of an acidic effluent calculated according to a 'safe' dilution formula adopted by the International Maritime Organisation (IMO) was being achieved in practice. At the time of the study 100 000 t of a liquid effluent (density (1.4×10^3 kg m^{-3}; pH < 2; organic content 3–7%; sulphuric acid 15–20%; ammonium bisulphate 40–45%) were discharged into an area of 150 km^2 with an average depth of water of 50 m, some 15 km out from the mouth of the River Tees in NE England (Figure 36.5).

Measurements of waste concentration were made by fluorimetry and pH which were determined both by *in situ* measurements made from a research vessel and in discrete samples which were analysed in an on-shore laboratory. Calibration of all measurement techniques was carried out in the on-shore laboratory using dilutions in clean off-shore sea water of the waste discharged that day. The research vessel was anchored by the stern near the middle of the disposal site and the instrument package deployed over the side. The dumping vessel then

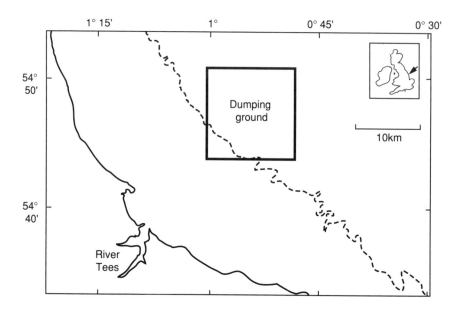

Figure 36.5   The location of the dumping ground off the River Tees (NE England). (Source Byrne *et al*., 1988)

made a series of traverses up current of the research vessel and perpendicular to the current direction so that the waste plume was carried to the research vessel by the current (Figure 36.6). A typical record from the *in situ* monitor is shown in Figure 36.7 from which several parameters such as the maximum concentration of the waste, the width of the slick and the distribution of waste in the slick can be derived. The results indicated that the *in situ* measurements of concentration were 3–10-fold lower than the IMO formula predicted. In addition the discrete samples produced lower concentrations by a factor of about two, reflecting the long time required to fill the bottles (around 10 s l^{-1}).

Case Study 2. Determination of Organochlorine Pesticides

The persistence of organochlorine pesticides in the environment is well known and the determination of these materials, especially at very low concentrations, in locations remote from sources can be subject to contamination during sampling. Sarkar and Sen Gupta (1989) have used an *in situ* sampler in a programme to determine organochlorine pesticides in coastal waters off the west coast of India at a depth of 20 m. The sampler pumps large volumes of water slowly ($50–200 \text{ ml min}^{-1}$) through a column containing either Amberlite XAD-2 resin or pre-cleaned polyurethane foam which retains any organochlorine compounds present in the water. After sampling about 150 l of water the instrument was retrieved, the column removed, sealed, stored on board and returned to the

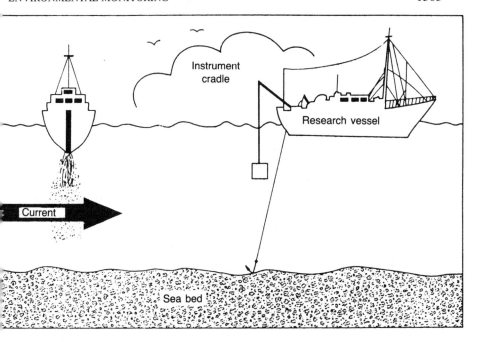

Figure 36.6   Diagram of the positions of vessels and instruments during the field study. (Source Byrne *et al.*, 1988)

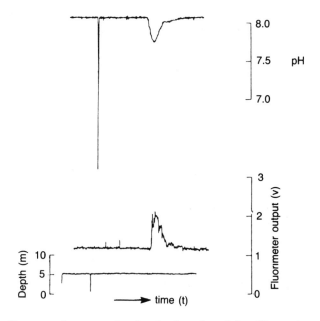

Figure 36.7   Chart recorder output showing the detection of the effluent plume by both the 'Aquatracka' fluorimeter and the pH probe. (Source Byrne *et al.*, 1988)

laboratory. The organochlorine pesticides were solvent eluted from the column. After concentration and purification of the extracts they were analysed by gas chromatography with an Ni^{63} electron capture detector. Concentrations of aldrin, dieldrin and total DDT sampled using XAD-2 resin were 1.42–9.8, 2.10–50.98 and 15.81– 444 ng l^{-1} respectively.

Case study 3. Speciation of Heavy metals in River Sediments

Knowledge of the speciation of heavy metals in waters and sediments is important from the aspect of the bioavailability and hence toxicity of the metals. The metal content of sediments reflect the quality of the water system and can be used to assess the presence of those metals which do not remain soluble after discharge. In addition sediments can act as sources of these materials as a result of changes in environmental conditions such as pH or redox potential.

Pardo *et al.* (1990) used voltammetric methods to determine the metal content for five different categories in sediment samples from the Pisuerga river in the industrialized town of Valladolid, Spain. They used the sequential extraction proposed by Tessier *et al.* (1979) to determine (a) adsorptive and exchangeable (b) bound to carbonates (c) bound to reducible phases (d) bound to organic matter and sulphides and (e) residual metals. Figure 36.8 shows a sample trace of the determination of zinc, cadmium, lead and copper by differential pulse anodic stripping voltammetry (DPASV). The results indicated that the polluting potentials of cadmium and lead were the greatest with 68 and 53% respectively of the total metal appearing in the first three, more mobile phases. Zinc showed a similar distribution, confirming its environmental similarity to cadmium.

36.4.3 Sampling and Sensing of Air Pollutants

General methodology

Monitoring of air pollutants can be conveniently divided into two main areas— the sampling and analysis of particles and the sampling and analysis of gases. Each of these two areas can have either passive or active collection systems, and each of these systems can be further subdivided into two types—direct or continuous measurement and collection followed by analysis in the laboratory. The sampling and sensing of air pollutants have been described in detail in several texts (Harrison, 1986; Noll and Miller, 1977; Stern, 1976; Stern *et al.*, 1984) and will only be summarized here.

There are essentially four components in any airborne pollutant sampling system:

● an inlet manifold which transports material to the collecting medium;

● a collecting medium which may be a filter for collecting particles, a liquid or solid adsorber for collecting gases, or a chamber to contain an aliquot of air for analysis;

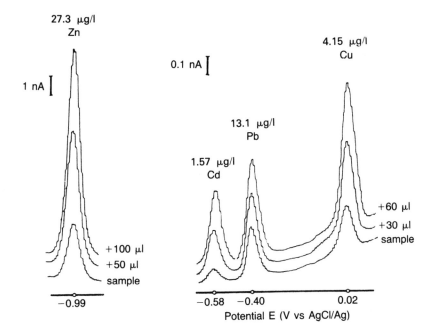

Figure 36.8 Determination of Zn, Cd, Pb and Cu by DPASVC. (Differential Pulse Anodic Stripping Voltammetry) Concentration of the standards: Zn = 15.0 mg/l, Cd = 2.5 mg/l, Pb = 40 mg/l, Cu = 3.0 mg/l

● a flow measurement device to determine the volume of air being sampled—this could be a rotameter, critical orifice or mass flow meter;

● a pump to draw air into the sampling system.

Particles and gases may be sensed by active or passive methods, using either collection and then analysis, or direct sensing methods.

Particles

In the *active* and *collect and analyse* procedure for measuring particulates the most common method for the collection of particles is filtration. Table 36.9 lists different types of filter media, their advantages and disadvantages, and the kinds of application for which they are suitable.

The sampling of particulate matter from stationary sources is important from the point of view of assessing the performance of arrestment equipment and compliance with emission standards. One example of a standard procedure (BSI, 1971) covers sampling from stacks for particles larger than 1 μm. The equipment includes a pitot static tube, inclined gauge manometer, thermometer, probe tube, flow meter, flow-rate control valve, connecting tubing, air pump and a particle collection component. One important aspect of stack sampling for particles is to ensure that the air flow in the sampling probe is such that the velocity at the face of the probe equals that immediately upstream in the stack. This is known as iso-kinetic sampling, and its importance can be seen in Figure 36.9. Particles travelling in a uniform gas stream have their inertia directed along the streamlines (Figure 36.9a). A probe inserted into the gas flow will be modified depending on whether there is no flow in the probe (Figure 36.9b), lower flow (Figure 36.9c), equal flow (Figure 36.9d) or higher flow (Figure 36.9e) compared to the flow in the stack. The sampling of particles in ambient air masses is seldom performed under isokinetic conditions owing to the large variations in wind speed and direction. The efficiency of ambient aerosol sampling devices is dependent upon the particle size, inlet geometry and wind conditions. High-volume samplers, which draw air at $1-2$ m^3 min^{-1} through a 20×25 cm glass fibre filter, have been commonly used for the assessment of ambient aerosol concentrations, especially in the USA.

Table 36.9   Advantages and disadvantages of different filter media

Filter type	Advantages	Disadvantages
Cellulose fibres	— Cheap — Suitable for light reflectance methods	— Poor collection — Trace impurities prelude use when chemical analysis of sample required
Glass fibres	— High efficiency — Can withstand high temperature — Low resistance to air flow	— Fragile — High trace element back-ground — Tendency for artefact formation (e.g. sulphate from SO$_2$)
Membrane	— Particles collected on surface so ideal for microscopical examination — High collection efficiency — Low levels of background impurities — Do no cause artefact formation of sulphates and nitrates	— High resistance to air flow — Brittle — Expensive

Once the particulate material has been collected it may be examined by a number of techniques such as microscopy, density gradient fractionation, light relection from the filter, X-ray diffraction, X-ray fluorescence. The filter deposit may also be extracted into either an aqueous or non-aqueous phase for the determination of metal, inorganic (e.g. sulphate, nitrate chloride) or organic (e.g. hydrocarbons) components using some of the analytical techniques described earlier for water pollutants.

In the *active* and *direct sensing techniques* analysis of airborne particles is exemplified by nephelometers in which scattered light is measured, the amount of scattered light being proportional to the mass loading of atmospheric particles for a typical atmospheric aerosol, and by the use of β-attenuation gauges where particles collected on a filter in a given time attenuate the electrons produced by a β-source, the degree of attenuation being a function of the mass of particles on the tape. The nephelometer technique is capable of giving an instantaneous measure of aerosol concentration, whereas the β-attenuation method will only produce an average concentration over the sampling period.

The *passive, collect and analyse* methods centre around use of a variety of deposit gauges which can be used for monitoring the deposition of particulate matter and these were probably the earliest instruments used for air pollution measurement. Harrison (1986) has reviewed various designs, examined the advantages and disadvantages and identified some applications of these gauges, particularly their use in identifying local emissions of dusts.

Gases

There are a number of active, collect and analyse methods for gaseous pollutants. They are briefly summarized in Table 36.10. It is important there should be no interference from particles often also present, and it is common for these to be removed from the air stream, usually by filtration. An example of the use of this technique for the collection and analysis of gaseous pollutants is the monitoring of sulphur dioxide (SO_2).

Analysis of absorbing solutions can take place by solution spectrophotometry or ion chromatographic techniques. The analysis of gaseous organic pollutants can be achieved by gas chromatographic methods.

Many of the *continuous direct sensing* methods employed for the direct sensing of gaseous airborne pollutants involve optical measurements. Two examples are now discussed.

In *chemiluminescence* use is made of chemical reactions that release energy in the form of light, called chemiluminescence. The radiation can be detected by photomultiplier tubes, the magnitude of the response being proportional to the gas concentration. Nitric oxide(NO) reacts in this way with ozone (O_3) producing excited NO_2 molecules which lose their energy either by light emission or by collisions with other molecules. Atmospheric NO_2 can be determined in the same instrument by first of all passing it through a heated stainless steel tube which decomposes the NO_2 to NO. The total NO + NO_2 (NO_x) is subsequently measured and the NO_2 obtained by difference. Other examples of this technique include the determination of ozone by reaction with ethylene, and the determination of

a.

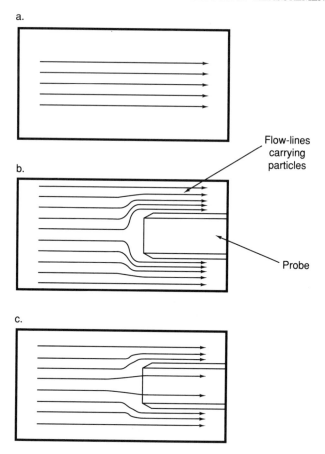

b.

Flow-lines
carrying
particles

Probe

c.

Figure 36.9   Streamline pattern around a sampling probe in a uniform flow field with different flow rates into the probe (See text)

gaseous sulphur compounds by burning them in a hydrogen-rich flame and monitoring the chemiluminescence produced from the radiative combination of atomic sulphur atoms.

Fluorescence can be used to monitor molecules which absorb radiant energy at one wavelength and re-emit it at another. The gas flows continuously through an optical cell and is continuously irradiated by a pulsed UV source with the fluorescence emission being detected by a photomultiplier tube placed at 90° to the excitation beam. Sensitivities of one part per billion can be achieved for SO_2.

The *remote sensing* technique is most important in atmospheric monitoring for measuring concentrations at heights in the first few kilometres above ground level (the lower troposphere) where the dispersion of pollutants takes place and is controlled by meteorological factors. Plumes can be measured where they still have no impact at ground level, and it is useful for measuring flux of emissions

d.

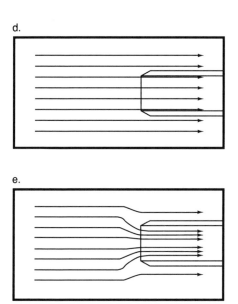

e.

Fig 36.9 (Cont.)

either from industrial sources or in total urban plumes. It has advantages too in ground-level measurements where an area can be rapidly scanned by a single

Table 36.10 Methods for collecting gaseous pollutants

Method	Comments
Adsorption	Adsorber must • have high relative surface area • have correct affinity for materials of interest • not react chemically with gases • be strong • have known retention capacity
Absorption	Pollutant of interest must be soluble in absorbing medium—this can be helped by allowing irreversible chemical reaction with absorbing solution. Absorption efficiency can be improved by increasing gas–liquid surface area and contact time. Absorption efficiency must be determined to avoid oversampling
Condensation	Gas stream cooled to below boiling point or freezing point of gases of interest.
Grab sampling	Direct sampling and isolation in an impermeable container. No concentration of sample possible, therefore sensitive analytical method required. Sample container must be assessed for reactivity with gas of interest

remote sensor as opposed to deploying a network of point samplers. There are two methods which can be adopted:

- either use solar radiation, and monitor modifications produced by pollutants,

- or introduce radiation and measure modifications along the light path either with instruments at both ends, or with a reflector at one end and transmitter and receiver at the other.

Correlation spectroscopy, LIDAR (LIght Detection And Ranging) and differential optical absorption spectroscopy are optical techniques used in remote sensing, and for further details the reader is referred to Varey (1986).

Another technique, utilized in the continuous monitoring of gaseous pollutants, particularly in the workplace environment, makes use of the energy liberated during the catalytic oxidation of a combustible gas. The detector is a miniature calorimeter or pellistor which consists of a coil of platinum wire supported on a refractory bead and surrounded by a precious metal catalyst. The adsorption of gases on the solid surface of semiconductor materials also provides a means of detecting toxic and flammable gases.

Passive samplers for the collection of gaseous pollutants depend upon the diffusion properties of the gas to be measured. Diffusion takes place through a stagnant air layer and the gas is absorbed in a specific reagent. One of the limitations of this technique is that because of the lower transfer rates compared to dynamic systems, coupled with insensitivity of analytical methods, it can generally be used only for long-term sampling or sampling where concentrations are high such as in workplace situations. An example is the determination of NO_2 using an acrylic tube (74 mm × 10 mm) open at one end and a triethanoloamine-coated stainless steel mesh at other. It was this particular example that Palmes and Gunnison (1973) used when they codified the factors controlling uptake rate in the application of Fick's laws of diffusion and demonstrated that the atmospheric concentration averaged over the exposure period can be determined. The adsorbed NO_2 is determined spectrophotometrically. These low-cost methods are extensively used in workplace monitoring (Berlin *et al.* 1987), in addition to being used in ambient sampling programmes to determine weekly average concentrations (DoE, 1989).

Case studies of air pollutant measurement

Case Study 1. Diurnal Variation of Ozone Concentrations in Athens

This study by Güsten *et al.* (1988) used continuous ozone monitors located at five sites in Greater Athens to elucidate the complex interaction of coastal wind circulation patterns and the generation and destruction of atmospheric oxidants in that city. Athens, a coastal city, is located in an area of complex topography and is surrounded on three sides by mountains over 1000 m high. About 40% of Greek industry and 50% of all vehicles registered in Greece are concentrated in the relatively small (450 km^2) area of the Athens basin.

Because of the high emissions of oxidant-forming precursors (NO_x and hydrocarbons) and the large number of cloudless days, high levels of ozone in excess of the US Air Quality Standard of 120 parts per billion are reached frequently in summer. Ozone was monitored continuously at five sites (see Figure 36.10) from June to September 1984 using continuous monitors, four based on UV absorption and one using the chemiluminescent reaction of ozone and ethylene. The average diurnal variation of the hourly ozone concentrations for the five sites of Figure 36.10 are shown in Figure 36.11. The situation at the Immitos site (1000 m above MSL), which shows a lack of any diurnal variation, is completely different from the other four sites in that on average it does not experience the ozone generated in the Athens basin below. The island site at Aegina, some 30 km from Athens, experiences diurnal variations in ozone concentration and this is attributed to transport by the land breeze of primary pollutants from Greater Athens which, under the conditions of strong insolation, high temperatures and light winds, react to generate ozone.

Case Study 2. Relation between Lead in Air and Lead in Petrol

Reduction in the lead content of petrol in the UK following the introduction of legislation required to implement EC Directive 85/210/EEC at the beginning of 1986, has been used to assess the relationship between lead in petrol and in outside air in urban environments (Pattenden and Branson, 1987).

Measurements of airborne lead concentrations were made at four sites in parts of London (Borough of Brent) and at four sites in Manchester. Samples were collected by drawing air vertically upwards at about 7 l min^{-1} through a Whatman 40 cellulose filter using a continuously operating pump. The total air flow passing through the filter was determined by a gas meter. Equipment was located about 5 m above ground level on the outside of buildings, generally in residential areas and not (with one exception) close to busy main roads. Filters were changed on a monthly basis and the lead was determined by X-ray fluorescence. Here the sample was bombarded with X-rays and absorption occurred followed by the emission of secondary X-rays. The intensity of these secondary emissions, by comparison with a suitable standard, was used to quantify the metal present. Calibration standards were prepared by dropping known volumes of standard solutions onto blank filters and allowing them to dry.

The monthly average concentration of lead in air at the London and Manchester sites for the two-year period from June 1984 is shown in Figure 36.12, which also shows (top line) the consumption rate of lead in petrol. The results in Figure 36.12 indicate that the rate of response of the airborne lead concentration was rapid and that the response time was probably less than the one-month time resolution of the measurements.

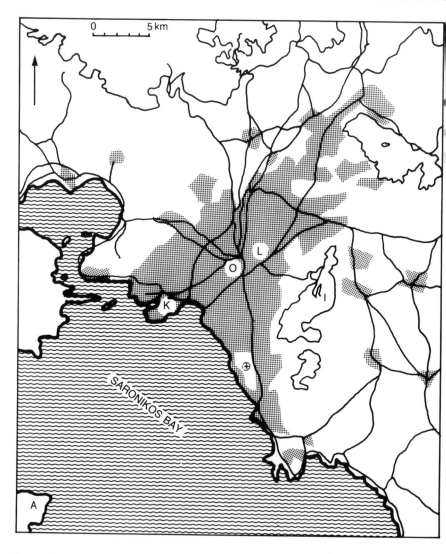

Figure 36.10   Map of Greater Athens with the O_3 monitoring sites. O, National Observatory of Athens (107 m above MSL); L, hill of Lykabetos (~ 200 m above MSL); I, Mount Immitos (~ 1000 m above MSL); K, Kastella (Piraeus) on the shoreline; A, island of Aegina (~ 200 m above MSL). Residential areas (dotted) and 500 m height contours are indicated

Case Study 3. Chloride Aerosols in the Atmosphere

A study of chloride aerosols at two sites (one urban—Leeds; one rural—Haverah Park) in central Northern England by Willison *et al.* (1989) used automatic dichotomous samplers to collect two aerosol size fractions. One fraction, designated fine, consisted of aerosols < 2.5 μm and the other fraction, designated coarse, consisted of aerosols in the range 2.5–15 μm. By collecting these two fractions it was possible to distinguish between aerosols derived from combustion sources(fine) and those derived from natural sources (coarse) such as sea spray and wind-blown dusts. Samples were collected on polypropylene-backed teflon filters every 24 hours. Filters were weighed before and after exposure on a microbalance, and were analysed for water-soluble ions as follows: chloride, sulphate and nitrate by ion chromatography; calcium and magnesium by atomic

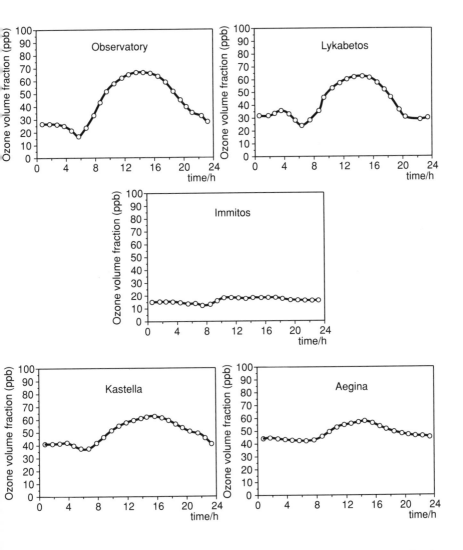

Figure 36.11 Average decimal variation of hourly O₃ volume fractions at different sites in greater Athens. (Source Güsten *et al.*, 1988)

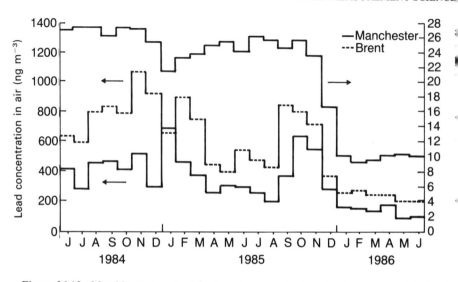

Figure 36.12 Monthly average lead particulate concentrations in air at residential sites in Manchester and Brent, London, and monthly lead consumption rates in petrol in the UK. (Source Pattenden and Branson, 1987)

absorption spectrophotometry; sodium and potassium by atomic emission spectrophotometry; and ammonium by the colorimetric phenol–hypochlorite reaction.

The relationships between the chloride, magnesium and sodium concentrations in both fractions of aerosol at the Leeds site is shown in Figure 36.13, which also shows the relationship between these elements in bulk sea-water. As can be seen there is good correlation for the coarse fraction but not so for the fine fraction. The contribution made by the main components of marine aerosols can be estimated by taking the measured sodium and magnesium concentrations and combining them with calculated chloride and sulphate concentrations derived from the bulk seawater relationships of

$$[Cl] = 0.95([Na + [Mg])$$

$$[SO_4] = 0.098([Na + [Mg])$$

and assuming that these relationships apply to aerosols. The contributions are shown in Table 36.11 and indicate that, even in the central UK, marine aerosols, of which chloride is the major component, contribute 10–20% to the total coarse particle mass. Furthermore, the average marine contribution to total chloride concentration at Leeds was about 64%.

Case Study 4. Nitrogen Dioxide Distribution in Street Canyons

The EC Directive on air-quality standards for nitrogen dioxide specifies that monitoring should be performed in areas predominantly affected by pollution

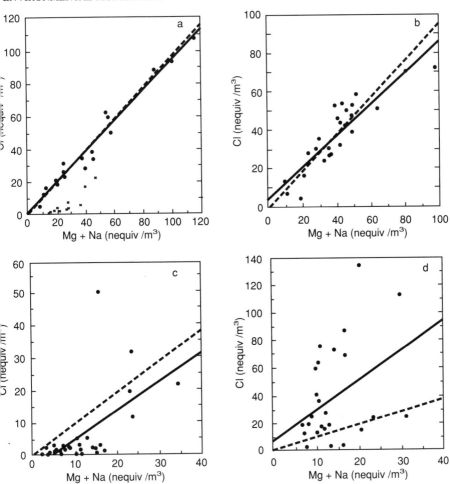

Figure 36.13 Relationship between Cl^- and $Mg^{2+}+Na^+$ in aerosols at Leeds: (a) coarse aerosol, August/September 1983; (b) coarse aerosol, December 1983; (c) fine aerosol, August/September 1983; (d) fine aerosol, December 1983. The relationship between these elements in bulk seawater is also shown (———). (Source Willson et al., 1989)

from motor vehicles and be in the vicinity of roads carrying heavy traffic with sites being selected to cover street canyons and intersections. Despite these apparently specific site descriptions there is considerable debate as to where precisely to sample.

Laxen and Noordally(1987) have measured NO_2 concentrations at over 40 locations in two canyon-like streets in central London by deploying passive diffusion tube samplers. The tubes were attached to windows or street furniture with the open end of the tube in freely circulating air wherever possible. A precision of about 8% was achieved based on replicate exposure of tubes.

Table 36.11   Measured and calculated concentrations ($\mu g\ m^{-3}$) of the main marine aerosol components and the marine aerosol contribution (%) to fine and coarse particles for the period August/September and December, 1983 (from Willison et al. (1989)

	Total mass	Na$^+$ m	Mg^{2+} m*	c†	Cl$^-$ m	c	SO$_4^{2-}$ m	c	Calculated marine aerosol contribution to total mass (%)
Leeds									
Coarse—									
summer	19.5	0.61	0.13	0.07	1.01	1.11	0.97	0.15	10
Winter	14.3	0.69	0.12	0.09	1.32	1.27	0.77	0.18	16
Fine—									
summer	30.3	0.21	0.02	0.02	0.20	0.38	7.38	0.05	2
Winter	29.1	0.27	0.03	0.01	1.36	0.44	4.85	0.06	3
Haverah Park									
Coarse—									
summer	11.0	0.42	0.08	0.05	0.66	0.77	0.45	0.11	12
Fine—									
summer	26.0	0.26	0.03	0.03	0.26	0.47	6.67	0.07	3

* m = measured
† c = calculated

The sampling locations and results at different times for one site are shown in Figure 36.14. At this site the buildings rise to 25 m on either side of the street which carries about 48 000 vehicles/day with approximately 13% being Diesel-engined powered. The results indicate concentrations decreasing from the kerb to the building facade with the highest concentrations being along the central reservation. The concentrations are fairly uniform along the street but are higher closer to and upstream of traffic lights. The concentrations decrease with height, Figure 36.15, by about 15–30% to become close to the local background near the top of the canyon. This study has demonstrated that if concentrations are close to the limit value for NO$_2$ then relatively small locational differences could be crucial in determining whether or not the standard is exceeded.

36.5   STANDARDIZATION AND CALIBRATION

There is need for clearly formulated measurement practice and protocols so that industry can discharge its responsibilities towards compliance with regulations. National regulatory bodies of, for example, EC countries have to develop a uniform approach to assessing compliance with community directives. Validation of air- and water-quality measurements requires an overall approach which includes:

• the development of sampling and analytical methods;

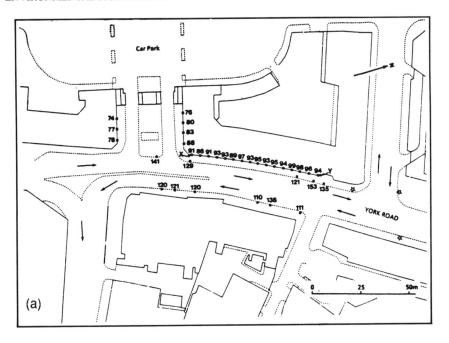

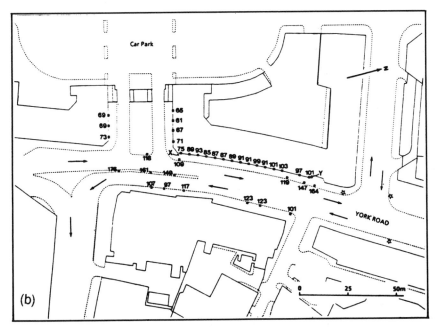

Figure 36.14   NO₂ concentrations ($\mu g/m^3$) at the York Road site: (a) 28 January to 4 February 1985; (b) 17–24 June 1985. The stars represent traffic lights. X and Y are the vertical profile locations. (Source Laxen and Noordally, 1987)

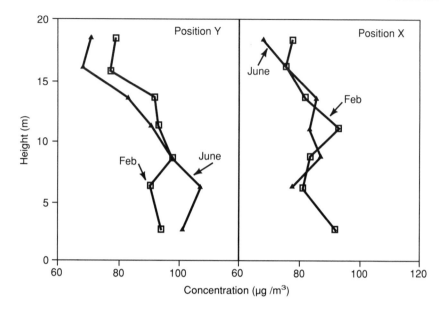

Figure 36.15   NO$_2$ variations with height at two locations in York Road (see Figure 36.14 for locations X and Y). (Source Laxen and Noordally, 1987)

- the provision of standards and certified reference materials;
- ensuring compatibility between measurements made by different laboratories.

In the EC the standardization and calibration initiatives are coordinated by the European Community Bureau of Reference (BCR). National laboratories and organizations within the member states contribute to expert working groups of the BCR.

In the UK the National Measurement System overseen by the Department of Trade and Industry provides a framework for standardized measurement not only in the field of environmental monitoring but also in all aspects of the economy. For air-quality matters Warren Spring Laboratory (WSL) undertakes the development and evaluation of sampling and analytical methods, and is also an accredited calibration centre for the measurement of important air pollutants such as sulphur dioxide, oxides of nitrogen, ozone and carbon monoxide and is available to laboratories who require their measurements to be traceable to international standards. WSL also coordinates inter-laboratory comparability exercises, particularly in the analysis of rainwaters (e.g. Heyes, 1988).

The Standing Committee of Analysts provides analytical protocols and assesses various techniques for the analysis of waters and associated materials. The British Standards Institution (BSI), in addition to providing product standards, also provides standards for environmental sampling and analytical techniques.

In the USA the Environmental Protection Agency (EPA) provides reference methods for environmental monitoring. Friedman, (1990) summarizes the different organizations within the EPA involved with the development and approval of testing methods and the steps required to issue a method.

The American Society for the Testing of Materials (ASTM) issues standard methods of sampling and analysis of environmental pollutants as does the Intersociety Committee of the American Public Health Association.

36.6 FUTURE TRENDS

Developments in analytical techniques

There is a need for environmental chemists to be able to analyse smaller and smaller quantities of pollutants in order to trace environmental pathways and to quantify human exposure.

Search for techniques that link chromatography (both gas and liquid) to atomic absorption, atomic fluorescence or plasma emission detectors is a currently expanding research area. As yet no commercial instruments are available. This is being applied to the determination of organometallic species. These species are important not only from the point of view of increased emissions from increased industrial production (e.g. organotins) but also from their production in the biogeochemical cycles of trace metals in sediments and the aquatic environment. Harrison and Rapsomanikis (1989) examines the state of the art and addresses application to organometallic compounds of tin, germanium, lead, arsenic, antimony, mercury and selenium.

Development of in-situ monitors

There are increasing requirements for continuous *in situ* monitoring to be able to reduce manpower resources involved in collecting samples and analysing them in the laboratory. This is particularly so in the case of water-quality monitoring where there are only a few techniques which are robust and reliable enough for continuous operation. In addition to developing robust and reliable methods for water-quality monitoring, there is also a need to devise techniques with the required sensitivity for direct measurement.

The use of biological systems such as fish monitors, which are being used in monitoring drinking water sources, is becoming more widespread (Gruber and Diamond, 1988).

This problem of having the necessary sensitivity for continuous systems is not so acute for air-quality monitoring, where there are already continuous analysers which will monitor directly ambient concentrations. However, there will be developments of existing principles such as spectroscopic techniques and remote sensing making use of laser light sources and fibre optics (Mohebati and King, 1988).

36.7 SOURCES OF FURTHER ADVICE

The reader is referred to the following textbooks and journals for further information on environmental monitoring.

Books

Air Pollution, Volume 3, Third edition. Stern, A. C. (ed.) (1976) Academic, New York.

The Chemical Analysis of Water-General Principles and Techniques, Second Edition. Hunt, D. T. E. and Wilson, A. L. (1986) The Royal Society of Chemistry, London.

Examination of Water for Pollution Control-A Reference Handbook
   Volume 1 Sampling Data Analysis and Laboratory Equipment
   Volume 2 Physical, Chemical and Radiological Examination
   Volume 3 Biological, Bacteriological and Virological Examination
   Suess, M. J. (ed) (1982) Published by Pergamon Press on behalf of the World Health Organisation Regional Office for Europe.

Handbook of Air Pollution Analysis, Second edition. Harrison, R. M. and Perry, R. (eds) (1986) Chapman and Hall, London.

Reclaiming Contaminated Land Cairney, T. (ed.) *(1987) Blackie, London.*

Journals

Analytical Chemistry American Chemical Society, USA.

Atmospheric Environment Pergamon Press, UK.

Environmental Pollution Elsevier, UK.

Environmental Science and Technology American Chemical Society, USA.

Environmental Technology Selper, UK.

Journal of the Air Pollution Control Association Air Pollution Control Assoc., USA.

Science of the Total Environment Elsevier, The Netherlands.

Water Research Pergamon Press, UK.

REFERENCES

ASME (1957). *Test Code for Determining Dust Concentrations in a Gas Stream*, Power Test Code, 27, American Society of Mechanical Engineers.

ASTM, American Society for Testing and Materials (1981). 'Standard Test Method for Particulates Independently of Particulates and Collected Residue Simultaneously in Stack Gases'. D3685–78, Part 26 in *ASTM Annual Book of Standards*, Philadelphia.

Berlin, A., Brown, R.H. and Saunders, K.J. (eds.) (1987). *Diffusive Sampling-An Alternative Approach to Workplace Air Monitoring*, Royal Society of Chemistry, London.

BSI British Standards Institution, (1961). *British Standard Simplified Methods for Measurement of Grit and Dust Emissions from Chimneys (Metric Units)*, BS 3405:1961, London.

BSI British Standards Institution, (1971). *British Standard Simplified Methods for Measurement of Grit and Dust Emissions (Metric Units)*, BS 3405:1971, London.

Byrne, C.D., Law, R.J., Hudson, P.M., Thain, J.E. and Fileman, T.W. (1988). 'Measurements of the dispersion of liquid industrial waste discharged into the wake of a dumping vessel', *Water Res.*, **22**, 1577–84.

CEC, Council of the European Communities (1980a). 'Directive on air quality limit values and guide values for sulphur dioxide and suspended particulates (80/779/EEC)', *Official Journal*, **L229**, 30 August.

CEC, Council of the European Communities (1980b). 'Directive relating to the quality of water intended for human consumption (80/778/EEC)', *Official Journal*, **L229**, 30 August.

CEC, Council of the European Communities (1982a). 'Directive on limit values and quality objectives for mercury discharges by the chloralkali electrolysis industry (82/176/EEC)', *Official Journal*, **L81**, 27 March.

CEC, Council of the European Communities (1982b). 'Directive on a limit value for lead in air (82/884/EEC)', *Official Journal*, **L378**, 31 December.

CEC, Council of the European Communities (1983). 'Directive on limit values and quality objectives for cadmium discharges (83/513/EEC)', *Official Journal* **L291**, 24 October.

CEC, Council of the European Communities (1984) 'Directive on limit values and quality objectives for mercury discharges by sectors other than the chloralkali electrolysis industry (84/156/EEC)', *Official Journal*, **L74**, 17 March.

CEC, Council of the European Communities (1985a). 'Directive on air quality standards for nitrogen dioxide (85/203/EEC)' *Official Journal*, **L87**, 27 March.

CEC, Council of the European Communities (1985b). 'Directive on the assessment of the effects of certain public and private projects on the environment (85/337/EEC)', *Official Journal*, **L175**, 5 July.

DOE, Department of the Environment (1987a). *Acid Deposition in the United Kingdom 1981–1985*, United Kingdom Review Group on Acid Rain Second Report, Warren Spring Laboratory.

DOE, Department of the Environment (1987b). *Ozone in the United Kingdom*, United Kingdom Photochemical Oxidants Review Group Interim Report, Harwell Laboratory.

DOE, Department of the Environment (1988). *The Effects of Acid Deposition on the Terrestrial Environment in the United Kingdom*, United Kingdom Terrestrial Effects Review Group First Report, HMSO, London.

DOE, Department of the Environment (1989). *The Effects of Acid Deposition on Buildings and Building Materials in the United Kingdom*, United Kingdom Building Effects Review Group Report, HMSO, London.

DOE, Department of the Environment (1990a). *Acid Deposition in the United Kingdom 1986–1988*, United Kingdom Review Group on Acid Rain Third Report, Warren Spring Laboratory.

DOE, Department of the Environment (1990b). *Oxides of Nitrogen in the United Kingdom*, United Kingdom Photochemical Oxidants Review Group Second Report. Harwell Laboratory.

EPA, Environmental Protection Agency (18 August 1977). Standards for Performance for New Stationary Sources, *US Federal Register*, Part II, **42** (160), 41753–89.

Friedman, D. (1990). Testing methodology in environmental monitoring, *Environ. Sci. Technol.*, **24**, 796–8.

Gruber, D.S. and Diamond, J.M. (1988). *Automated Biomonitoring—Living Sensors as Environmental Monitors*, Ellis Horwood, Chichester.

Güsten, H., Heinrich, G., Cvitas, T., Klasinc, L., Ruscic, B., Lalas, D.P. and Petrakis, M. (1988). 'Photochemical formation and transport of ozone in Athens, Greece', *Atmos. Environ.*, **22**, 1855–61.

Harrison, R.M. (1986). 'Analysis of particulate pollutants', in *Handbook of Air Pollution Analysis* (2nd edn), Harrison, R.M. and Perry, R. (eds.) Chapman & Hall, London.

Harrison, R.M. and Rapsomanikis, S. (1989). *Environmental Analysis using Chromatography Interfaced with Atomic Spectroscopy*, Ellis Horwood, Chichester.

Heyes, C.J. (1988). *Interlaboratory Comparison of Precipitation Analysis within the National Materials Exposure Programme—January 1988*, LR 695(AP), Warren Spring Laboratory.

Hunt, D.T.E. and Wilson, A.L. (1986). *The Chemical Analysis of Water-General Principles and Techniques* (2nd edn), Royal Society of Chemistry, London.

Laxen, D.H.P. and Noordally, E. (1987). 'Nitrogen dioxide distribution in street canyons', *Atmos. Environ.*, **21**, 1899–1903.

Lord, D.W. (1987). 'Appropriate site investigations', in *Reclaiming Contaminated Land*, Cairney, T. (ed.), Blackie, London.

Mancey, K.H. and Allen, H.E. (1982a). 'Design of measurement systems', in *Examination of Water for Pollution Control*, Vol. 1, Suess, M.J. (ed.) Pergamon Press, Oxford.

Mancey, K.H. and Allen, H.E. (1982b). 'Automated monitoring and analysis', in *Examination of Water for Pollution Control*, Vol. 1, Suess, M.J. (ed.) Pergamon Press, Oxford.

Mitchell, N.T. (1982). 'Radiological examination', in *Examination of Water for Pollution Control*, Vol. 2, Suess, M.J. (ed.), Pergamon Press, Oxford.

Mohebati, A. and King, T.A. (1988). Remote detection of gases by diode laser spectroscopy, *J. Modern Optics*, **35**, 319–24.

Moore, D.J. (1986a). 'Planning and execution of an air pollution study', in *Handbook of Air Pollution Analysis* (2nd edn), Harrison, R.M. and Perry, R. (eds.) Chapman & Hall, London.

Moore, D.J. (1986b). 'Air pollution meteorology', in *Handbook of Air Pollution Analysis* (2nd edn), Harrison, R.M. and Perry, R. (eds.) Chapman & Hall, London.

Noll, K.E. and Miller, T.L. (1977). *Air Monitoring Survey Design*, Ann Arbor Science, Ann Arbor.

Palmes, E.D. and Gunnison, A.F. (1973). 'Personal monitoring device for gaseous contaminants', *Am. Ind. Hyg. Assoc.* **32**, 78–81.

Pardo, R., Barrado, E., Perez, L. and Vega, M. (1990). 'Determination and specification of heavy metals in sediments of the Pisuerga River', *Water Res.*, **24**, 373–9.

Pasquill, F. and Smith, F.B. (1983). *Atmospheric Diffusion* (3rd edn), Ellis Horwood, Chichester.

Pattenden, N.J. and Branson, J.R. (1987). 'Relation between lead in air and in petrol in two urban areas of Britain', *Atmos. Environ.*, **21**, 2481–3.

Pio, C.A. (1986). 'General sampling techniques', in *Handbook of Air Pollution Analysis*, (2nd edn), Harrison, R.M. and Perry, R. (eds.) Chapman & Hall, London.

Sarkar, A. and Sen Gupta, R. (1989). 'Determination of organochlorine pesticides in Indian coastal water using a moored in-situ sampler', *Water Res.*, **23**, 975–8.

Schulte, H.F. (1976). 'Radionuclide surveillance', in *Air Pollution* (3rd edn), Vol. 3, Academic Press, New York.

Standing Committee of Analysts, Department of the Environment (1984). 'High performance liquid chromatography, ion chromatography, thin layer and column chromatography of water samples 1983', *Methods for the Examination of Waters and Associated Materials*, HMSO, London.

Stern, A.C. (ed.) (1976). *Air Pollution* (3rd edn), Vol. 3, Academic Press, New York.

Stern, A.C., Boubel, R.W., Turner, D.B. and Fox, D.L. (1984). *Fundamentals of Air Pollution* (2nd edn), Academic Press, New York.

Tessier, A., Campbell, P.G.C. and Bisson, M. (1979). 'Sequential extraction procedure for the speciation of particulate trace metals', *Analyt. Chem.*, **51**, 844–51.

Thompson, K.C. (1980). 'Ultra-violet and visible solution spectrophotometry and colorimetry 1980 version—an essay review', *Methods for the Examination of Waters and Associated Materials*, HMSO, London.

Varey, R.H. (1986). 'Remote monitoring techniques', in *Handbook of Air Pollution Analysis* (2nd edn), Harrison, R.M. and Perry, R. (eds.) Chapman & Hall, London.

Warren Spring Laboratory (1972). *National Survey of Air Pollution 1961–1972*, Vol. 1, HMSO, London.

Willison M.J., Clarke, A.G. and Zeki, E.M. (1989). 'Chloride aerosols in central northern England', *Atmos. Environ.*, **23**, 2231–9.

Wilson, A.L. (1982). 'Design of sampling programmes', in *Examination of Water for Pollution Control*, Vol. 1, Suess, M.J. (Ed.), Pergamon Press, Oxford.

Wilson, D.C. and Stevens, C. (1981). *'Problems Arising from the Redevelopment of Gas Works and Similar Sites*, AERE Harwell Report R 10366.

PART 2

Developing Technologies

Chapter

37 M. R. HASKARD

Microelectronics in Instrumentation

Editorial introduction

The fastest-growing technology used to manufacture sensors is the microelectronic form. These developments have taken place over several decades being characterized by intense academic and industrial research that has not always provided production results as fast as were expected. Microelectronic sensors are, however, now finding application in automobiles in literally billions. This indicates they are a maturing sensor form that will soon become common place in all applications where low cost and small size are needed. Within this technology the youngest form is the biosensor; Chapter 40 deals with those.

37.1 INTRODUCTION

The term 'microelectronics' was coined during the 1950s. However, that does not mean that such devices did not previously exist. The prefix 'micro' is from the Greek *mikros*, meaning small or exceptionally little. Certainly, there were many attempts to make miniature circuits for the armed services during the Second World War, for hearing aids and so forth. Nor should one assume that the integrated circuit (IC), accepted as the cornerstone of microelectronics, is a recent idea. German vacuum tube manufacturers had, by 1926, integrated into a single glass envelope several vacuum tubes with resistor–capacitor networks to form an audio amplifier.

Today, microelectronics is an all pervasive technology and can be found in every manufacturing industry. It is, therefore, not surprising that it forms an important part of today's instrumentation systems. Figure 37.1 shows one example where on a single silicon chip there is a complete instrument system

Handbook of Measurement Science, Volume 3
Edited by P. H. Sydenham and R. Thorn
© 1992 John Wiley & Sons Ltd

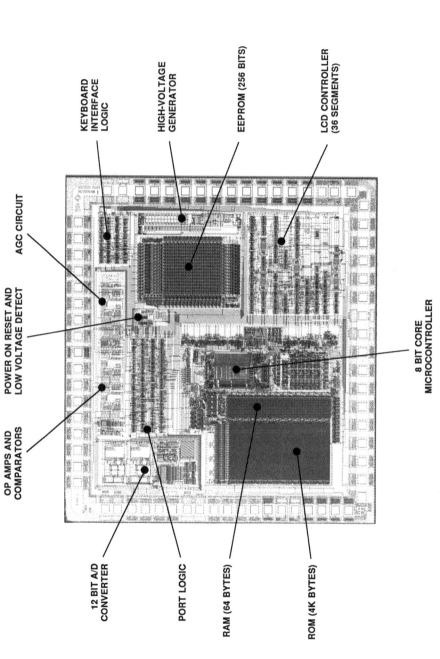

Figure 37.1 A complete system on a chip. A medical measurement and control system which includes digital, analog, memory and 8-bit microcontroller circuitry. (Courtesy of SIERRA Semiconductors, California, USA)

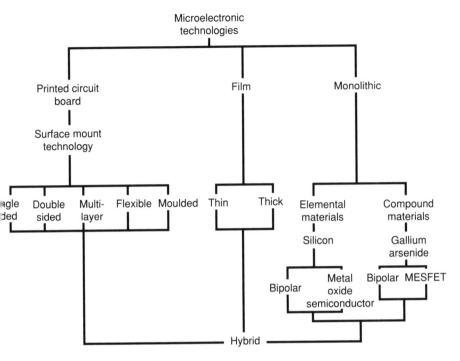

Figure 37.2   Microelectronic technologies

comprising operational amplifiers, 12-bit analog-to-digital (A/D) converter, an 8-bit microcontroller, memory consisting of 64 bytes of RAM, 4 kilobytes of ROM and 256 bits of EEPROM, a keyboard interface and an AGC circuit. The flexibility, reliability and ability to provide high performance comes about because microelectronics is not a single technology, but a range of technologies, that can be used separately or merged together to provide a solution. Figure 37.2 shows these technologies in tree diagram form.

Each of the three main types of microelectronics, namely printed circuit board, film and monolithic can provide two services, components and interconnections. Printed circuit board technology is only capable of producing a few components such as inductors, capacitors and strip line components. Its major function is interconnection. Miniature components are reflow soldered, or chip and wire bonded, to the board surfaces. Film technologies also have an important interconnection role to play, but provide a greater range of low-cost components, principally the passive components such as resistors, inductors, capacitors and microstrip. Again, discrete components and integrated circuits are mounted on the substrate using reflow soldering, conducting epoxy or chip and wire bonding methods. Until recently, the monolithic process had been solely used to produce complex active circuits of both digital and analog form. However, silicon substrate technology is now being developed to interconnect silicon integrated circuits. Since the same technology is being employed, there are no thermal mismatches, elimination of waste area taken up by mechanical transformation or going through bonding pads and packing densities can be extremely high (Dettmer, 1988).

With the passage of time all of these technologies are becoming easier to use, principally through improvement in computer aided design methods. Not only is the software more user friendly and fully integrated, but the range of software has expanded allowing schematic entry, electrical and thermal analysis, automatic layout, generation of net lists and test vectors and so forth. Part of this process has been the improvement in the library cells, so that the need to generate both primitive and complex circuits is eliminated. Libraries contain layout and electrical information on simple cells like logic gates and amplifiers through to complex cells such as A/D converters and microprocessors.

While each of these technologies can be used individually, it is more usual that two or more be merged to provide the optimum manufacturing solution to any problem. Such a solution is called a hybrid. All three technologies may be employed in a complex system, for example, monolithic chips mounted on single in-line thick film daughter boards which are in turn mounted end-on to a printed circuit motherboard. Which technologies are used depends very much upon the

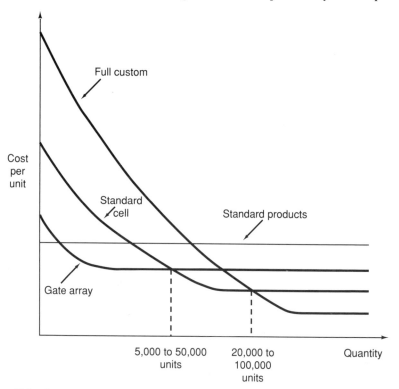

Figure 37.3   Cost comparison for several integrated circuit manufacturing processes showing their dependence upon product quantity

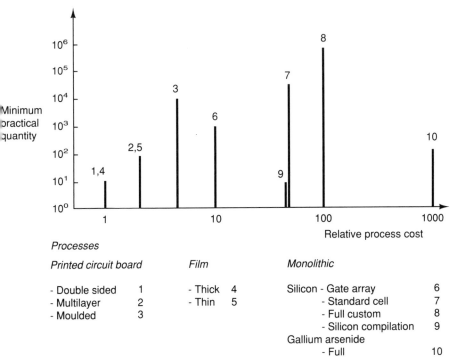

Figure 37.4 Comparison of relative costs for the major microelectronic technologies

system complexity and the number of systems required. For example, in the monolithic area there are many approaches and the correct one depends upon the product quantity. This is illustrated in Figure 37.3. Similarly, thin film circuits are more complex than thick film, and moulded and multi-layer printed circuit boards more expensive than double sided. This is illustrated in Figure 37.4.

Each of these technologies will now briefly be considered.

37.2 MICROELECTRONIC TECHNOLOGIES

37.2.1 Introductory Remarks

As previously discussed there are three main manufacturing technologies, but they can be broken down into further groups. In this section only the major methods will be discussed. They are all planar processes in that the three-dimensional construction is divided down into a number of layers or planes. The design process is to define the geometry or layout for each of these planes, while manufacture involves the building up or processing of each of these layers to form the final device or circuit.

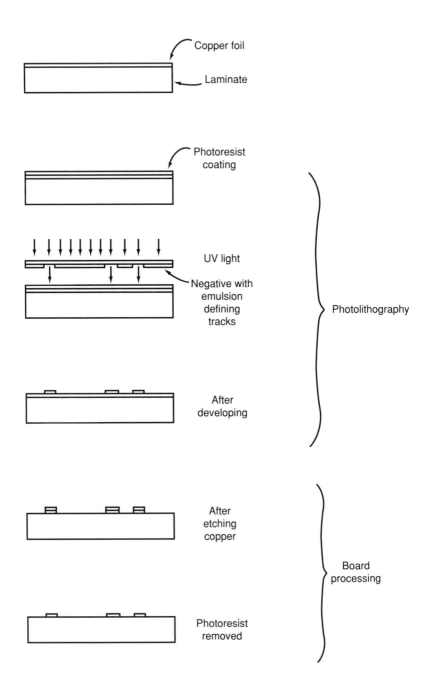

Figure 37.5   Photolithography steps and etching of a printed circuit board

37.2.2 Printed Circuit Boards

Originally invented in 1943 during the Second World War in the United Kingdom, the principal advantages of the PCB technology were that it produced consistent electrical parameters and was ideal for mass production of electrical and electronic products (Dummer, 1978). It consists of a layer of copper foil (half or one ounce weight) fixed to one or both sides of an insulating laminate such as synthetic rosin bonded paper or epoxy glass.

The conducting track layout is transferred to the board by a photolithographic method illustrated in Figure 37.5. The board is coated with a photo-resist material, either by roller coating, dipping, spraying or spinning and then exposed to ultraviolet light through a photographic negative. The photo-resist that receives the light polymerizes and in development of the resist does not dissolve away.

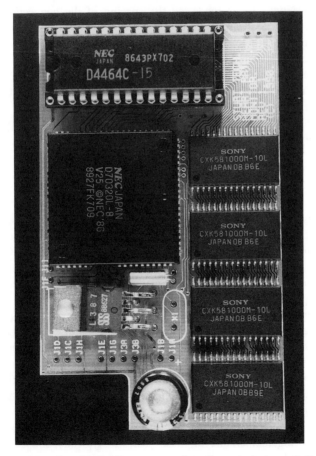

Figure 37.6 Example of an SMT AT computer of credit card size, actual dimensions are 87 mm × 55 mm. (Courtesy of Sturt Technologies, Adelaide, Australia)

Thus, where tracks are to remain the boards are protected by a layer of photo-resist. The board is next etched in an appropriate solution, such as ferric chloride or ammonium persulphate. The photo-resist is then removed, the board cleaned, coated to protect the track from oxidizing, guillotined to size and then drilled. Since the development of this simple process additional steps have been added to allow coating of the copper with tin lead solder, solder masking so that the soldering only occurs on the pad areas, gold plating of connector pins, plated through holes and vias and more than two conductor layers (multi-layer boards). In all cases track widths and spacing are typically greater than or equal to 125 μm (5 mil) with 250 μm (10 mil) the preferred figure.

In the late 1980s surface mount technology (SMT) was introduced, where, as the name implies, components are mounted direct onto the tracks (one or both sides) using special surface mount packaged components and reflow soldering. In the case of very large-scale integration (VLSI), silicon chip packages may have several hundred connections to be made, so the spacing of the connections have been reduced over the years from 2500 to 1250, to 1000 to 625 and now to 500 μm. This has forced board manufacturers to reduce track width and spacing resulting in very complex boards. Figure 37.6 shows an example of a microprocessor board the size of a credit card having the power of an AT personal computer.

Surface mount technology is a process that can be highly automated. It is therefore suitable for high-volume production runs, providing sophisticated, small-size, highly reliable products. Haskard (1991) gives an in-depth account of this assembly process.

37.2.3 Thick-film Technology

Thick film is the name given to the method of producing circuits using the screen printing process. Like the printed circuit board it has its origins in the 1940s where it was developed to produce low-cost proximity fuses for bombs (Cadenhed and DeCoursey, 1985). Special conductive, resistive and dielectric pastes (or inks) are screen printed onto insulating substrates, the pattern on the screen determining where each of the pastes is to be printed. In between each printing there is normally a settling time, a drying time and finally a furnace firing in accordance with a predetermined time temperature profile. Some pastes allow co-firing. Figure 37.7 shows an example of a thick-film module consisting of a pressure sensor and processing circuitry to provide both a digital and analog output.

There are three basic thick film families—high, medium and low firing tempera-ture types. In the first, palladium or ruthinium based pastes are employed in conjunc-tion with a noble metal such as silver or gold to produce conductors and resistors. Borosilicate glass frit is also added and this material forms the basis for dielectric pastes. The firing temperature of these pastes is typically 850°C. The substrate material, normally alumina of 94–96% purity, is able to withstand this temperature. Components are printed to about a tolerance of ±20% and, where necessary, are laser

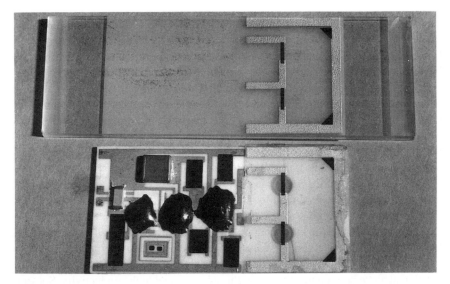

Figure 37.7 A thick-film pressure sensor and processing circuitry. (Courtesy Microelectronics Centre, University of South Australia)

trimmed to ±1%. These pastes provide the most stable properties of all three types and have found wide acceptance in the electronics industry.

The medium temperature family of pastes are similar in composition to those just described but designed to fire at about 600°C. Special substrates based on low carbon steel coated with porcelain or a screened dielectric layer are employed. This family is not as popular as the two other kinds.

The final family requires firing at temperatures of only 150°C and can be printed on many plastic materials including printed circuit boards. It fits nicely into already established production technology. Two types of paste exist, the thermoplastic type used mainly for membrane switches, while the second, thermosetting plastic, allows a complete range of conductor, resistor and dielectric pastes, based principally on silver, carbon and an epoxy type material.

To accommodate the new VLSI packages with large lead counts on fine pitches, multi-layer processes have been developed using either the conventional or complementary approaches. Layouts are undertaken using computer aided design workstations to specified geometric rules. Track widths and spacings are typically greater than, or equal to 0.5 mm with similar minimum resistor and capacitor dimensions. An in-depth account of thick film is given in Haskard (1988).

37.2.4 Thin-film Technology

While the name suggests that the only difference between this type of microelectronic circuit and that just discussed is the thickness of the material laid down, such is not the case. Today, thin film principally refers to the vacuum deposition of electronic circuits. Thick-film circuits are typically 15 μm thick, whereas thin film can be as low as a few tens of Ångström but, with plated steps, can exceed thick-film thicknesses.

Early pioneers of this technology were Thomas Edison, in 1884, and W.R. Gove, in 1952, developing the evaporation and sputtering processes respectively (Cadenhed and DeCoursey, 1985). While the basic ingredients of the process were formulated last century, it was not until midway through this century that the understanding and technologies developed sufficiently so that they could be used to manufacture repeatable and reliable circuits.

There are many thin film systems, but only two will be discussed here. The first uses nichrome for resistors, gold for conductors and silicon monoxide as the dielectric for capacitors and crossovers. All are evaporated using an electron beam gun. The second approach is to use tantalum-based materials, a process developed by Bell Laboratories. The dielectric for capacitors is formed by anodizing the deposited tantalum. The interconnection material is complex, consisting of multiple layers of metal to achieve both good adhesion to the substrate and high conductivity. Substrate materials are either special low alkaline content glasses or polished alumina of approximately 99% purity. Both the processes mentioned employ either masking and/or photolithographic methods to define the required circuit areas. The thin-film process is more expensive than thick film, but offers better resolution and therefore finer geometries and higher quality passive components.

37.2.5   Bipolar integrated circuits

From the invention of the bipolar transistor by Shockley and his team in 1948, it took only 10 years for the development of the integrated circuit. In America, Texas Instruments produced an integrated flip-flop while, in the United Kingdom, Plessey made an integrated relaxation oscillator. When Fairchild developed their Planar process integrated circuits took a giant leap forward.

Silicon bipolar technology has, until recently, been the main process in every area, be it digital, analog or high-voltage industrial. Because of this, there has been a tremendous amount of design effort and accumulation of bipolar technology knowledge. It is a stable, well-documented process, giving a reliable product. Various splinter processes were developed, allowing the selection of one which would best suit a customer's requirement. Thus transistor–transistor logic (TTL) and analog/integrated injection logic I^2L or merge transistor logic (MTL) are examples of such processes, the latter allowing the integration of analog and digital circuits on a single chip. All this can be done at low power levels. One process that is worthy of special mention is emitter-coupled logic (ECL), which provides moderate packing densities of circuits, but at very high speeds (Topham, 1987; Hunt and Saul, 1988). Transistors can have cut-off frequencies of several tens of gigahertz, therefore, frequency counters operating at 10 GHz and A/D

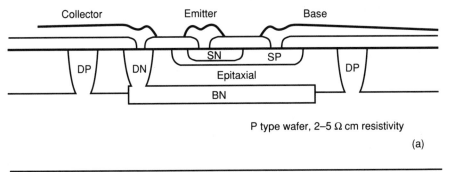

P type wafer, 2–5 Ω cm resistivity

(a)

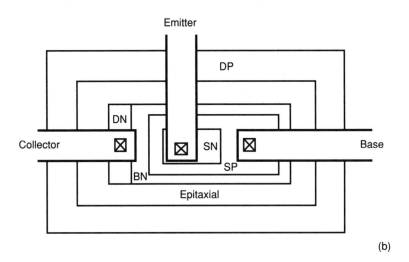

(b)

Figure 37.8   Standard NPN transistor: (a) diagrammatic cross-section; (b) top view

converters with bandwidths of 500 MHz are possible. Recently gallium arsenide circuits, based on MESFETs, have become available with similar performance figures, but often at higher costs.

The basic silicon bipolar process has as its standard component a vertical NPN transistor (Figure 37.8). Both vertical and lateral PNP transistors can also be implemented, but in the basic processes they have inferior properties. Such transistors are made in a wafer of single-crystal silicon, by a series of processing steps including epitaxial growth, diffusion or ion implantation (to control the level of impurity atoms and thus make an N or P region) and one or more layers of metallization to interconnect the transistors so fabricated to form the circuits.

The deep P layer (DP) forms the isolation area in the epitaxial layer in which the transistor is made. The deep N (DN) makes contact to the buried N (BN) layer to ensure good saturation resistance while the high-resistance epitaxial region is used to produced a high breakdown voltage transistor. The shallow P (SP) and shallow N (SN) regions form the base and emitter areas respectively.

Simple resistors made using the shallow N or shallow P diffusion or capacitors made by reverse biasing junctions can be employed in a circuit. An overglaze layer may be added to protect the circuit. Thus the number of layers and associated masks for the lithographic method can be many. For an analog/I^2L process the number of layers for a double metal process are as shown in Table 37.1.

Because of size, number of components and parasitic capacities, it is not possible to design integrated circuits using breadboarding techniques. Consequently, circuits must be carefully designed and simulated using sophisticated computer aided design methods. This is possible only when suitable models, including physical and manufacturing parameters, are available. This often presents a problem when a new design, such as a sensor, is being designed, for neither a model nor the appropriate parameters for that model exist.

Table 37.1 Number of layers for analog/I^2L double metal process

Mask Number	Mask symbol	Mask type	Comments
1	BN	Buried N plus	
—	—	Epitaxial	No mask needed
2	DP	Deep P	
3	DN	Deep N	
4	SP	Shallow P	
5	SN	Shallow N	
6	CO	Contact holes	
7	CE	Contact holes	Schottky diodes
8	$M1$	Metal 1 layer	
9	$V1$	Via M1 to $M2$	
10	$M2$	Metal 2 layer	
11	GZ	Over glaze	Protection

37.2.6 Metal Oxide Semiconductor Integrated Circuits

Since the advent of the integrated circuit its complexity has doubled approximately every year (Moore's law). In the case of microprocessors the number of transistors employed has increased by an order every five or six years (Barron, 1990; Nass, 1990) and is projected to reach 100 million by about the year 2000 (Figure 37.9). This growth has been sustained in recent years through the metal oxide semiconductor (MOS) process.

While there are a variety of processes the most common today is the complementary MOS (or CMOS) bulk process. It employs complementary enhanced MOS transistors (N and P channel types) that allow circuits with low static current drains, reasonable speeds (up to 100 MHz) and high packing densities.

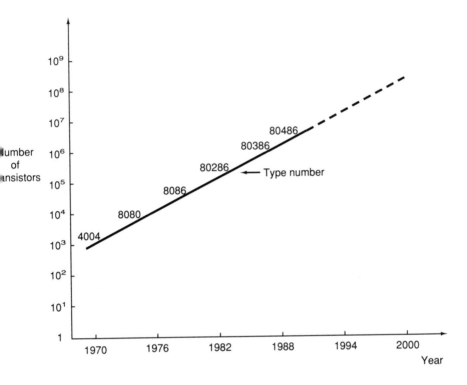

Figure 37.9 Actual and projected complexity of microprocessor chips with time

Up to one quarter of a million gates can be accommodated on a single die (that part of a wafer forming one of its many identical circuits). This CMOS process is suitable for both digital and analog circuits so that total systems can be placed on a single chip as illustrated earlier in Figure 37.1.

Like the bipolar process, MOS transistors are made simultaneously in a wafer of single crystal silicon or gallium arsenide and employs similar steps. One additional step is the chemical vapour deposition of polysilicon used as the gate material for the transistors. A cross section of a CMOS inverter is shown in Figure 37.10. As seen from this figure the total number of masks required for the process is at least 10. The essential ones for a CMOS bulk process are listed in Table 37.2.

Other CMOS based processes of interest are:

(1) silicon on sapphire (SOS) process which has improved speed because the parasitic capacitors associated with a conducting substrate are eliminated;

(2) BICMOS, which combines the bipolar and CMOS technologies to obtain the best from both processes.

Table 37.2   Masks needed for CMOS bulk process

Mask Number	Mask symbol	Comments
1	AA	Active area—transistor areas
2	PW and/or NW	P and N well areas
3	PI	P implant—source/drains
4	NI	N implant—source/drains
5	PY	Polysilicon for gates
6	CO	Contact cuts
7	M1	Metal 1 interconnect
8	V1	Vias between metal layers
9	M2	Metal 2 interconnect
10	GZ	Overglaze—protection

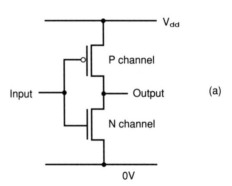

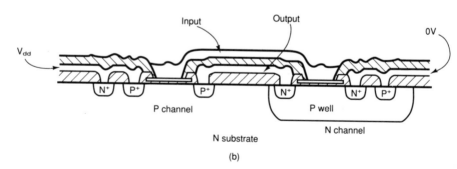

Figure 37.10   Cross-section of a simple CMOS inverter. The wells and substrates are connected to zero and positive supply rails respectively to reverse bias them ensuring isolation: (a) schematic circuit; (b) diagrammatic physical cross-section

37.2.7 Hybrid Circuits

Hybrid technology combines two or more of the electronic manufacturing processes already discussed. Since the monolithic process produces devices containing a large number of active devices and both the film and printed circuit board technologies are ideal for interconnection, the two types of processes complement each other. For example, a thick film circuit may provide the passive components and interconnection for several VLSI chips to form a compact instrument system. For larger systems, several such boards may be mounted vertically, single in line (SIL), on a printed circuit card to form a mother–daughter board system. Other components like power-supply bypass capacitors may be mounted onto the mother printed circuit board. This is all illustrated in Figure 37.11. For more detail see Haskard (1988).

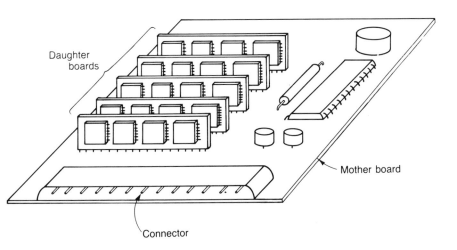

Figure 37.11   Mother–daughter board system giving a small densely populated microelectronic packaging system

37.3 MICROELECTRONICS BUILDING BLOCKS FOR AN INSTRUMENT SYSTEM

Microelectronics, and particularly the monolithic form, has significantly influenced the development of all electronic systems, including instrument systems. Most instrument and control systems interface into the real or physical world, which is analog and, therefore it is not surprising that until recently instrument systems were dominantly analog in nature.

Analog electronic systems employed few components, allowed the use of linear circuit and system theory to design them and thereby resulted in systems that operated in real time, at low cost. The systems employed analog computing

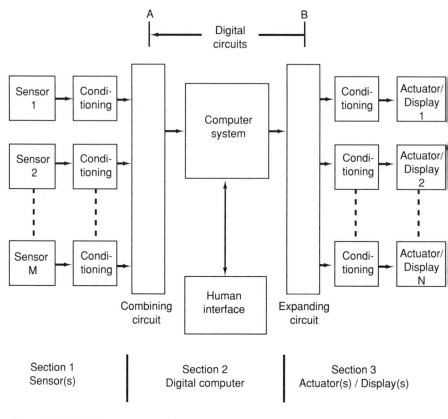

Figure 37.12   Basic components of an instrument or control system

techniques, namely operational amplifiers, integrators, multipliers and summation circuits. Although simple in concept there were many disadvantages, including noise and instability. Further, each system was designed for a dedicated purpose.

Digital systems tended to be far more complex and, therefore, expensive and the theory of discrete signal systems had not been developed to the same extent as continuous signal systems. The coming of the integrated circuit and, in particular, the microprocessor changed all of this. The monolithic process has the advantage that simple circuits can be easily replicated and integrated together to form complex systems with the cost of manufacture of these complex systems frequently being no more than that of a discrete simple device or circuit. This property, coupled with the poor absolute tolerance achievable with a monolithic process, meant that complex low-cost digital, rather than analog circuits, started to appear in the early 1970s.

A decade later, microprocessor and personal computers were common so that discrete signal analysis and simulation methods developed rapidly. Use of digital computers in instrument and other systems not only allowed the characteristics of a system to be placed under software control, but also gave the advantages of improvements in process power and human interface.

Today, the thrust of instrument system design is to convert signals into a digital form as quickly as possible, allowing as much of the signal processing to be undertaken in digital form. The conversion can even occur in the sensor itself, so that they can be interfaced directly into a computer-based system. A typical instrument or control system can be divided into three sections (Figure 37.12) namely:

(1) sensors and conditioning circuits;

(2) digital computer;

(3) actuators/displays.

The computer system employed in the central section may vary considerably in complexity. It could be a simple microprocessor chip where the interface to the human operator is a small liquid crystal display and keypad. A printer or floppy disk drive may be employed for extracting data from the system. At the other end of the spectrum a personal computer may be used with full keyboard, display, hard disk drive and many other features. Plug-in instrumentation cards exist for many of the common personal computer systems, including multiplexers, analog amplifiers, A/D and D/A converters. Software packages are commer-

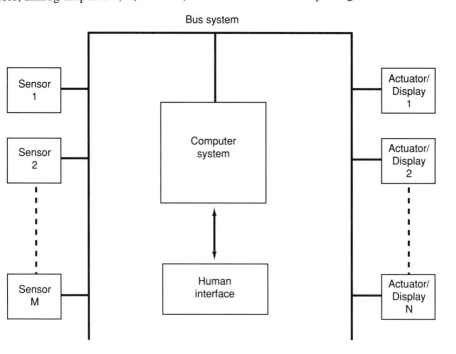

Figure 37.13 Computer-bus organized instrument and control system

cially available for these systems to perform analysis on any data and to display results as graphs, tables, pie charts and statistical tables.

Today, the emphasis on instrument system design is to try and extend the boundaries between the analog and digital interface (shown as point A and B in Figure 37.12) by shifting these points to the left and right respectively as far as possible. Further, if the sensor (conditioning block) output can be digital then all sensors (actuators) can be placed on the computer bus and treated as any other peripheral, Figure 37.13. In Figure 37.12 the output from the sensor conditioning unit is either an analog voltage, an RS232 interface, a 4–20 MA current loop or other forms. In the bus system shown in Figure 37.13, a standard instrumentation bus such as the IEEE 488 or a simpler two-wire bus such as the proprietary Inter Integrated Circuit (IIC) or microline type used. For further details of instrumentation bus systems see Chapter 42.

No matter which of the two instrument systems is used, microelectronic components are employed to implement them. Blocks such as instrument and operational amplifiers, comparators, A/D and D/A converters, bus drivers, multiplexers and so forth are all available as standard integrated circuit building blocks. They can be connected into a system using film or printed circuit board technologies. This approach has the advantage that the most appropriate technology may be selected for each block and combined into a total system, again using the most appropriate technology. No significant compromises are necessary. Perhaps the only penalty is the lack of commercial security in that other organizations can readily duplicate the system since standard products have been used.

Microelectronics has also had a further impact on instrument systems in that all of the technologies are suitable for mass production and can result in comparably low-cost products. This is needed if microelectronic-based instrument and control systems are to be included in many consumer products. This includes domestic products such as irons for pressing clothes, light switches for detecting people's presence and controlling the level of illumination, and automobiles which can have numerous systems to monitor and control engine performance and car-handling ability. It has been estimated that in a single year more pressure sensors would be used in motor vehicles than the total number that have been ever manufactured in the world since pressure transducers were first invented. The increased range of low-cost microelectronics sensor systems is now flowing on into the professional field so that it is becoming easier to expand and apply the concept in other areas. Examples are:

- *Agriculture.* Measurement of parameters such as depth and pressure on cultivator tines, grain moisture level, body weight and fat levels of animals, automatic feed facilities.

- *Horticulture.* Control of glass houses, soil moisture level and automatic watering.

- *Medical.* Personal instruments for monitoring glucose levels, heart rate and body temperature.

● *Building.* Energy measurement and control in buildings, building movement monitoring in earthquake areas.

In developing these systems it has been realized that the same principles that apply to computer networks could also be applied to instrument systems, that is

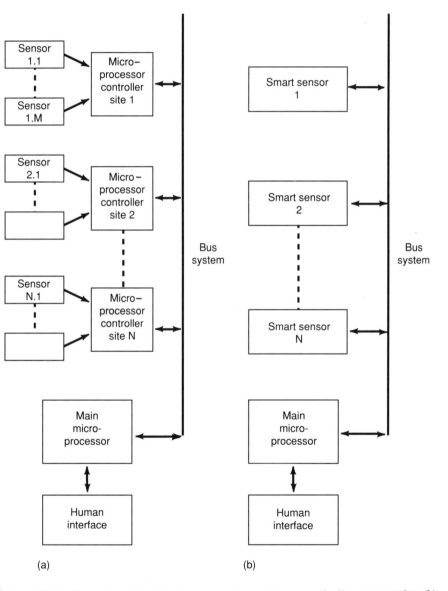

(a)   (b)

Figure 37.14   Examples of distributed sensor systems: (a) sensors feeding smart nodes; (b) sensors contain smart features

they are predominantly sensor and acuator networks. Just as computer peripherals today are smart peripherals and the network intelligence is distributed, so the thrust with instrument systems is to distribute microprocessors throughout. Two examples are shown in Figure 37.14. In case (b) there is a move to put more intelligence right at the sensor interface.

37.4 INTELLIGENT SENSORS

37.4.1 Introduction

The concept of intelligence in sensors has been promoted for at least a decade (Giachino, 1986; Haskard, 1986; Brignell, 1989). An example is the magnetic field detector produced in 1984 by combining a silicon Hall effect sensor with a standard microprocessor chip using thick-film technology (Cooper and Brignell, 1985). The term *smart sensor* is often used in the literature, but due care must be taken to understand what authors mean. Often a sensor used with any form of integrated electronics is called a smart sensor. It does not necessarily mean the integration of a microprocessor in the sensor.

Sensor types can be categorized as shown in Figure 37.15. Simple sensors incorporate some level of electronics that allow corrections to be made typically to offset, range (gain) and temperature. The analog interface is typically an output

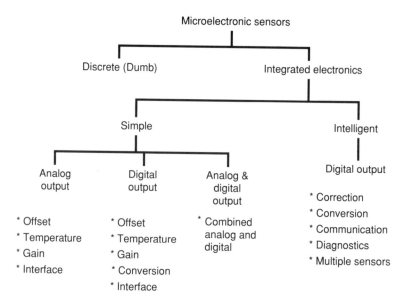

Figure 37.15   Classification of sensor types in terms of abilities

voltage or a 4–20 mA current loop. With digital output sensors, output can be as simple as a digital frequency, an RS232 interface or full bus interface. With many sensors containing integrated electronics the actual output from the sensor proper is analog and is converted to a binary signal with an on-board A/D converter. It is, therefore, often possible to provide both a digital and analog output from the sensor.

With the inclusion of a microprocessor on the sensor chip, a range of operations can be carried out under software control. These include

- correction;

- conversion;

- communication;

- control.

There is a range of corrections that can be applied, including offset, range, linearity and temperature effects. In the case of temperature, two options are available. The first is to undertake what is common practice for the simple integrated electronic sensor, namely, to build into the circuit temperature compensation. The second approach is to include a temperature sensor, so that correction can be made under software control for any temperature at which the sensor chip is operating. Combination of both methods is possible.

Conversion of the signal-to-digital form does not necessarily have to be achieved using the microprocessor. A simple oscillator, with a sensor incorporated in it causing frequency changes, is an example where A/D conversion methods are not necessary. The advantage of such an approach is the reduced silicon area. However, the inclusion of an on-chip A/D converter under microprocessor control may not consume a great amount of silicon area and has the advantage of flexibility in that multiple sensors may be multiplexed into the converter, a sensor being either on, or off, of the chip. Methods of achieving A/D conversion will be discussed later.

Without exception, intelligent sensor chips incorporate some bus connection systems. They have a programmable address set up using hard wire pins or preset using an on-chip ROM. Typically this may be achieved using fusible links, EPROM or EEPROM technologies. The unit may be allowed to operate only as a slave, that is, pass information to the control microprocessor when it is interrogated, or may be allowed to be a master/slave unit, so that it can request information from the control processor such as updated calibration information.

Control is the function of the microprocessor to monitor and ensure that the total system is operating correctly. This is achieved by running a diagnostic program. While seen as a giant step forward in improving sensor reliability, practical implementation is difficult. How to check that the various elements of the sensor chip are functioning correctly is a problem. The sensor element, converter, communication system as well as the microprocessor itself all need to be checked and confirmed to be operating correctly without greatly increasing the complexity of the system. A typical strategy used is to let the central microprocessor initiate a diagnostic check. This may be immediately prior to every reading, once a day or at any other preset time. The procedure is as follows:

(1) Self-test requested

(2) If the sensor responds within a certain time then the communication section is operating correctly.

(3) The on-chip microprocessor is tested by two simple arithmetic/logical/shift operations. The second is to make sure that a register has not stuck at the correct result by accident.

(4) Test the A/D converter by calibrating the error.

(5) Test the sensor proper by taking a reading and checking that it is in the linear range.

(6) Respond to the central microprocessor with the results from (3)–(5) above.

(7) The central processor can request a further sensor reading and compare it with the results from (5) above to have a better appreciation as to whether or not the sensor proper is operating correctly.

    The three elements of the intelligent sensor that need further comments are the microprocessor, converter and sensor element. These are now discussed in more detail with the proviso that a fuller discussion on sensors will be given later. This section concentrates on the sensor chip circuitry and its ability to handle multiple sensors.

37.4.2 Microprocessor Designs

The major thrust in conventional microprocessor design is speed and complexity to give improved performance. In the case of an on-chip microprocessor for sensor use, neither of these are required. In the first place the majority of physical and chemical reactions are relatively slow by electronic standards, so that sensor readings need only be taken at low rates. Secondly, the silicon area of any intelligent sensor chip must be kept minimal if it is to be an adequately low-cost device. Even the minimal complexity microprocessor occupies considerably more area than any of the other parts on the chip, be it the sensor or converter. Research into a minimum area microprocessor using a direct decoded instruction set of 16 instructions shows that a processor occupied an area 2 orders more than a sensor (including some electronics). Figures for a 2 μm CMOS process are shown in Table 37.3 (Haskard, 1990).

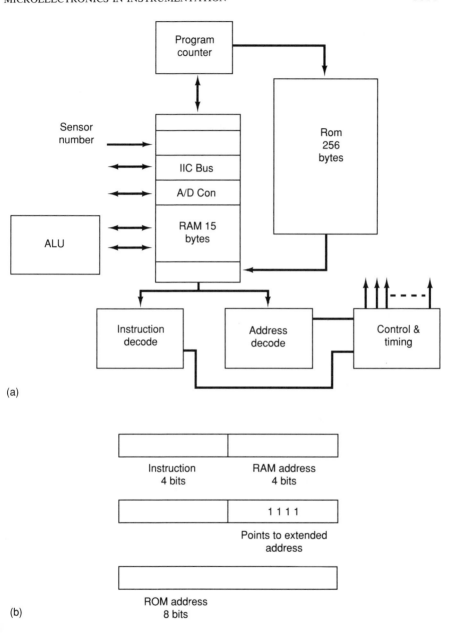

Figure 37.16 Trial minimum area microprocessor for an intelligent sensor: (a) system architecture; (b) register format

The simple microprocessor is shown in block diagram form in Figure 37.16, while Table 37.4 gives the silicon area required for each block.

Table 37.3   Silicon area for components of an
intelligent sensor

Function Block	Area (mm^2)
Hall sensor with amplifier	0.025
Temperature sensor with amplifier	0.06
A/D converter	0.13
8-bit serial microprocessor	4.6
8-bit parallel microprocessor	6.5

The initial approach taken (George *et al.*, 1988) was to employ a microprogrammed machine, but the resulting area was excessive. Any design must provide
a correct balance between the central core of the microprocessor (ALU, control
and timing blocks) and memory. A microprogrammed machine gives programming flexibility so that each software instruction can perform a complex operation
resulting in programs consisting of few instructions. Thus the program memory
size is reduced, but there is now the additional microcontroller memory and its
complexity. A simple direct decoded instruction set machine reduces the central
core of the processor, but increases the memory required for the program, for
each instruction performs a relatively simple operation. The balance point appears to be a directly coded instruction set machine with a reasonable set of well-
thought-out instructions. The design of the microprocessor should follow conventional VLSI practices, namely using tightly packed regular structures that can

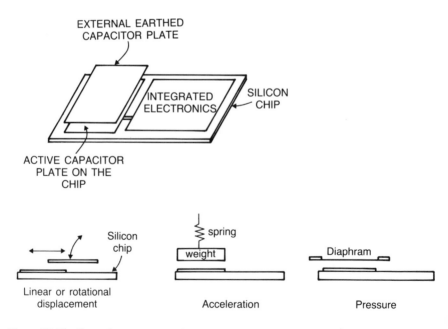

Figure 37.17   General purpose capacitance based sensor, with example applications.

be replicated many times. Thus RAM and ROM type structures rather than random logic areas should be employed.

Table 37.4 Silicon area required to produce the minimum area processor in 2 μm double metal CMOS process for both serial and parallel versions.

Function block	Serial processor		Parallel processor	
	Transistor count	Area(mm^2) count	Transistor count	Area(mm^2) count
Arithmetic Unit	80	0.1	626	0.75
Instruction decode	128	0.1	128	0.1
Accumulator and multiplexer	82	0.1	152	0.11
Program counter and parallel load	288	0.2	288	0.2
RAM decode	128	0.1	128	0.1
RAM (R_0 to R_{13})	980	1.1	1568	2.06
ROM	4096	1.67	4096	1.67
Timing	296	0.27	202	0.2
Communication channel		0.9		1.3
Total areas		4.6		6.5

The balance between the power or authority of the control microprocessor and the sensor chip microprocessor needs careful consideration. For example, in a truly distributed system the sensor processor has considerable power. This usually means large memory and, therefore, large silicon area. Holding information in the central processor and passing it to the sensor as and when required can reduce sensor chip memory, but puts more strain on the bus interface.

The technology used to produce the sensor is also important. Conventional CMOS technology is restrictive in that initial and subsequent sensor calibration information has to be stored dynamically and fed from the central microprocessor to each sensor on 'power up'. If EPROM or EEPROM technology is used then this information can be stored in the sensor. Such technology allows the sensor software to be upgraded and changed as and when the need arises.

Overall, the area occupied by the on-chip microprocessor will always be considerably larger than the other function blocks. Consequently, any intelligent sensor design should be developed to be as flexible as possible so that its applications can be extended to increase quantities produced, thereby reducing costs. Intelligent sensors should preferably be of a general type so that the on-chip sensor can sense a range of parameters. This may be by using several different sensor types on a chip, a capacitor type sensor that can be employed to measure a range of mechanical parameters (see Figure 37.17) and certainly the option of accepting inputs from off chip sensors.

37.4.3   Conversion techniques for intelligent sensors

The production of a digital output from an analog sensor can be achieved in one of the following ways (Haskard and May, 1987) They are:

(1)   A/D conversion;

(2)   voltage to frequency circuits;

(3)   sensor arrays.

These will be considered in turn. Their basic operation is explained as this then enables description of how they need modification for intelligent instrumentation use.

There are three approaches normally used in integrated circuit technology to produce A/D converters. They are shown in Figure 37.18.

The first class employs D/A converters, combined with a comparator and a digital register. The register is cycled in some way, its output connected to the D/A converter so that this result is compared in the comparator to the analog input. When the comparator switches, the two analog signals are identical so the register holds in digital form the equivalent of the analog signal. In the simplest case, a binary register can be counted up in staircase fashion. However, for an N-bit register this can take up to 2^N clock pulses. With the successive approximation method, starting from the most significant bit, each bit in the register is set and the output from the D/A converter compared with the analog input. Should the D/A converter output be too large, then the register bit just set is reset to zero as the next significant bit is set to 1. However, should the converter output be too small then only the next significant output bit is set. In this way only $N + 1$ clock cycles are needed, the extra one to initially reset the register and clear the previous result.

The D/A converters used in these two systems are based on having a number of identical resistors, capacitors or transistors. The principles are illustrated in

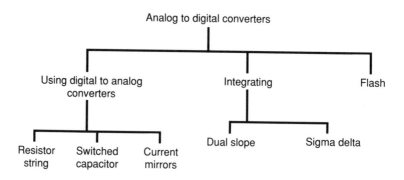

Figure 37.18   Integrated circuit, analog-to-digital converter classifications

Figure 37.19. The resistor string method is shown in Figure 37.19(a). As the binary register count increases the series switches allowing tapping further up the string, giving a larger analog output voltage. For an N-bit converter, there are 2^N resistors in the string. For a three-bit case illustrated there are eight resistors.

The switched capacitor type depends upon charge redistribution when capacitors are shorted together. For the three-bit case illustrated in Figure 37.19(b), the capacitors are weighted C, $2C$ and $4C$ with $4C$ being the most significant bit. The capacitors are switched to V_{ref} or zero volts depending on the value of a_i in the register. On the phase one clock pulse these voltages are applied to the capacitors so that the resulting charge is

charge on capacitor $C = a_0 C V_{ref}$
charge on capacitor $2C = 2a_1 C V_{ref}$
charge on capacitor $4C = 4a_2 C V_{ref}$

On the clock pulse phase two, all capacitors are shorted together to give the output. The charge is averaged over the total capacity of $7C$ giving

$$\text{analogue output voltage} = (4a_2 C + 2a_1 C + a_0 C)V_{ref}/7 \qquad (37.1)$$

If a fourth dummy capacitor of value C is added so that on phase one of the clock it is switched to zero volts and then on phase two is switched in parallel, it contributes no charge but shares the charge. We then obtain

$$\text{analogue output voltage} = (a_2/2 + a_1/4 + a_0/8)C V_{ref} \qquad (37.2)$$

In the current mirror method, Figure 37.19(c) two identical transistors are connected so that the second transistor mirrors the current in the first. Consider the bipolar transistors shown in Figure 37.20(a). The emitter current of each transistor is given approximately by

$$I_e = I_{CEO}\left(e^{\frac{qV_{be}}{kT}} - 1\right) \qquad (37.3)$$

where I_{CEO} is collector/emitter leakage current.

k is Boltzmann's constant
q is charge on an electron
T is absolute temperature.

Thus the two identical transistors, sharing the same base emitter voltage, have equal emitter currents. If they are high-gain transistors then base currents are negligible and collector currents equal emitter currents giving $I_1 = I_2$. In practice the non-finite current gains (β) produce an error so that

$$I_1/I_2 = 1 + 2/\beta \qquad (37.4)$$

where $2/\beta$ is the error term.

Three and four transistor circuits are used to reduce this error. However, this basic Widler circuit illustrates the D/A converter principle. If we replace transistor Q_2 by two transistors in parallel, both transistors being identical to Q_1, the current flow is now $I_2 = 2I_1 N$ transistors in parallel, will give current NI_1 (Figure 37.20(b)). It can now be understood by referring to Figure 37.19(c) that the output of the register switches, or does not switch, into resistor R, binary weighted currents I, $2I$ and $4I$.

All circuits given in Figure 37.19 are illustrative only and in practice are normally employed with an operational amplifier to provide buffering and reduce the effects of stray reactances.

Under the heading of integrating, there are two important types of A/D converters. The first, dual slope offers a very high immunity to noise and gives an extremely stable performance. The second employs very simple circuits that can be readily made in integrated circuit form. Figure 37.21(a) shows, in block diagram form, the dual slope converter.

The integrator is first switched to the analog input signal V_a for a specified number of clock cycles N. Thus the integrator output voltage is

$$V_i = -NTV_a \qquad (37.5)$$

where T is the clock period.

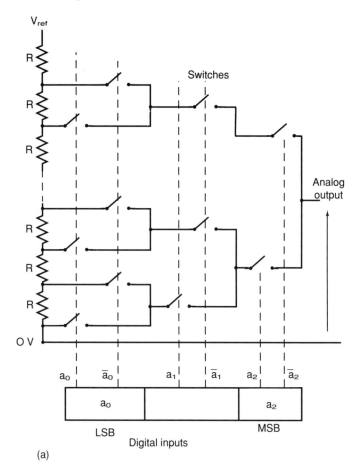

(a)

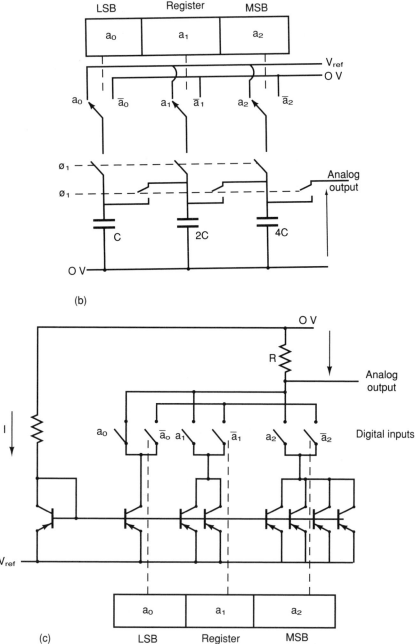

Figure 37.19 Circuit schematics illustrating the principles of commonly used integrated circuit digital to analog converters: (a) resistor string; (b) switched capacitor; (c) current mirror

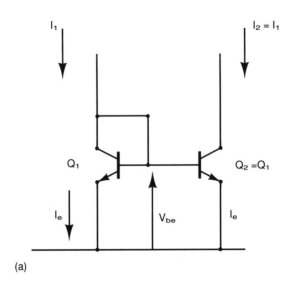

(a)

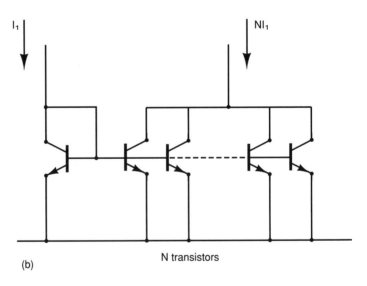

(b)   N transistors

Figure 37.20   Basic current mirror circuits: (a) Widlar; (b) current scaling through use of a parallel transistor combination

After N cycles the control switch to the integrator is changed to the reference voltage whose polarity is opposite that to that of the analog input voltage. The integrator output now integrates down towards zero. The counter counts the number of clock pulses X. Once the integrator output reaches zero volts, the comparator closes the 'AND' gate stopping counting. Thus we have

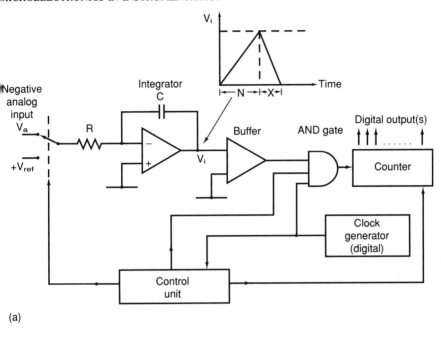

(a)

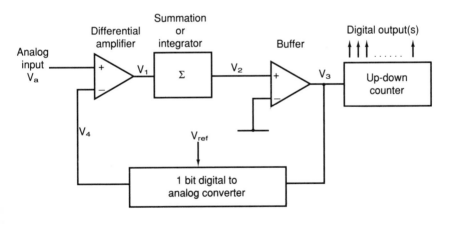

(b)

Figure 37.21 Integrating converter: (a) dual slope; (b) Sigma Delta or over sampling type

$$V_i = -XTV_{ref} \qquad (37.6)$$

Equating to equation (37.5), we obtain

$$V_a = XV_{ref}/N \qquad (37.7)$$

By scaling the term V_{ref}/N, the counter output can be made direct reading.

The Sigma Delta system of Figure 37.21(b) gets its name from the first two blocks, that is the output V_2 is the sum of the difference of the analog input voltage and the output from the simple, one-bit, D/A converter. The equations describing the operation of this converter are:

$$
\begin{aligned}
V_1 &= V_a - V_4 \\
V_2 &= \Sigma V_1 \\
V_3 &= 1 \quad \text{if} \quad V_2 > 0; \quad \text{else} \quad 0 \\
V_4 &= V_{ref} \quad \text{if} \quad V_3 = 1; \quad \text{else} \quad -V_{ref} \qquad (37.8)
\end{aligned}
$$

If we examine the case for a constant input voltage of 0.2 volts and a reference voltage equal to 1 volt then it can be seen from Table 37.5 that the process repeats itself after five cycles. The average output from the D/A converter is, for those five cycles, 0.2 volts, that is, the input voltage. Further, if the number of one and zero states of voltage V_3 are counted then we can achieve the same results provided that the number of samples is taken as 2×10^N where N is the number of significant bits (decimal bits in this example).

Table 37.5 Examples showing the operation of the Sigma Delta digital to analog converter

Sample No.	V_1	V_2	V_3	V_4	Comments
1	0.2	0.2	1	1	
2	−0.8	−0.6	0	−1	← *
3	1.2	0.6	1	1	
4	−0.8	−0.2	0	−1	Average output
5	1.2	1.0	1	1	$((-1+1-1+1+1)/5$
6	−0.8	0.2	1	1	$= 0.2$
7	−0.8	−0.6	0	−1	← *
8	1.2	0.6	1	1	
9	0.8	−0.2	0	−1	

For an analog input voltage of 0.2 volts and a reference voltage of 1 volt (see Figure 37.21(b)), the operation is illustrated in more detail in Figure 37.22. The output into the counter is shown firstly for 20 samples giving an average count of 4 or $4/20 = 0.2$ volts. If 200 samples are taken then the answer can be resolved to 2 significant figures. If for example the input voltage is 0.19 volts rather than precisely 0.2 volts then the output as shown in the figure indicates that there are now 38 excess 1 pulses giving 38/200 or 0.19 volts output. In all cases the analog input voltage is held constant during the sampling time. Hence an alternative name for this converter is 'over sampling'.

The final converter type covered here gives the fastest conversion time and hence its name—flash converter. An N-bit converter consists of 2^N comparators coupled to a logic block to derive the output. This is illustrated in Figure 37.23.

Input voltage V_a = 0.2 volts
Output signal V_3 sequence for 20 samples   10101101011010110101
giving 4 excess 1 pulses = 4/20 = 0.2 volts

Input voltage V_a = 0.19 volts
Output voltage V_3 sequence for 200 samples   10101101011010110101
01101011010110101101
01101011010110101101
01011010110101101011
01011010110101101011
11010110101101011010
11010110101101011010
10110101101011010110
10110101101011010110
10101101011010110101

There are 38 excess 1's giving 38/200 equal 0.19 volts.

Figure 37.22 Illustrative case for analog voltages of 0.2 and 0.19 V respectively and the pattern of 1s that are fed into the up–down counter

The comparators that are connected to reference voltages that are less than the analog input voltage, switch, so the logic block simply searches for the uppermost comparator to switch. Because the basic system employs many comparators various schemes are available to reduce the number, for example, using half the comparators and sampling the bottom half of the range and then the upper half (Cole, 1987). Alternatively, comparators can be omitted and interpolation methods employed (Van de Grift et al., 1987).

Having described the basic operation of A/D conversion discussion now is directed to their further modification to provide self-calibration.

In the area of intelligent sensors, the speed of the flash comparator is not required. What is needed is a simple converter for which self-calibration is possible. These are normally the dual slope and the switched capacitor types. Figure 37.24 illustrates how self-calibrating methods may be employed. The dummy capacitor C added to the circuit given in Figure 37.19(b) is employed as the standard reference capacitor. Firstly, it is compared with the other capacitor of value C (Figure 37.19(b)) and the out-of-balance charge noted and used to provide a correction figure during conversion. Next the two single C capacitors are connected in parallel and then compared with the single $2C$ capacitor as shown in Figure 37.24(c). Again the out-of-balance charge is noted and stored in the microprocessor memory as an error term. Continuing this process, capacitors are paired, checked and errors noted. The error for each bit is stored digitally and then applied after a conversion through a series of arithmetic additions or subtractions.

A simple serial D/A converter, requiring two capacitors that can be made self-calibrating, is shown in Figure 37.25.

The D/A converter (Figure 37.25(a)) relies upon the fact that if a capacitor with voltage V across it is shunted with an identical value capacitor having no charge or

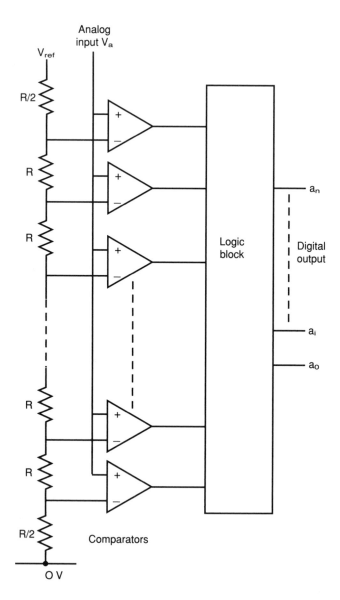

Figure 37.23   Flash analog-to-digital converter

voltage, then the voltage is halved. Operation of the circuit is as follows. Capa-citor C on the right-hand side is initially discharged through switch S_2. Data is fed in serially, starting with the least significant bit. The left-hand capacitor C is charged to either V_{ref} or 0 volts, depending on the value of a_0 being a 1 or 0 respectively. Thus the total charge on the capacitor after feeding in the first digit is

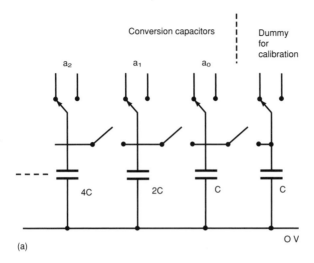

(a)

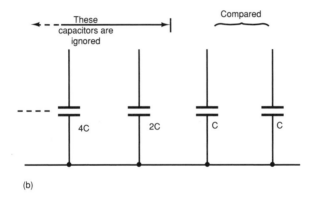

(b)

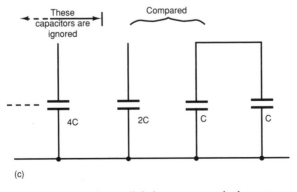

(c)

Figure 37.24 A self-calibrating analog-to-digital converter method, see text

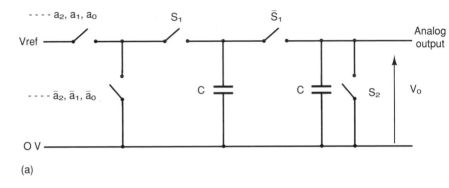

(a)

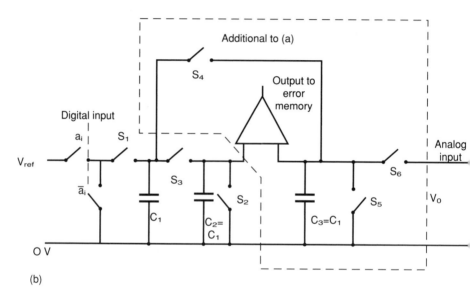

(b)

Figure 37.25   A simple two-capacitor digital-to-analog converter: (a) basic converter; (b) circuit with self-calibrating features.

$$Q_0 = a_0 V_{\text{ref}} C + 0 \qquad (37.9)$$

When switch S_1 is opened the two capacitors are shorted ($\overline{S_1}$ is closed) so that the
output voltage is

$$V_0 = a_0 V_{\text{ref}}/2 \qquad (37.10)$$

The process is repeated with a_1 as the input. Note that the capacitor C on the
right-hand side is now not shorted ($\overline{S_1}$ open) and retains its charge $a_0 C V_{\text{ref}}/2$. The
charge now in the two capacitors is

$$Q_1 = a_1 V_{\text{ref}} C + a_0 C V_{\text{ref}}/2 \qquad (37.11)$$

Opening switch S_1 and shorting the capacitors ($\overline{S_1}$ closed) produces

$$V_0 = V_{\text{ref}}(a_1/2 + a_0/4) \tag{37.12}$$

In this way the analog output voltage is built up.

Figure 37.25(b) shows the A/D converter with self-calibrating facilities added. The serial converter is on the left-hand side. The analog voltage is fed into capacitor C_3 via switches S_5 and S_6 and compared with the D/A converter output. The microprocessor controls the switches and can be programmed to make the converter function either as a simple staircase or a successive approximation type. The staircase type requires a shorter program and therefore requires less memory space. The additional switch S_4 allows capacitor C_1 and capacitor C_3 to act as a serial D/A converter so that known signals can be fed into both sides of the comparator. Using this method the comparator offset voltage and linearity of the converter (how well C_1 and C_2 are matched) can be derived.

37.4.4 Voltage to Frequency Circuits

Oscillator circuits are a simple way of converting an analog signal to a digital one. The basic circuit is a voltage-controlled oscillator such that the input analog signal controls the frequency of oscillation. One such circuit is a ring oscillator where an odd number of modified inverter circuits are connected in a cascaded loop or ring (Figure 37.26). The modified inverter circuit simply has a series resistor with a shunt capacitor on the input. The frequency of oscillation of the ring is primarily determined by the number of stages and the RC time constant. If one or more of the series resistors is constructed using a MOS transistor, then by varying the transistor gate potential, the effective resistance and hence the frequency of the ring oscillator can be varied. To achieve temperature stability two ring oscillators are employed, the second to monitor frequency variations due to ambient temperature changes.

With an integrated resistive or capacitive sensor on a silicon chip it is possible to use the sensor directly as part of the modified inverter stage RC time constant, to give a direct sensor output as frequency.

37.4.5 Multiple Sensor Arrays

Because the intelligent sensor is under microprocessor control, the chip can be made with multiple sensors on it. Several approaches are possible. In the first instance several sensors of the same type can be included and majority logic decisions taken to determine the value of the parameter. In this way if a sensor is damaged or contaminated the system can still function. An alternative scheme is to add a range of different sensor types so that the chip is of a general nature. The chip is programmed or conditioned to accept the output from the sensor type required. In some cases the output from several sensors can be used, for example to measure temperature, acceleration and revolutions of an engine. Yet another alternative is to use several similar sensors conditioned to perform slightly differently. An example is an ion selective chemical sensor each sensor being conditioned to detect a different type of ion.

A problem that can occur with sensor arrays arises from sensor cross sensitivity. In the case of the array of chemical sensors above, the output from each sensor may be 'fuzzy' and complex processing is required to decipher the true ion concentration. An alternative to using a microprocessor is to include a neural network, either on chip or within the central processor.

37.5 MICROELECTRONIC SENSOR TYPES

37.5.1 Introduction

Perhaps the question that needs to be answered is why use microelectronic sensors, for sensors can be manufactured by many different processes, each tailored to provide a sensor of optimum performance from the most suitable materials. The microelectronic manufacturing process offers a restricted range of materials and processing steps. However, against this, all of the processes are standard, well characterized, allow a range of production quantities typically from a few hundred to many millions of units, minimise costs, produce units that are small in physical size, reliable and with repeatable performance. All of these characteristics are required for sensor production. The design methodology developed for microelectronic circuits is also directly applicable to sensor design. Further, the numerous microelectronic manufacturing companies make second sourcing easier. All of these advantages do not eliminate the need for other sensor types, but simply establishes a class of sensor called microelectronic sensors.

Where microelectronic sensors have integrated electronics the inclusion of such electronics can make up for the non-ideal materials and processes. Increased sensitivity, temperature stabilization and improved noise immunity are all possible.

A further question needing consideration is which microelectronic technology to use, for any of the technologies shown in Figure 37.2 can be employed. Three factors are important in making a decision on which technology to use namely, appropriateness, cost effectiveness and process availability. The first of these points will be addressed in this section as each major technology is examined. Cost has already been covered while process availability depends very much upon location.

Sensors made from each of the microelectronic technologies can be subdivided into three categories, namely, those that make use of:

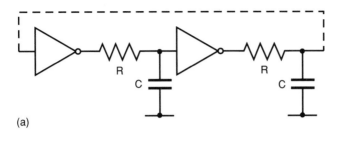

(a)

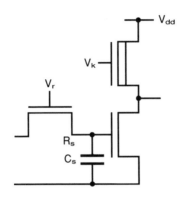

(b)

Figure 37.26 A ring oscillator: (a) basic circuit; (b) detailed single section. The number of stages must be odd, only two are shown

(1) material properties, for example the piezo-resistance of a thick film paste or photo conduction of silicon;

(2) circuit techniques, the capacitor sensor of Figure 37.17 being an example;

(3) the technology as a manufacturing method to produce a low-cost unit. The sensor material is added as a post-processing stage.

   In the next section emphasis is given to the first category.

37.5.2 Printed Circuit Board Technology in Sensor Use

With the event of surface mount technology very small, complex and robust sensors can be made. They usually fall into categories (2) and (3) above, for there are few material properties of the copper and laminate that can be utilized. Figure 37.27 shows a simple conductivity cell (Jarmyn *et al.*, 1987) using standard

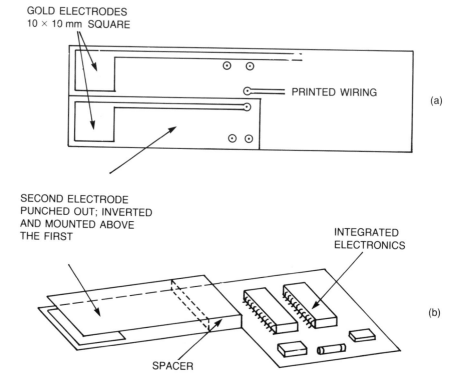

Figure 37.27   Liquid conductivity meter based on printed circuit board technology: (a) printed circuit card; (b) assembled sensor

printed circuit gold-plated electrodes. On-board integrated electronics allowed manufacture of an inexpensive hand-held instrument.

37.5.3   Film Technology in Sensor Use

Sensors are frequently based on the properties of film and particularly the properties of thick film paste (Table 37.6).

Table 37.6   Examples of sensors based on thick film properties

Input	Mechanism	Sensor example
Mechanical	Piezo-resistance	Strain gauge, pressure transducer
Thermal	Temperature coefficient	Thermistor, platinum resistance thermometer
	Seebeck effect	Gas sensor
Physical	Absorption	Humidity monitor

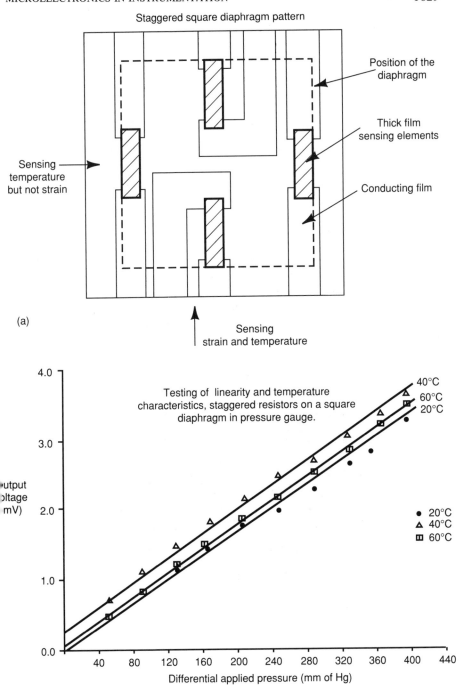

Figure 37.28 Simple pressure sensor using polymer thick film resistors, screen printed onto a polyester film diaphragm: (a) construction; (b) results

Some paste manufacturers now produce special pastes to exploit these mechanisms. An example is in its use in temperature sensors. Standard conductor pastes have thermal/resistance coefficients in the range of 1000–5000 ppm/°C, while resistor pastes are typically ±100 ppm/°C. Special pastes that are now available with regulated temperature coefficients include:

(1) thermistor paste with temperature coefficients −5000 to −10000 ppm/°C;

(2) posistor pastes with temperature coefficients +2000 to +4000 ppm/°C;

(3) nickel resistance thermometer paste with temperature coefficients of approximately +5500 ppm/°C

(4) platinum resistance thermometer paste with temperature coefficients of 3500 ± 200 ppm/°C.

The piezo-resistance effect, where in the resistance of the paste changes linearly with strain, can be used not only for simple strain gauges, but also for pressure measurement. Diaphragms can be made from special stainless steel overprinted with a dielectric, the gauge resistors and conductors then being printed on top of the insulating dielectric (White, 1989). Alternatively, low-firing-temperature pastes can be printed onto plastic membranes to produce low-pressure sensors.

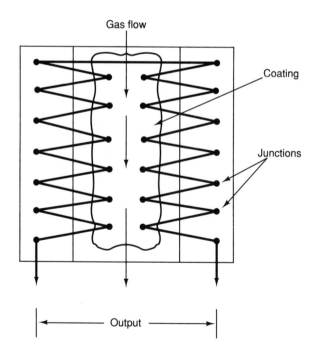

Figure 37.29   Gas sensor using the Seebeck effect to monitor temperature change

Figure 37.28 gives an example, showing both the layout and output voltage percentage change against pressure and temperature. The four resistors are printed so that only two have strain applied to them, the others providing temperature compensation. The resistors are connected as a simple wheatstone bridge.

Where two dissimilar metals join and there is a hot and cold junction a voltage is produced by the Seebeck effect (Beam, 1965). Thick and thin film materials show this property. The difficulty with ascertaining the Seebeck voltage for thick film materials is that the pastes are not pure metals, but mixtures. Consequently thermo-voltages must be determined by measurement. Table 37.7 gives some of the findings and shows that low resistor pastes give higher voltage outputs than conductor pastes.

Table 37.7 Seebeck voltages for various thick film paste combinations

Paste combination	Average e.f.m. ($\mu V/°C/$junction pair)
Ag–Pd (ESL 9635) Resistor 100 ohm/square (ESL 2812)	18.19
Au–Pd (ESL 6835 A) Resistor 100 ohm/square (ESL 2812)	16.2
Au–Pd (ESL 6835 A) Ag (DP 6320)	9.4
Ag–Pd (ESL 9635 A) Ag (DP 6320)	8.0

Using the gold palladium and 100 ohms/square resistor paste a thermopile was made to detect chemical thermal reactions, for a simple carbon dioxide detector (Esmail Zadeh et al., 1987). To provide thermal isolation of the hot junction on the alumina substrate, a glass dielectric thermal insulator was printed onto the substrate as shown in Figure 37.29. The central junctions were coated with various carbonate materials to test their sensitivity to the presence of carbon dioxide. Using zinc carbonate, carbon dioxide levels of less than $\frac{1}{4}\%$ could be detected.

A capacitor sensor to sense humidity can be made if a porous paste or polymide dielectric layer is used. Thick or thin film construction can be employed. Silver should not be used in the construction because of migration in the humid atmosphere. Either gold or low resistance carbon polymer electrode materials can be used. The construction is illustrated in Figure 37.30. The dielectric constant of the absorbing material is in the range of 3–6, whereas for water it is 78.5 at 25°C, giving a large variation in capacitance for normal humidity ranges.

Like printed circuit board technology, thick film and thin film processes can be used to construct sensors which are not based on the properties of the film material. Examples of this are

(1)   variable conductive gas sensors;

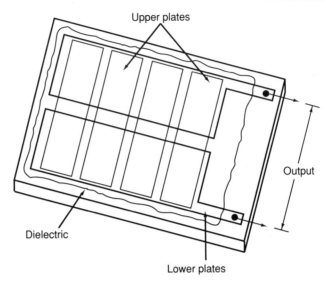

Figure 37.30   Capacitor humidity sensor

(2)   oxygen sensors;

(3)   surface acoustic wave (SAW);

(4)   ion-selective sensors.

In the first instance inter-digitized fingers are deposited on top of a platinum paste resistive heater. The heater can be on the opposite side of the substrate or on the same side, but electrically insulated by a thin dielectric layer. The construction is illustrated in Figure 37.31(a). The fingers are coated with various materials sensitive to particular gases. These may be tin oxide based with additives to give specific selectivity for the hydrocarbon gases or metal phthalocyanines (PCN) for gases such as nitrous oxide and carbon monoxide. The sensitivity of the material varies with temperature as shown in Figure 37.31(b). Most of the materials show poor selectivity, but by using several sensors heated to different temperatures improved selectivity can be obtained. Post processing is critical and again either conventional computers (Shurmer, 1990) or neural networks can be employed.

The oxygen sensor is based on using zirconia substrates as a solid electrolyte cell (Moseley and Tofield, 1987), with a thick film printed platinum paste heater and electrodes as shown diagrammatically in Figure 37.32.

While there are three possible modes of operation, namely concentration cell, fuel cell and ion pump cell, only the first will be considered here. A sample containing a partial pressure of oxygen P_s is admitted to the sample chamber while the reference chamber contains oxygen having a partial pressure P_r. The potential across the electrodes (E) is can be derived from the Nernst equation

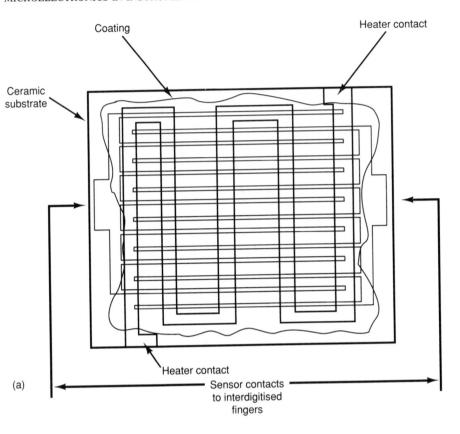

Figure 37.31   Simple conductivity gas sensor: (a) construction;

$$E = \frac{RT}{4F} \ln(P_s/P_r) \qquad (37.13)$$

where R—universal gas constant

     F—Faraday constant

     T—the temperature of the cell in K.

Bulk and surface acoustic wave (SAW) sensors depend upon the properties of piezoelectric materials such as quartz and lithium niobate. The sensors are micro balances, detecting mass changes as low as a nanogram (Wlodarski *et al.*, 1990). In the case of the SAW device illustrated in Figure 37.33 two sets of interdigitized thin-film fingers (in gold or aluminium) both launch and detect a Raleigh wave along the surface of the material. Frequencies used range from 30 to 1000 MHz, with the dimensions of the fingers determining the frequency, (Wohltjen, 1984; Czarnecki, 1988).

In one form the SAW device is connected in a feedback loop to produce a simple oscillator, whose frequency change is caused by a mass change:

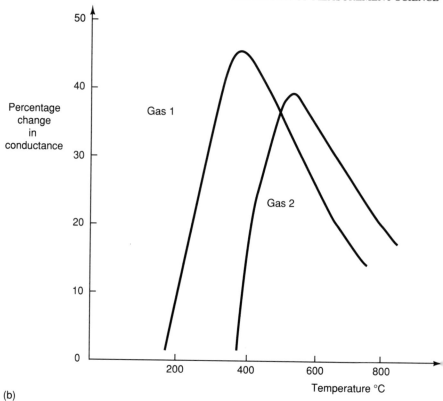

(b)

Figure 37.31   Simple conductivity gas sensor: (b) response with temperature to two different gas types

$$\Delta f = (K_1 + K_2) f_0^2 h\rho' - K_2 f_0^2 \left[\frac{4\mu'}{Vr^2} \left(\frac{\lambda' + \mu'}{\lambda' + 2\mu'} \right) \right] \qquad (37.14)$$

where f_0— unperturbed resonant frequency
  h— coating thickness
  ρ'— coating density
  λ'— Lane constant for the coating
  μ'— shear modulus of the substrate
  V_r— wave velocity of the substrate
  K_1, K_2 — constants dependent upon physical properties of material.

The first term is due to the mass loading, while the second results from changes in the properties of the coating film. Normally this term is small and can be ignored.

The device, as it stands, has a temperature coefficient and can, therefore, be employed as a temperature sensor. In all other applications a second sensor is required to compensate for the temperature effects. SAW sensors, can be used to

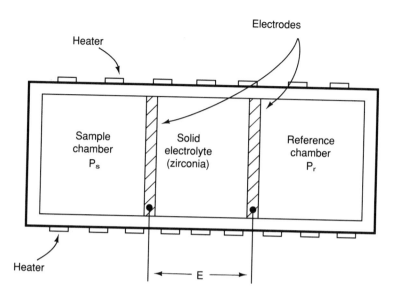

Figure 37.32   Simplified construction of an oxygen ion conductivity solid electrolyte cell

measure both physical and chemical parameters. In the latter case a coating is applied to the surface of the SAW device. Examples of coatings are polymers that absorb moisture or specific gases so that there are mass changes resulting in a change in output frequency.

Finally, a sensor to determine the concentration of ions in a solution is described. When ion sensitive and reference electrodes are inserted into a solution containing specific ions, as shown in Figure 37.34 (a), the output potential ΔE can be derived from the Nernst equation.

$$\Delta E = \frac{2.303RT}{Z_iF} \log_{10} C_i \qquad (37.15)$$

where C_i— ionic concentration
   R— universal gas constant
   T— temperature in °K
   Z_i— charge on the ion of interest
   F— Faraday constant.

The electrodes can readily be made using standard carbon polymer thick-film pastes as illustrated in Figure 37.34 (b) and (c), (Hoffman *et al.*, 1984; Shoubridge *et al.*, 1985). High-resistivity pastes, for example, 100 $k\Omega$/square, are used as the lower resistance pastes may have silver added to reduce their resistivity. The ion selective coating may be mixed with the carbon polymer paste and screen printed or vacuum deposited. Figure 37.35 gives typical results for a thick film system used to monitor pH and various heavy metal ions.

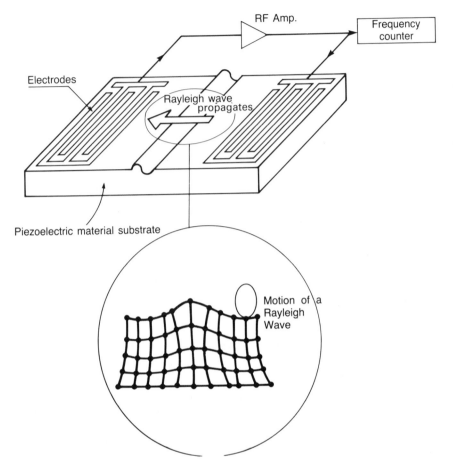

Figure 37.33   Construction of a SAW sensor

37.5.4 Silicon Technology Sensors

Semiconductor materials display many properties that make them suitable for sensors. (Middelhoek and Audet, 1989) Table 37.8 show the important properties for silicon. Other mechanisms can be added for different materials. For example, gallium arsenide displays piezoelectric properties which have been exploited for sensors (Fricke *et al.*, 1990).

Photodetectors, due both to photoconductance and photovoltaic effects constitute perhaps one of the simplest types of sensor, and are therefore in common use. Further, it is simple to manufacture large arrays of such devices including charge-coupled device arrays, that can be employed in a solid state TV camera. In this case the electron–hole pairs generated change the charge in the wells of

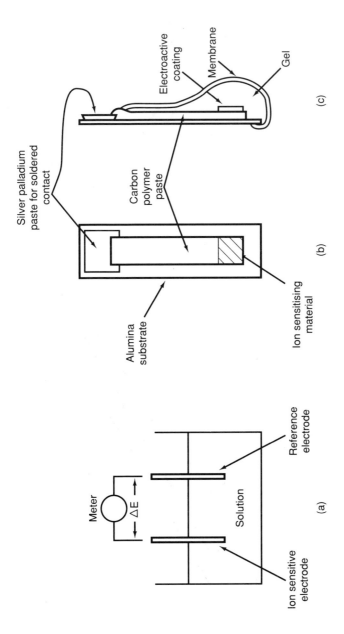

Figure 37.34 Ion sensitive sensors: (a) principle of operation; (b) thick-film sensor; (c) thick-film reference electrode

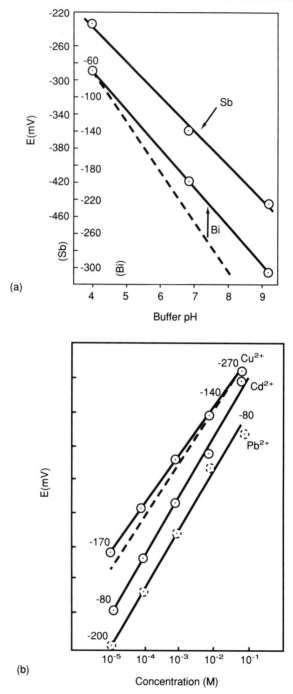

Figure 37.35   Ion concentration measurements using thick-film electrodes for (a) pH and (b) three heavy metals. The dotted line is the theoretical Nernstian slope

Table 37.8   Selective properties of silicon and corresponding sensor types

Input	Mechanism	Sensor device
Radiation	Photoconduction	Photodiode/transistor
	Photovoltaic	Photodiode
Mechanical	Piezo-resistance	Strain gauge
		Pressure transducer
	Piezo-junction	Pressure transducer
Thermal	Electron hole pair generation	Thermal diode/transistor
	Fermi level change	
Magnetic	Hall effect	Hall device (current or voltage)
		Magnetotransistor
Electric	Modulating conduction	Field-effect transistor
Chemical	Modulation conduction	Ion selective/chemical selective
		field-effect transistor

the array so that on readout the charge is dependent upon the intensity of light falling on each well. The equation for photogeneration is (Green, 1982)

$$I_L = qLWG(L_e + L_h + \omega) \tag{37.16}$$

while the open circuit voltage is

$$V_L = \frac{kT}{q} \ln\left(\frac{I_L}{I_o} + 1\right) \tag{37.17}$$

where G—generation rate of electron–hole pairs
$\qquad$ L_e, L_h—diffusion length of electrons and holes
$\qquad$ ω—width of the depletion region
$\qquad$ L and W—dimensions of the diode
$\qquad$ T—temperature in K
$\qquad$ q—charge on an electron
$\qquad$ k—Boltzmann's constant.

The spectral response of the device is dependent upon the band-gap voltage of the material and for silicon this peaks at 0.8 μm wavelength, Figure 37.36. Care must be taken in the manufacture of such devices, for if the thickness of the field oxide and other transparent layers above the semiconductor material surface are not controlled then optical interference can produce nulls in the spectral response. This is illustrated in Figure 37.36.

Thermal effects are closely related to photo effects in that the incoming energy generates electron–hole pairs. Thus the leakage current of diodes and transistors increases with temperature, approximately doubling every 8°C rise in temperature. In the case of MOS transistors the gate threshold voltage V_{gs} is also temperature dependent (Blauschild et al., 1978):

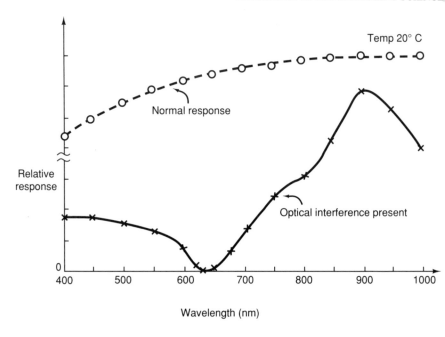

Figure 37.36   Spectral response of a silicon photosensor with and without optical interference effects

$$\frac{dV_{gs}}{dT} = -\alpha + \sqrt{\left(\frac{L}{C_{ox}\,W}\right)\frac{d}{dt}\sqrt{\frac{1}{\mu_0}\left(\frac{T}{T_0}\right)^{3/2}}} \tag{37.18}$$

where α—gate threshold temperature coefficient
L and W—device dimensions, length and width respectively
C_{ox}—gate capacitance per unit area
μ_0 and T_0—reference mobility and temperature values
T—operating temperature in °K.

Perhaps the most common method of making an accurate temperature sensor is to employ two transistors having different emitter current densities (Timko, 1976). Consider the circuit shown in Figure 37.37. Transistor Q_2 has an emitter area N times that of Q_1. From the Ebers and Moll model (Ebers and Moll, 1954) for a transistor we have the emitter current is, using parameters as defined in Figure 37.3.7.

$$I_e = I_0 \left(e^{\frac{-q V_{be}}{kT}} - 1 \right) \tag{37.19}$$

or

$$V_{be} \approx \frac{kT}{q} \ln\left(\frac{I_e}{I_0}\right) \tag{37.20}$$

Now

$$\Delta V = V_{be1} - V_{be2}$$

$$= \frac{kT}{q}\left[\ln\left(\frac{I_{e1}}{I_o}\right) - \ln\left(\frac{I_{e2}}{NI_o}\right)\right]$$

Because transistor Q_2 has an emitter area, and hence leakage current N times that of transistor Q_1, and I_{e1} is equal to I_{e2},

$$\Delta V = \frac{kT}{q}\ln(N) \qquad (37.21)$$

where k—Boltzmann's constant
        T—Temperature in K
        q—charge on an electron.
Further

$$\Delta V = IR$$

hence

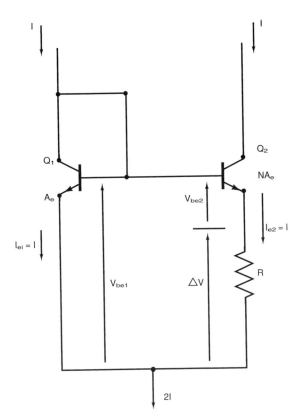

Figure 37.37   Principle of a bipolar transistor thermal sensor

$$I = \frac{kT}{q} \ln (N) \qquad (37.22)$$

The current flowing out the circuit is twice this value. It is proportional to temperature and by the correct selection of N and R can be made direct reading—typically 273 μA is equivalent to 273 K.

A large number of semiconductor devices are capable of detecting the presence of a magnetic field, (Zieren and Duyndam, 1982; Brini and Kamarinos, 1981; Kordic, 1986). The most common is the Hall effect device shown in Figure 37.38(a).

The Hall voltage V_H is given by

$$V_H = WR_H(\bar{J} \times \bar{B}) \qquad (37.23)$$

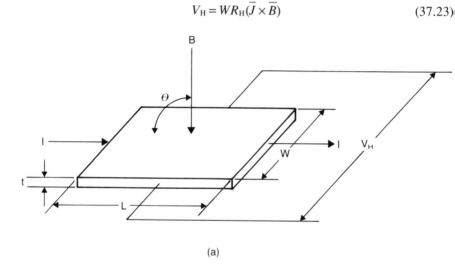

(a)

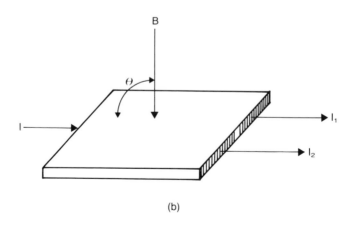

(b)

Figure 37.38   Hall-effect sensors: (a) conventional type; (b) modified current version

where R_H—Hall coefficient
      J—current density given by

$$\bar{J} = \bar{I}/Wt \tag{37.24}$$

Thus

$$V_H = \frac{R_H}{t}(\bar{I} \times \bar{B}) \tag{37.25}$$
$$= R_H IB \sin\theta/t$$

To maximize the output, the semiconductor should have a high resistance, and therefore usually the epitaxial material in a bipolar process or the channel region under the gate for an MOS process are used.

This latter case has the advantage that thickness t is also small, thereby increasing the sensitivity. In the ideal case V_H is zero for no magnetic field. Unfortunately, fabrication tolerances and stresses induced when mounting the device upset the balance and an offset occurs. This offset can be corrected by employing two devices on the one chip, mounted at 90° to each other.

An alternative, current version, is shown in Figure 37.38(b). Here the current flowing is split into two currents I_1 and I_2. The difference between them is dependent upon the magnetic field. Thus

$$\Delta I = I_1 - I_2 = R'_H I \sin\theta \tag{37.26}$$

where R'_H—Hall coefficient.

Examples of sensors based on this approach are Davies and Barnicoat (1970) and Cooper and Brignell (1984).

The MOS field effect transistor is a device where the current flowing between the source and drain depends upon the electric field along the channel length. In a normal device, the gate potential establishes the field. However, if the gate metallization is omitted, then external fields may be employed to control the current flow. For low source-to-drain voltages MOS transistors operate in a linear mode. Since most devices are made symmetrical they are, therefore, linear in both first and third quadrants of operation. Ignoring second-order effects the channel conductance g_{ds} for an external field E (Figure 37.39) is

$$g_{ds} = E(\mu\varepsilon W/L) \tag{37.27}$$

where W and L —width and length of the device
      μ—carrier mobility in the semiconductor
      ε—permittivity of the effective gate.

The equation shows that the strength of an external electric field can be measured by monitoring the source drain conductance (Horenstien, 1985).

A similar situation arises when a MOS transistor is used to measure the presence of ions in a solution, that is the field produced by the ions is used to modulate the channel conductance. This concept was first proposed by Bergveld

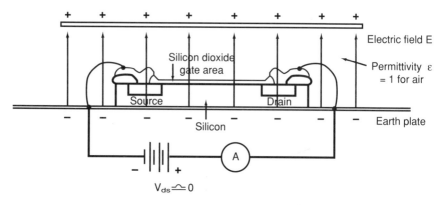

Figure 37.39   MOS sensor to measure electric fields

(1970). He showed that the flat band voltage can be chemically influenced by ion interaction with the gate electrode. Following this, it was shown that this is also true for interaction with gases (Lundstrom, 1975).

With the ion selective field effect transistor (ISFET) the applied gate voltage is assumed to be constant and the threshold voltage V_t variable. Barabash *et al.* (1987) have derived an expression for this threshold voltage and examined its temperature characteristics. For the system shown in Figure 37.40 the normal MOS equation for the non-saturation region is

$$I_d = \mu \, C_{ox} \frac{W}{L} \left[(V_g - V_t) V_{ds} - \frac{V_{ds}^2}{2} \right] \tag{37.28}$$

and

$$V_t = \varepsilon_{ref} + \Delta \varphi_{ij} - \varphi_{0t} + \chi_{el} + \varphi_F - \left(\xi + \frac{E_G}{2} \right) \Big/ q - (Q_{SS} + Q_B)/C_{ox} \tag{37.29}$$

where I_d—MOS drain current
  μ—carrier mobility in the semiconductor
  W and L—device width and length respectively
  C_{ox}—gate capacitance per unit area
  V_g and V_{ds}—applied gate and drain source voltages
  ε—absolute potential of the reference electrode
  $\Delta \varphi_{ij}$—liquid junction potential difference
  φ_{0t}—value of the insulator–electrolyte interface with respect
        to the electrolyte bulk at threshold voltage
  χ_{el}—electrolyte dipole potential
  φ_F—potential of the Fermi energy level of silicon with respect
        to the mid-band
  ξ—electron affinity of silicon
  E_G—energy band gap of silicon

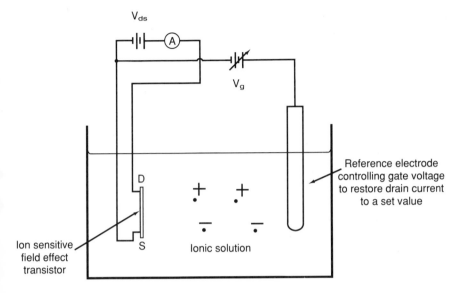

Figure 37.40 Simple ISFET/reference electrode system

q—electron charge

Q_{SS}—equivalent Si/SiO_2 interface charge per unit area

Q_B—silicon depletion charge per unit area.

The term χ_{el} is a function of the ionic strength I, such that

$$\chi_{el}(25°C) = 50(1 - e^{(0.86 \log I)}) \tag{37.30}$$

The forward voltage drop V_f of a diode is pressure dependent, due to the change in bandcap energy E_G with pressure (Wortman and Monteith, 1969; Wlodarski, 1984). This allows a pressure sensor to be formed. It can be shown that

$$V_f = \left(\frac{\partial E_G}{\partial p}\right)\frac{\Delta p}{q} \tag{37.31}$$

where Δp—change in applied pressure to the diode.

The pressure sensitivity of this forward voltage $(\partial V_f/\partial p)$ is small, typically 10^{-11} V/P$_a$, is negative for silicon and is temperature dependent. However, the property is used for high-pressure sensors (ten of megapascals) such as in fuel injector systems.

Another mechanism shown in Table 37.8 for forming a sensor piezoresistance, and its application to strain gauges, accelerometers, flow meters and pressure transducers. It finds use because of the ability to micromachine silicon through plasma and chemical etching (Institute of Physics, 1986; Delapierre, 1989; Linden et al., 1989; Gilles, 1989; Moser et al., 1990). Such is the skill being acquired

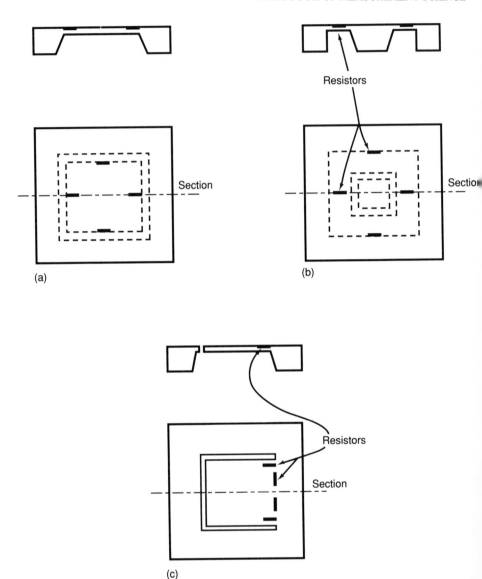

Figure 37.41   Selection of micro machined sensors: (a) pressure sensor; (b) accelerometer; (c) cantilever flow meter. In all cases four piezo resistors are connected as a bridge to monitor movement

in this area that it is now possible to make miniature silicon actuators including electrostatic motors and other parts such as micro gears, valves, nozzles etc. (Bart *et al.*, 1988; Tarrow *et al.*, 1990). Figure 37.41 shows, in cross-section, some sensor structures. Each has four piezoresistors (two for temperature compensa-

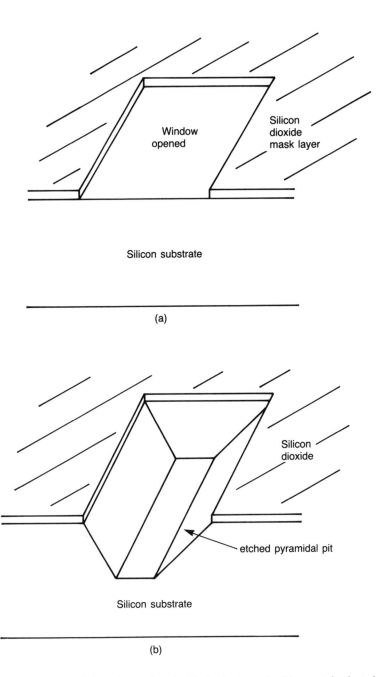

(a)

(b)

Figure 37.42 The shape of the anisotropic etched holes is determined by crystal orientation: (a) shows the masking of a 100 wafer while (b) shows the final etched shape, a pyramidal pit with 111 side walls

tion) connected in a wheatstone bridge to measure the deflections. Optical methods can also be employed to measure deflections.

Micromachining can be achieved by several methods, normally classified as wet or dry (plasma) etching. Photolithography is used to define the areas to be etched. The etching can be classified as isotropic, that is there is no directional preference, or anisotropic, where the etch is dependent upon the silicon crystographic orientation. The latter is the most common method employed for wet etching. Commonly used anisotropic etches are potassium hydroxide and etholene diamine/pyrochaprschol (EDP). For both substances the etch in the 111 plane is much slower than in any other plane. For example, the etch rate in the 100 plane for potassium hydroxide is some 400 times faster than for the 111 plane. Another important property of anisotropic etches is that their rate depends upon doping type and density. With potassium hydroxide the etch rate in p-type boron doped material is reduced by a factor of 20 when the dopant exceeds $7 \times 10^9 \, cm^{-3}$. Electrochemical wet etching methods are also employed to improve etch rate control. Figure 37.42 illustrates the effect of anisotropic etching.

Considerably improved precision can be obtained using dry or plasmic etching. The processes are isotropic and involve expensive plasma equipment. The reactive ion etching (RIE) process is the most satisfactory method at present. This is illustrated in Figure 37.43. Gas pressure is low to achieve a glow discharge plasma (approximately 10^{-1}). The substrates are placed in a radio-frequency field so that a self-bias potential of a few hundred volts develops that causes the ions to bombard the substrate. Gas types used in the plasma include CF_4, CF_3Cl and C_2F_5Cl.

From the examples of microelectronic sensors given, it can be clearly seen that the diversity of the technology allows a wide range of sensor types to be produced. They are small in size, low cost for large volumes giving repeatable and reliable performance.

37.6 PROBLEMS

While the microelectronic technologies allow the production of a wide range of sensors there are several problems that must be overcome before the technology can be fully exploited. They are:

(1) packaging of the sensor;

(2) production volumes;

(3) lifetimes.

By far the greatest difficulty is how to package the sensor. Conventional integrated circuit packaging methods can be used in some instances. For example, ceramic packages pass infrared radiation and magnetic fields. Quartz windows

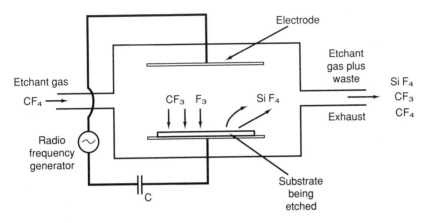

Figure 37.43   Principles of reactive ion etching using carbon tetrafluoride gas

Figure 37.44   An intelligent MOS power transistor (right-hand side) with signal processing on the left. A BICMOS manufacturing process was used. (Courtesy of Hitachi, Japan)

allow visible and ultraviolet light transmission. However, many sensors must operate in hostile environments. For ion-selective sensors, this may be an acidic or alkaline solution, for pressure sensors on a vehicle engine temperatures may reach 200°C. Thus, to accommodate the range of environments new and sometimes costly packaging methods are needed.

The packaging of sensors must be different to integrated circuit packages for a second reason. Most integrated circuits are designed to be mounted on a printed circuit card. This card is needed to assemble the complete system. In the case of a sensor often with its own internal electronics, no other components are needed at the sensor site and therefore the use of a printed circuit board is an unnecessary expense. The package therefore should be self-contained, allowing mounting of the sensor and associated plug to connect it to the remainder of the system.

The advantage of many microelectronic manufacturing methods is that they are suited to high-volume production. At present those high volumes have not been realized widely enough. Perhaps the increasing number of sensors being employed in automobiles may start to change this.

Finally, there is the uncertainly of lifetime. All of these sensor types are relatively new and while they make use of well-known processes they are exploiting them in new and different ways. As a consequence there is no record of experience to provide reliability and long-term performance figures. Consequently, there is often a reluctance to employ them.

37.7   APPLICATION AREAS

While sensors of various types are required for any measurement or control system, the application of microelectronic sensors is driven by those areas requiring large volumes at low cost. These are principally the consumer products area. Part of this driving force is the advent of the microprocessor. As microprocessors are applied to new consumer product areas they will automatically bring with them the need for microelectronic sensors. The two go hand in hand. What are some of these areas?

Perhaps the present greatest growth area is the automobile area. Engine and vehicle control is being augmented using microprocessors, so sensors are needed to measure mechanical and chemical parameters. This includes existing sensors like speed, mileage and fuel tank level, but will grow to include many others. For example, exhaust gas levels, fuel mixture levels, wheel slip and air velocity. Passenger comfort will also become an important factor with accurate control of air temperature and humidity, window optical density to match brightness and many in-built safety monitor features. Navigational aids will become standard so that the vehicle position using of navigational satellite will be displayed or projected maps. Speed zone signs will be recognized and the car navigational system will give appropriate warnings. Little imagination is required to see that more and more low-cost, high-volume, sensor types will be needed as vehicle manufacturers improve their products. Figure 37.44 shows an intelligent power transistor made specifically for the application of electronic control to automobiles.

As microprocessor chips continue to drop in price they will become part of even the simplest of consumer products. Light switches will become microprocessor controlled to activate whenever people enter or leave rooms and will operate light(s) to achieve the correct illumination in a room, no matter what time of day. Sensors will be built in to detect movement and light levels. Energy conserving homes will become important and sensors will be used to monitor and therefore to control the state of the dwelling and the features within, be it washing machines, cookers, refrigerators or central heating and cooling. Protection of homes is a further growth area in most large cities. Detection of the presence of unwanted people and preventing them from stealing and doing bodily harm to the occupants is a need. Microprocessor systems with appropriate sensors will become standard items in homes.

With such sensors available for the consumer market it is to be expected that they will be adapted and applied to industry and the professional areas. Their application will only be limited by one's imagination, for microprocessors with integrated sensors will become the nuts and bolts of many industrial systems. We have already seen that a sensor for one parameter can often be modified to measure another parameter. The Hall-effect sensor, designed to measure the strength of a magnetic field, can be used to achieve many operations. Examples are:

(1)   to measure pressure through mounting a permanent magnet onto a diaphram;

(2)   to monitor whether a doorway is open or closed by placing a magnet on the door frame;

(3)   to note the presence or absence of a large magnetic material mass such as a car due to disturbance of a magnetic field.

Consequently, it will become standard practice to take simple microelectronic devices with microelectronic sensors and prepackage them to perform a whole new range of functions and, therefore, products.

The measurement and instrumentation fields are at present feeling the impact of the microprocessor. In the not too distance future there will be further changes as new microelectronic devices, including smart sensors, become available as standard products for all to use.

37.8   SOURCES OF FURTHER ADVICE

The reader is referred to the following textbooks and journals for further information on microelectronics in instrumentation.

Books

Harashima, F. (ed.) (1989) *Integrated Micro-Motion Systems* Elsevier, Amsterdam.

Jones, B. E. (ed.) (1987) *Current Advances in Sensors* Adam Hilger, Bristol.

Madou, M. J. and Morrison, S. R. (1988) *Chemical Sensing with Solid State Device* Academic New York.

Middelhoek, S. and Van Spiegel, J. (eds) (1987) *Sensors and Actuators* Elsevier Sequoia Lausanne.

Muller, R. S. *et. al.* (eds) (1990) *Microsensors* IEEE Press, New York.

Journals

Electronic Letters IEE, UK.

IEEE Transactions on Electronic Devices IEEE, USA.

Journal of Solid State Circuits IEEE, USA.

Measurement Science and Technology (formerly *Journal of Physics E: Scientific Instru* ments), Institute of Physics, UK.

Microelectronics Journal UK.

Sensors and Actuators A Elsevier Sequoia SA, Lausanne.

REFERENCES

Barabash, P.R., Cobbold, R.S.C. and Wlodarski, W.B. (1987). 'Analysis of the threshold voltage and its temperature dependence in electrolyte–insulator–semiconductor field effect transistors', IEEE Trans, **ED-34**, 1271–82.

Barron, I. (1990). 'Lots of transistors', *Keynote Address Microelectronics Conference '90* Adelaide, Australia, 2–4 July.

Bart, S.F., Lober, T.A., Howe, R.T., Lang, J.H. and Schlecht, M.F. (1988). 'Design consideration for micromachined electric actuators', *Sensor and Actuators*, **14** 269–92.

Beam, W.R. (1965). *Electronics of Solids*, McGraw-Hill, New York.

Bergveld, P. (1970). 'Development of an ion sensitive solid state device for neurophysiological measurements', *IEEE Trans. Biomed. Eng.*, **17**, 70–1.

Blauschild, R.A., Tucci, P.A., Muller, R.S. and Meyer, R.G. (1978). 'A new NMOS temperature stable voltage reference', *IEEE J.*, **SC-13**, 767–74.

Brignell, J.E. (1989). 'Smart sensors', in *Sensors: A Comprehensive Survey*, Vol. 1, Chapter 12, Gopel, W., Hesse, J. and Zemel, J.N. (series editors). VCH Verlagsgessellschaft mbH, Weinhiem, Germany.

Brini, J. and Kamarinos, G. (1981). 'The unijunction transistor is a high sensitive magnetic sensor', *Sensors and Actuators*, **2**, 149–54.

Cadenhed, R.L. and DeCoursey, D.T. (1985). 'The history of microelectronics, Part 1', *Inst. of Hybrid Microelectronics*, **8**(3), 14–30.

Cole, B.C. (1987). 'Crystal builds the first 12-bit flash converter chip', *Electronics*, **60**(25), 67–9.

Cooper, A.R. and Brignell, J.E. (1984). 'A magnetic field transducer with frequency modulated output', *J. Phys. E. Sci Instrum.*, **17**, 3146–7.

Cooper, A.R. and Brignell, J.E. (1985). 'Electronic processing of transducer signals: Hall effect as an example', *Sensors and Actuators*, **7**, 189–98.

Czarnecki, L. (1988). 'SAW filters for timing recovery at data rates up to 2.5G bits', *Siemens Components*, **13**, 186–90.

Davies, L.W. and Barnicoat, G.P. (1970). 'Integrated Hall current element', *Int. Conf. on Microelectronics, Circuits and System Theory*, 18–21 August, Sydney, Australia, 32–3.

Delapierre, G. (1989). 'Micro-machining: a survey of the most commonly used processes', *Sensors and Actuators*, **17**, 123–38.

Dettmer, R. (1988). 'The silicon PCB', *IEE Review*, **34**, 411–3.

Dummer, G.W.A. (1978). *Electronic Inventions and Discoveries*, Pergamon International Library, Oxford.

Ebers, J.J. and Moll, J.L. (1954) 'Large signal behavior of junction transistors' *Proc. IRE* **42**, 1761–72.

Esmail Zadeh, R., Davey, D., Mulcahy, D.E. and Haskard, M.R. (1987). *A Carbon Dioxide Alarm Module*, South Australian Institute of Technology, School of Electronic Engineering Report YW-87-8.

Fricke, K., Schweeger, J., Würfl, J. and Hartnagel, H.I. (1990). 'Piezo-electric pressure sensors based on GaAs', *14th European Workshop on Compound Semiconductors Devices and Integrated Circuits*, 8–10 May, Cardiff, Wales.

George, M., Haskard, M.R., Koh, S.N. and Kong, R.Y. (1988). 'Intelligent sensor with serial bus interface', *Microelectronics Conference '88*, 16–18 May, Sydney, Australia, 127–34.

Giachino, J.M. (1986). 'Smart sensors', *Sensors and Actuators*, **10**, 239–48.

Gilles, D. (1989). 'Micromachining: a survey of the most commonly used processes', *Sensors and Actuators*, **17**, 123–38.

Green, M. (1982). *Solar Cells: Operation, Principles, Technology and System Applications*, Prentice-Hall, Englewood Cliffs, NJ.

Haskard, M.R. (1986). 'General purpose intelligent sensors', *Microelectronics J.* **17**(5), 9–14.

Haskard, M.R. (1988). *Thick Film Hybrids: Manufacture and Design*, Prentice-Hall Englewood Cliffs, NJ.

Haskard, M.R. (1990). 'An experiment in smart sensor design', *Sensors and Actuators*, **A24**, 163–9.

Haskard, M.R. (1991). *Electronic Circuit Cards, Including Surface* Mount Technology, Prentice Hall, Englewood Cliffs, NJ.

Haskard, M.R. and May, I.C. (1987). *Analogue VLSI Design: nMOS and CMOS*. Prentice Hall, Englewood Cliffs, NJ.

Hoffman, C.R., Haskard, M.R. and Mulcahy, D.E. (1984). 'Carbon-filled polymer paste ion-selective probes', *Analytical Letters*, Pt. A, **17**, 1499–1509.

Horenstien, M.N. (1985). 'A direct gate field effect transistor for the measurement of DC electric fields', *IEEE Trans*, **ED-32**, 716–17.

Hunt, P.C. and Saul, P.H. (1988). 'Process and circuit innovation in silicon technology', *Plessey Research and Technology Review '88*, 59–66.

Institute of Physics (1986). *Silicon Based Sensors*, Inst. of Physics—Short Meeting Series No. 3, Bristol, UK.

Jarmyn, M., Haskard, M.R. and Mulcahy, M.E. (1987). 'The design of portable conductivity meters for agricultural applications', *Aust. J. Instr. and Control*, 2(3), 18–21.

Kordic, S. (1986). 'Integrated silicon magnetic field sensors', *Sensors and Actuators*, 10, 347–78.

Linden, Y., Tererz, L., Tiren, J. and Hok, B. (1989). 'Fabrication of three dimensional silicon structure by means of doping—selective etching, *Sensors and Actuators*, 16, 67–82.

Lundstrom, I. (1975). 'A hydrogen sensitive Pd gate MOS transistor', *J. App. Physics*, 46, 3876–81.

Middelhoek, S. and Audet, S.A. (1989). *Silicon Sensors*, Academic Press, London.

Moser, D., Parameswaran, M. and Baltes, H. (1990). 'Field oxide microbridges, cantilever beams, coils and suspended membranes in SACMOS technology', *Sensors and Actuators*, A21–3, 1019–22.

Moseley, P. T. and Tofield, B.C. (1987). *Solid State Gas Sensors*, Adam Hilger, Bristol, UK.

Nass, R. (1990). '100 million transistors on a chip', *Electronic Design*, 38(11), 16.

Shoubridge, P.J., Haskard, M.R. and Mulcahy, D. E. (1985). 'Carbon-filled polymer paste reference electrode', *Analytical Letters*, Pt. A, 18, 1457–63.

Shurmer, F. (1990). 'The fifth sense', *IEE Review*, 36, 95–8.

Tarrow, L.S., Bart, S.F., Lang, J. H. and Schlecht, M. F. (1990) 'A LOCOS process for an electrostatic microfabricated motor', *Sensor and Actuators*, A21–3, 893–8.

Timko, M.P. (1976). 'A two terminal integrated circuit temperature transducer', *IEEE J.*, SC-11, 784–8.

Topham, P.J. (1987). 'High speed integrated circuits using heterojunction bipolar transistors', *Plessey Research and Technology Review '87*, 55–61.

Van de Grift, R.E.T., Rutten, I.W.J.M. and Van der Veen, M. (1987). 'An 8-bit video ADC incorporating folding and interpolation techniques', *IEEE J.*, SC-22, 944–53.

White, N.M. (1989). 'An assessment of thick film piezoresistors on insulated steel substrates', *Hybrid Circuits*, 20, 23–7.

Wlodarski, W. (1984). 'Semiconductor transducers for measurement of rapidly varying pressures in fuel pumps of diesel engines', *TENCON '84 Conference on Industrial Electronics and Applications*, 17–19, April, Singapore, 17–19.

Wlodarski, W., Mulcahy, D.E., Haskard, M.R. and Shanks, R.A. (1990). 'Surface acoustic wave sensors', *Microelectronics Conference '90*, Adelaide, Australia, 2–4 July, 147–52.

Wohltjen, H. (1984). 'Mechanism of Operation and Design Considerations for Surface Acoustic Wave Device Vapour Sensors', *Sensors and Actuators*, 5, 307–35.

Wortman, J.J. and Monteith, L.K. (1969), 'Semiconductor mechanical sensors', *IEEE Trans*, ED-16, 855–60.

Zieren, V. and Duyndam, B.P.M. (1982). 'Magnetic field sensitive multicollector NPN transistors', *IEEE Trans.*, ED-29, 83–9.

Chapter

38 L. C. LYNNWORTH

Ultrasonics in Instrumentation

Editorial introduction

Ultrasonic principles are able to provide sensing solutions in many difficult applications. Early uses did not always provide enough measurement performance to be practical in industrial processes. Today, sometimes, with sophisticated constructions that make use of complex electronic circuitry, ultrasonic equipment is robust and metrologically acceptable and often the product of choice that can be used in the harshest of environments.

38.1 BASIC PRINCIPLES OF ULTRASONIC TRANSDUCERS

38.1.1 Transducers as Active and Passive Sensors—Effects of the Measurand on Generation, Propagation and Detection of Ultrasound

Transducers for launching or detecting ultrasonic waves utilize mechanical, electromagnetic, chemical, pneumatic, fluid dynamic or thermal phenomena. Transducers for ultrasonic instruments are mainly in the electroacoustic category. Here, electrical energy is converted to acoustic energy, when transmitting; acoustic or mechanical wave energy is converted to an electrical signal, when receiving.

An important concern in the design of ultrasonic transducers and measuring instruments is the optimization of the system with respect to reliable and accurate determination of the measurand—the parameter whose value is to be inferred or computed based on 'ultrasonic' observations. In an ultrasonic experiment one may observe transit time, amplitude of the received signal, oscillator frequency, oscillator strength, or terms related to these. If more than one parameter is varying, say

Handbook of Measurement Science, Volume 3
Edited by P. H. Sydenham and R. Thorn
© 1992 John Wiley & Sons Ltd

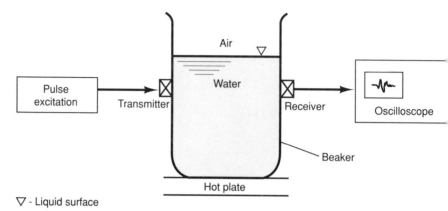

∇ - Liquid surface

Figure 38.1   Ultrasonic effects of heating water on which a separated pair of resonant pie-zoelectric quartz transducers are mounted

temperature and pressure, it may be necessary to observe more than one charac-teristic of the generated or received wave to resolve ambiguities. Generally speaking, one can say that each measurand (flow, temperature, pressure, ...) influences one or more of the following: generation, propagation and detection of ultrasound. For a fuller account of such influences and their measurement see Lynnworth (1989).

Consider the simple through-transmission arrangement of Figure 38.1. As the water is heated from room temperature until it all boils away, one could observe changes in generation, propagation and detection as listed in Table 38.1.

Table 38.1   Examples of heating effects on generation, propagation and detection of ultra-sound in water, in the laboratory test of Figure 38.1

| Temperature | Heating effects on: | | |
	Generation	Propagation	Detection
Increase from 20 to 74°C	Frequency increases	Speed of sound increases; absorption decreases	Signal arrives progressively earlier
75 to 99°C	Frequency increases	Speed of sound decreases; absorption decreases	Signal arrives progressively later
100°C	Frequency increases	Propagation is largely blocked by boiling and then by low-impedance air after all the water boils away	Signal weakened by scattering due to boiling and may be undetectable

Without further study it is not obvious as to the best way to proceed, where, say, the object is to measure water temperature. As will be seen in Section 38.7, however, ultrasonic transducers and instruments are available commercially for measuring temperature. Their operation is based on the temperature dependence of: (a) the transducer frequency; (b) the speed observed in the water; (c) the speed of sound in an intrusive sensor or probe; or (d) the speed of sound in the beaker material.

In discussing the laboratory experiment depicted in Figure 38.1, the influence of pressure has been disregarded. Actually the speed of sound depends on pressure in all three media involved: the solid (quartz crystal), the liquid (water) and the gas (air) remaining after all the water has boiled away. One might expect pressure effects to be negligibly small in an 'incompressible' liquid like water, and certainly negligible in a solid such as quartz. But this is not the case. For example, in the particular cut of quartz used by Hewlett-Packard in their quartz pressure gauge, a change of pressure of 69 Pa (0.01 psi) is resolvable as a difference in its resonant frequency. In the case of air, the influence of pressure on the speed of sound may be calculable if one can utilize the amplitude of the received signal as a measure of air pressure. Once pressure is known, air temperature can be calculated from the speed of sound.

The dependence of the speed of sound on temperature and density in a fluid such as oxygen at low temperature is given in Figure 38.2. This figure suggests that, over the range studied, the density of oxygen could be computed from measurements of speed of sound and temperature.

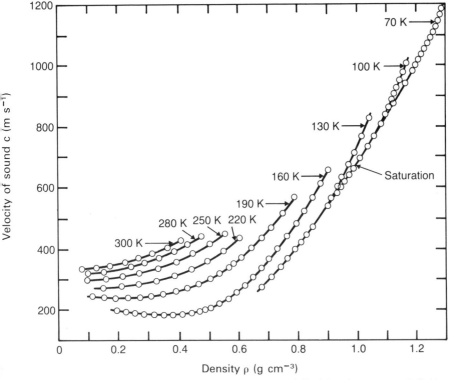

Figure 38.2 Velocity of sound versus density for saturated liquid and compressed fluid oxygen. (After Straty and Younglove, 1973)

38.1.2 Piezoelectric, Magnetostrictive, Electrostatic, and Electromagnetic Mechanisms

The *piezoelectric* effect, discovered by the Curie brothers over a century ago, involves a coupling between elastic and dielectric phenomena. Piezoelectricity is found only in materials that have no centre of symmetry. Quartz, lithium niobate, tourmaline and other crystals exhibit the effect, but probably most industrial transducers, as used in testing and process measurements today, use poled ceramics as the electroacoustic element. PZT (lead zirconate titanate) and LM (lead metaniobate) are examples of such materials. As a transmitter, piezoelectric materials expand (or contract), or exhibit shear, in response to a driving voltage. As a receiver, an electrical signal is generated in response to an incoming pressure or shear wave. These effects can be observed over a wide temperature range, from nearly absolute zero up to the Curie point or transformation temperature of the element, e.g. 573°C for quartz, 1210°C for $LiNbO_3$. The usable frequency range extends down below the ultrasonic band into the audio and infrasonic range, and up to the GHz range.

Magnetostrictive devices are usually limited to frequencies below a few hundred kilohertz by the electrical skin effect, although operation into the megahertz range has been reported. Ni, Fe, Co and their alloys are magnetostrictive, i.e., they twist or change their dimensions in response to an applied magnetic field. Conversely, in response to a torsional, tensile or compressional load, they generate a magnetic field that is typically sensed by a coil surrounding the magnetostrictor. As a stress wave propagates along a magnetostrictive wire, one can imagine a little bar magnet moving through the wire at the speed of sound. When that magnet passes through an encircling coil, a voltage is induced in that coil.

An *electrostatic* transducer may be thought of as a capacitor, one of whose electrodes is the surface of a conductive (or conductively clad) specimen. Coulomb forces are the basis of its transmitter action. As a detector, if the charge is constant, a change in gap capacitance leads to a corresponding change in voltage.

Electromagnetic (emat) transducers are based on Lorentz forces, and offer a non-contact way to generate and detect ultrasonic waves in electrical conductors but only in the presence of a biasing magnetic field (which can be pulsed). The transmitter force is between eddy currents and the biasing field. By proper design of the coil that generates the eddy currents, and the biasing field orientation, one controls the type of wave generated. In theory, there would appear to be no upper temperature limit on the use of emat principles in conductive solids or liquids, or perhaps even plasmas, but results to date appear to be limited by practical considerations to below approximately 1000°C.

For further details see Sachse and Hsu (1979) or Silk (1984).

38.1.3 Waveguides, Buffer Rods, Radiators, Reflectors

It is not always convenient to couple or connect the electroacoustic transducer directly to the medium in which propagation is desired. The medium may be too

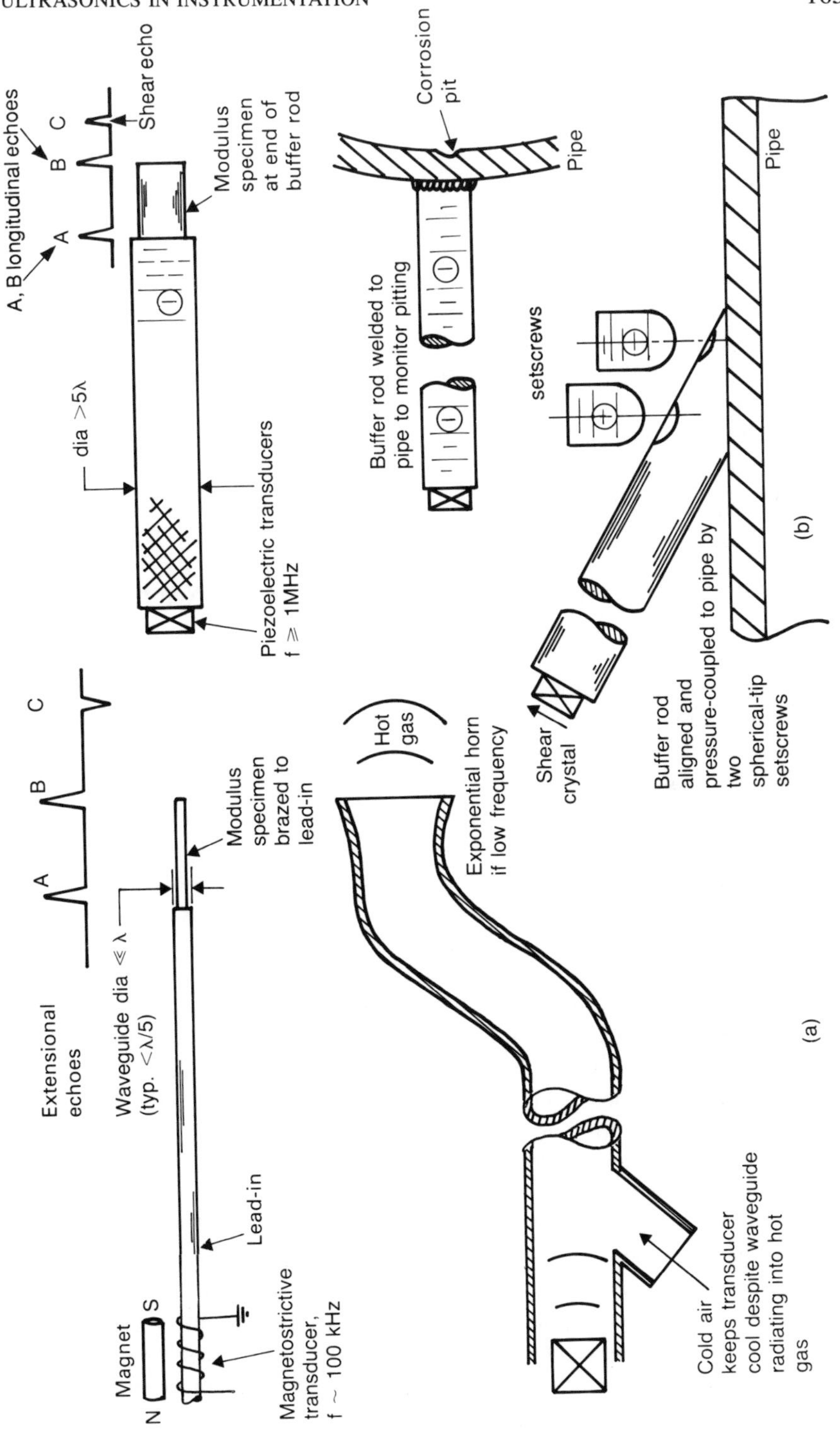

Figure 38.3 Examples of (a) waveguides and (b) buffer rods

hot, too distant, or of a chemical nature too hostile for the transducer. In such cases use is made of waveguides or buffer rods to convey the sound wave to its destination. In other cases it may be possible to increase the amount of energy radiated into the medium, e.g., liquid in a pipe, by arranging for the sound wave to travel some distance along the pipe. As the wave propagates it leaks or radiates into the liquid over an extended region. In other cases it may be desirable to aim a beam of sound in a direction other than perpendicular to the transducer's face; lenses and/or reflectors may provide the remedy in these instances.

Waveguides may be solid or hollow. A waveguide may be thought of as an attenuator, a filter and a delay line. Hence one is concerned about the loss, dispersion, phase shifts, multipaths and ringing in the waveguide. Buffer rods, or solid waveguides, are often threaded or knurled to reduce unwanted mode conversions that otherwise occur at the sidewalls when longitudinal waves are conveyed (Jen *et al.*, 1990a).

Examples of waveguides and buffer rods are given in Figure 38.3.

38.1.4   Container Walls as Sensors or Windows for Non-invasive Interrogation

When categorizing ultrasonic instrumentation principles and techniques it is convenient to organize them according to whether the sensor is the medium itself or an intrusive sensor. Here we use 'sensor' to mean the medium, material or structure in which the generation, reception or propagation of ultrasound is measured and in which the said generation, reception or propagation is influenced by the state, characteristics or value(s) of measurands in the path.

Referring again to Figure 38.1, if we think of water as its own sensor ('Propagation' column in Table 38.1) we must nevertheless acknowledge that the electroacoustic transducers legitimately might be called 'sensors'. By designating water as the sensor we can emphasize that the speed of sound in water is the basis of the way that we are obtaining information on the water's temperature. The electroacoustic piezoelectric quartz transducers shown in this particular illustration could be replaced by an external laser generation and detection system, focused on opposite sides of a black beaker, yet the speed of sound in water could still serve as the basis for determining water temperature.

Finally, one must not overlook what might appear to be an intermediate category, namely, the existing wall used as a sensor. As one example, ringing in the wall of a vertical standpipe is often used to tell if liquid is present at that level. In the common example of a clamp-on flowmeter the wall of the pipe provides an acoustically transparent window through which the interrogating beam gains access to the liquid in the pipe. The liquid is its own sensor. The pipe and suitable transducer assemblies make it possible to utilize that sensor without invading or physically penetrating the boundaries of the fluid, see Figure 38.4.

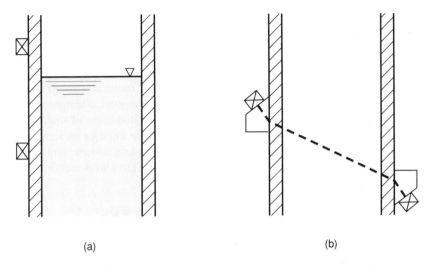

(a) (b)

Figure 38.4 Pipe wall used as (a) liquid level sensor, or (b) window for non-invasive flow measurement.

38.1.5 Limitations on Coupling, Propagation, Spatial Resolution and Correlations among Propagation and Measurands

In the field of ultrasonic non-destructive testing (NDT) a variety of couplants—liquids, oils, pastes, gels—are available for transmitting the interrogating wave from the transducer assembly to the part being tested. The couplant tends to match acoustic impedances as it replaces the low-impedance air gap that would otherwise block most of the transmission between a solid transducer and a solid test object. At temperature extremes the choice of couplants is greatly diminished. Accordingly in cryogenic or high-temperature situations one may resort to pressure coupling (sometimes aided by a compliant film of metal) or to solid bonds of epoxy, solder or ceramic cements. Coupling to liquid metals is sometimes facilitated by coating the probe so that it is wetted but not dissolved by the liquid metal.

Temperature extremes may be associated with severe sound speed gradients or other inhomogeneous conditions that block propagation. On the one hand, ultrasonic measurements have been made in the laboratory from nearly absolute zero to plasma temperatures of 20,000 K. On the other hand, practical measurements in cryofluids or merely in hot water near saturation conditions are often interrupted by 'flashing' or boiling, because of the severe mismatch in acoustic impedance between liquid and gas phases.

Another limitation in ultrasonic measuring systems is spatial resolution. Apart from speckle and wavelength-dependent limits, the propagation characteristics of the medium between the transducer and the region under study influence resolution in that region. As in optical systems, diffraction limits the lateral resolution. Depth resolution is usually limited by the time duration of the interrogating pulse, or its bandwidth. Refraction (Snell's law) limits the angle that a beam can be aimed relative to an interface. This means noninvasive interrogations may be unable to examine all areas inside a bounding pipe or vessel.

Yet another limitation applies to attempts to generalize correlations between variables such as, say, strength and the speed of sound, molecular weight and the speed of sound, or Young's modulus and the speed of sound. Such correlations may be quite useful when judiciously applied to a limited class of materials, or to specimens of similar alloys of like cross section. One must be alert, however, to avoid errors introduced by unwanted variables, such as moisture, temperature, pressure, boundary dimensions comparable to wavelength and dispersion.

38.2   DESIGN OF TRANSDUCERS

38.2.1   Material, Frequency, Dimensions, Vibration mode, Installation in a Measuring Chamber

In-depth accounts of piezoelectric, magnetostrictive and other electroacoustic transducer materials appear in the books by Mattiat (1971), Brockelsby et al. (1963), Silk (1984), and in a 129–page chapter by Sachse and Hsu (1979). These works emphasize piezoelectric materials, examples being quartz, PZT (lead zirconate titanate), LM (lead metaniobate) and $LiNbO_3$ (lithium niobate). Practical guidelines on material selection, sufficient for many design applications, are also available from manufacturers of transducer materials, e.g. Clevite, Staveley Sensors, Keramos, Valpey Fisher, Vernitron. The choice of materials sometimes depends on price, but more often on performance factors such as activity, sensitivity, dielectric constant, loss tangent, electrode material, mode purity, ability to launch specific mode(s), Curie point, thermal expansion coefficient, mechanical Q, available size and frequency–thickness constant.

Most people are familiar with the low-frequency percussion sounds of a big bass drum in contrast to the higher-frequency tone of a small snare drum. Similarly, the natural frequency of a small resonator is higher than that of a large resonator of the same material and proportions. The resonant frequency f_0 of a thickness mode disk depends on the disk's speed of sound c and thickness x. At half-wave resonance,

$$x = \frac{\lambda}{2} = \frac{c}{2f_0}$$

where λ = wavelength.

The frequency–thickness product is

$$f_0 x = \frac{c}{2}.$$

Thus, if the crystal's speed of sound is known, the thickness required to produce a given resonant frequency f_0 is readily calculated as

$$x = \frac{c}{2f_0}.$$

This calculation implies that the neutral plane lies exactly midway between the crystal's main faces. When crystals are bonded to other structural members, the latter's influence on f_0 and bandwidth must be taken into account. Transverse dimensions, when not much greater than λ, also influence f_0 and mode purity.

The most commonly used transducer vibration modes are compression and transverse shear, analogous to P and S waves in seismology. Radial, flexural (bending) and torsional modes are also used, depending on the interaction sought with the measurand, or to produce or detect a lower frequency with a smaller transducer than would be possible with an element used in its thickness (compression) mode or transverse shear mode (Kunkel *et al.*, 1990). Interdigital and wedge designs are commonly used to launch Rayleigh (surface) waves (Figure 38.5).

For laboratory measurements in non-corrosive, non-hazardous, non-conducting fluids the crystal often may be edge-supported right in the fluid. For industrial measurements it is usually the practice, however, to partly or totally surround the electroacoustic element by a metal housing. The housing, in turn, may be sealed by O-rings or other means to the chamber if leakage is to be avoided. The housing may be permanently or removably mounted in a flange or other port. The housing also serves as a radio-frequency interference shield and as a safety device to

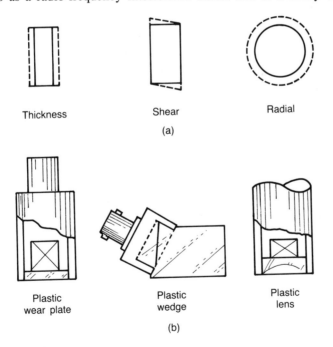

Figure 38.5   Examples of (a) thickness mode, transverse shear and radial mode crystals; (b) epoxy-sealed NDT transducers

prevent operator contact with high-voltage electrodes. Housed transducers are available from manufacturers of ultrasonic non-destructive testing (NDT) equipment. These transducers typically have radiating faces that are epoxy-sealed, and therefore are not usable in all fluid environments. Some components of natural gas and flare gas, for example, seem to have an uncanny ability to penetrate all the usual epoxies. Metal-sealed transducers are available from Massa and Panametrics. Designs from the latter firm include, if required, an internal λ/4 impedance-matching layer.

38.2.2   Special Considerations for Temperature and Pressure Extremes

Temperature extremes can cause failure of transducers in several ways: depolarization when the Curie point is exceeded; chemical reactions when hot; cracking or disbonding due to differential thermal expansion; couplant boils away or becomes embrittled at hot or cold extremes; transducer material becomes electrically conductive when hot; attenuation or speed of sound in transducer backing or adjacent wedge changes too much; soldered connections melt. In cryogenic applications it is desirable to minimize heat transfer to the cryofluid. High-purity gas applications may require that during vacuum bakeout the transducer neither outgases nor leaks. For use at high pressure, the transducer must neither deform nor leak. In a downhole tool for exploring oil, gas or geothermal reserves, transducers may be subjected simultaneously to high temperature and high pressure (e.g. $T = 100$ to $350°C$; $P = 20$ to 70 MPa). In testing red-hot steel, buffer rods or momentary contact may be able to isolate the transducer element from the high temperature of the product and the high pressure required for dry coupling.

38.3   DESIGN EXAMPLES

38.3.1   Transducers for Gases

In contrast to liquids or solids, gases are of low density ρ, low speed of sound c, and consequently of low acoustic impedance Z—see comparisons given in Table 38.2.

Table 38.2   Physical characteristics of sound in air, water and steel

Medium (20°C, 10^5Pa)	Density, ρ (kg/m^3)	Speed of sound, c (m/s)	Acoustic impedance Z (kg/m^{2}s)
Air	1.29	343	450
Water	998	1482	1.48×10^6
Stainless steel 316	7833	5760	45.1×10^6

This means that if a high-acoustic-impedance transducer is used, it is much more difficult to transmit ultrasound into or out of gas compared to liquids or solids. One seeks low-impedance transducers for gases. Audio equipment offers low-Z design suggestions, as do organs for speech and hearing.

Apart from low-mass membrane ultrasonic transducers (Section 38.5.2 and 38.5.3), which are appropriate for air, and tubular-waveguide-protected audio-frequency transducers for high-temperature gases (see Section 38.7.1), there are several housed piezoelectric disk designs to be considered. These may be categorized as unmatched and matched. Unmatched designs may encapsulate a resonator (thickness, radial, or flexural mode) in an impervious housing. Matched designs, for which more details have been published, include a single $\lambda/4$ matcher having an impedance as close as practical to the geometric mean of the impedances of the disk and the gas, and two-layer designs (Figure 38.6). The simpler one-layer design has proved adequate in hundreds of natural gas and flare gas flowmeters installed in pipelines, refineries and chemical plants between 1984 and 1990.

Acoustic isolation of transmitter and receiver, to prevent cross-talk, is often one of the most difficult design aspects of gas transducers used in pairs. Some

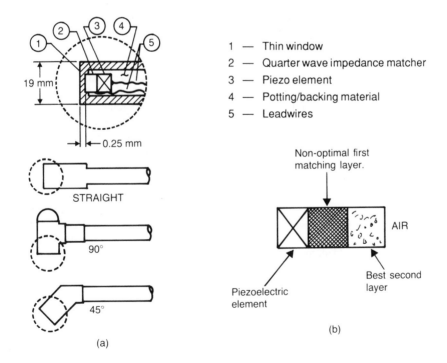

1 — Thin window
2 — Quarter wave impedance matcher
3 — Piezo element
4 — Potting/backing material
5 — Leadwires

19 mm

0.25 mm

STRAIGHT

90°

45°

(a)

Non-optimal first matching layer.

AIR

Best second layer

Piezoelectric element

(b)

Figure 38.6 Transducers for gases: (a) metallurgically sealed single-layer impedance-matched designs for flare gases according to a patent by Lynnworth *et al.* (1981); (b) two-layer unencapsulated design for air, after Khuri-Yakub *et al.* (1988).

approaches are illustrated in Lynnworth (1989, 1990a) utilizing membranes, intentionally alternated impedance mismatches, small contact areas, time delays or damping. If the same transducer is used as transmitter and receiver, ringdown becomes the key problem as far as coherent noise is concerned. Signals indicative of the measurand cannot be detected until coherent noise generated on transmission has subsided sufficiently or unless it can be subtracted.

38.3.2　Transducers for Liquids—Wetted, Clamp-on, Hybrid

In this section we introduce three categories of transducer designs for liquids, Figure 38.7, whose principal characteristics are compared in Table 38.3.

Table 38.3　Categories of transducer designs for liquids

Type	Advantages	Disadvantages
Wetted	Beam path is controllable; used in the most accurate ultrasonic methods; usable over wide frequency range e.g., $0.1 < f < 10$ MHz. May comprise part of replaceable module, e.g. avionic flowmeter	Requires penetration of pressure boundary; requires islation valve or empty pipe for installing and removing; cavities distort flow.
Non-wetted clamp-on, strap-on, snap-on, epoxy-on, braze-on, weld-on	Non-invasive, removable unless bonded; easy to retrofit, no cavities. In many cases, it would appear to be the safest way to measure flow in sense that the pipe integrity is not jeopardized.	Oblique beam path depends on liquid; tendency for acoustic short circuit noise when conduit wall is thick; usually $f < 2$ MHz. Limited usually to liquids; sometimes usable with gases at high pressure, or gases at atmosperic pressure in special circumstances only.
Hybrid	Combines advantages of wetted and non-wetted approaches. In some cases, the 'window' can be welded, moulded or cast into a valve or pipe fitting to accomodate the occasional or permanent attachment of the transducer.	Requires initial penetration of pressure boundary if window is installed as a retrofit; introduces cavities.

Wetted designs (Figure 38.7(a)) typically employ piezoelements resonant near 0.5, 1 or 2 MHz and may be narrowband or broadband depending on the backing impedance and on the window thickness. Low-impedance backings and thin windows lead to narrowband waveforms. Backings that match the element's

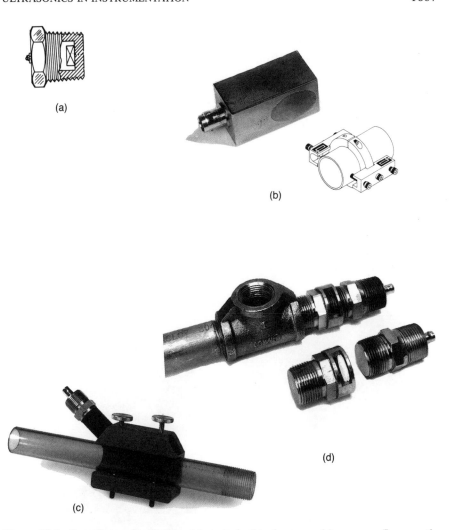

Figure 38.7 Transducers for liquids: (a) wetted; (b) clamp-on; (c) snap-on (Lynnworth, 1990a), (d) hybrid

impedance and thick metal windows lead to broadband waveforms. Tapered-pipe-threaded housings can be rated up to 20 Mpa (3000 psi).

Non-wetted clamp-on types (Figure 38.7(b)) are designed to fit in standard yokes or tracks. Tracks are available with scales so that the axial separation between transducers, as required in contrapropagation or tag flow measurements, can be set easily. For large pipes, magnets or wrap-around straps are commonly used to hold the yokes, tracks or transducers. For small pipes, below 50 mm ID, one or more transducers may be fixed temporarily or permanently within a given clamp-on or snap-on assembly. Such dual-transducer flowmeter assemblies, for

example, became available in the 1980s from Transonic Systems for soft biomedical tubing having internal diameters (ID) down to 1 mm, and from Panametrics in 1990 for metal tubing from 10 to 50 mm ID. Non-wetted *braze-on* flexural
transducers measured two-phase flow by the tag cross-correlation method in a 1
mm ID stainless steel tube, as explained in Section 38.6.5.

The two principal liquid-related uses of external non-wetted transducers have
been to measure liquid level and flow in pipes. The same or similar transducers
can often be used to measure thickness and integrity of pipes and other pressure
vessels—see Section 38.3.3.

Hybrid flowmeter transducers evolved first as a convenience for removability
and second, and independently, as a remedy to the occasional nuisance and
sometimes intolerable errors caused by uncertainty or drifting of the obliquely
refracted beam launched in the liquid by a clamp-on angle beam transducer. A
hybrid transducer consists of a flat-faced plug that can permanently maintain the
pressure boundary while defining one terminus of a known fixed measuring path
normal to the wetted face of the plug, and a removable non-wetted transducer

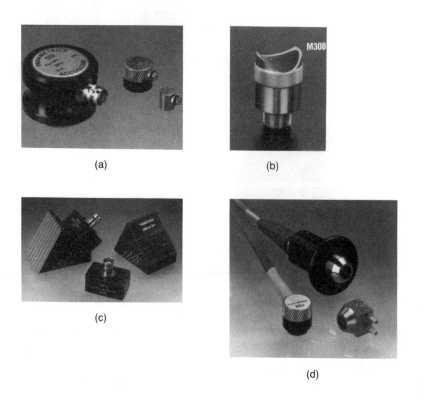

(a)

(b)

(c)

(d)

Figure 38.8   NDT transducers: (a) contact, flat; (b) wetted, focused; (c) angled; (d) dual-ele-
ment for corrosion testing. (Illustration courtesy Pamametrics).

that repeatably mates with said plug. A hybrid flowcell may be taken to mean a streamline flowcell with external transducers, which is flanged, screwed, welded or otherwise incorporated into a pipeline.

38.3.3 Transducers Used in Non-destructive Testing, Inspection and Evaluation of Solid Materials

Although the markets and the manufacturers may differ accordingly to one's interpretation of ultrasonic instrumentation (i.e. 'analytic' instrumentation, 'process' instrumentation, 'NDE/NDT' instrumentation), one would expect that the physics underlying the wave/measurand interaction would largely determine whether the design of transducers for such applications must differ or could be identical. In practice, many hand-held transducers for manual inspection of manufactured parts reflect ergometric design considerations; size, weight, surface contours and finish. Other design considerations include part accessibility, inspectability, transducer wear resistance and longevity. Because of the wide range of materials, geometries, dimensions, microstructures, bonds, quality levels and a market that has matured over the past half-century, a far greater variety of transducers exists for NDT applications than for any other single measurand. Most ultrasonic NDT transducers operate in the 1–10 MHz decade. Others are available down to 20 kHz and up to 100 MHz using bondable electroacoustic elements, and up to the GHz range using deposited piezoelectrics for acoustic microscopy, see Figure 38.8.

38.4 PROBLEMS OF PACKAGING ELECTRONICS AND TRANSDUCERS

38.4.1 General Comment

Packaging reflects function, appearance, tradition, materials, compatibility with the environment and with other instrumentation, available power, space, applicable codes, safety, portability, reliability, inputs/outputs and data displays, cost and other constraints.

Sometimes preamplifiers or other electronics are built into the transducer housing; sometimes the transducer is built into the electronics console; sometimes both are built into the same envelope, as in a downhole flow tool (Jacobson et al., 1992). What is perhaps relatively unique about packaging ultrasonic equipment is the need to keep coherent noise out of the receiver. Such noise comes from ringdown and cross-talk. As in electromagnetic communication equipment, damping and transmitter/receiver isolation have to be maintained while meeting other design objectives.

38.4.2   Environmental Problems

Examples of some environmental problems and solutions appear in Table 38.4.

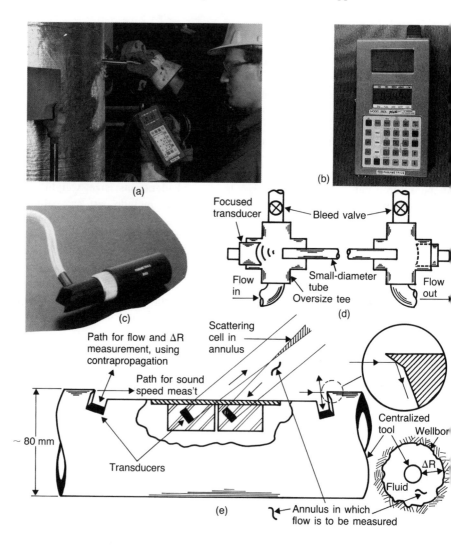

Figure 38.9 Examples of portable battery-powered electronics and/or transducers packaged compactly to meet constraint of one or more small dimensions: (a) hand-held corrosion gauge; (b) hand-held digital thickness gauge with keypad, two display screens and data-logging capability; (c) thickness-gauging transducer for small tubing, with vee-shaped aligner; (d) axial path offset flowcell with wetted transducers for contrapropagation measurements of flow in 1 mm ID SS tubing; (e) section of 80 mm OD downhole geothermal flow tool showing transducers for transmission and reflection measurements. These high-temperature, high-pressure transducers use piezoelectric elements of approximately 13 mm diameter.

Table 38.4   Environmental problems and solutions encountered in ultrasonic sensing

Problem	Solution
Explosive atmosphere	Use intrinsically safe design; purge electronics; avoid arcing contacts; use explosion-proof transducers; obtain safety certifications
Seismic vibration	Analyse and/or test to maximum shock and vibration levels over spectrum of interest (e.g. IEEE Standard 344-1971); minimize mass, stress, cantilevers, moments of inertia
High temperature	Use buffer rods, water jacket or momentary contact to isolate transducer from exposure to damaging temperature; use transducer with high Curie point; avoid different coefficients of thermal expansion; use Dewar and heat sinks to minimize heat transfer to electronics.
High pressure	Support transducer with incompressible backing
High nuclear fluence	Use inorganic transducer materials.

38.4.3   Space and Form Factor Constraints

Examples of some space and form factor constraints for ultrasonic instrumentation already developed or under development in 1990 are given in Table 38.5.

Table 38.5   Examples of space and form constraints

Application	Geometric constraints on electronics and/or transducer
Hand-held corrosion gauge	Desirable dimensions less than $100 \times 75 \times 10$ mm for electronics; transducer grippable by fingers, bare or gloved.
Space shuttle main engine flowmeter	All electronics on one board about 127×178 mm; transducer radial dimension less than 10 mm.
Downhole oil flow tool	Electronics and transducer must fit inside a 43 mm diameter envelope, but length can be 1–3 m.
Fluid management systems for hospital operating room	Equipment should fit on IV (intravenous) pole, preferably attachable into or installable into equipment already on that pole.

Examples of equipment commercially available in 1990 or designed at that time that satisfy requirements for small volume or at least one small dimension appear in Figure 38.9.

38.5   INFLUENCE OF NEW ELECTROACOUSTIC MATERIALS, COMPONENTS AND TECHNIQUES

38.5.1   General Comment

Referring to Section 38.3.1, it will be understood that air has been a major barrier to ultrasonically testing and measuring materials, or even testing air itself, i.e., measuring air flow, air temperature. Some of the new electroacoustic materials covered in this section find biomedical applications in body tissue and technical applications in liquids and in some gases other than air.

38.5.2   Polyvinylidene Fluoride and Copolymer Piezoelectric Sheet Materials

Available from several manufacturers in Europe, Japan and the US, polyvinylidene fluoride (PVDF or VF_2) is a long-chain semicrystalline polymer containing repeating units of $CH_2 - CF_2$. The material has excellent resistance to stress fatigue, abrasion, and cold flow. It is lightweight ($\rho = 1.78 \text{ g/cm}^3$), transparent and flexible. Its longitudinal wave speed in the thickness direction is 2200 m/s, leading to a characteristic acoustic impedance greater than water's by a factor of only 2.6. PVDF is readily fabricated in continuous sheets or complex shapes. Its non-polar crystal form (α form) can be converted to the polar and, therefore, piezoelectric (β) form by stretching, followed by poling in a high field while near 80°C. It is often supplied polarized and metallized. The copolymer $VF_2 - VF_3$ can be deposited from solution and then poled *in situ*, or it can be moulded into curved shapes; stretching is not required if the VF_3 content is above 20%. Construction details for a deposited-from-solution design due to Chan *et al.* (1989) are given in Figure 38.10(a).

38.5.3   Low-cost Electret Transducer Assemblies

Available from Polaroid since the early 1980s, this type of polymer film transducer is variously termed a capacitive, electret or electrostatic type. The housed device encloses a gold-plated Kapton element stretched over a grooved backing member. Details are given in Figure 38.10(b).

38.5.4   Composite Transducers

In attempts to secure some of the advantages of a low-characteristic impedance transducer material such as found in polymer transducers, yet still retain some of the advantages of ceramic transducers, several investigators have chosen the path of the composite. In some cases the acoustic impedance, bandwidth, and radiation patterns of the composite transducer can be controlled in a manner beyond what is possible in single-phase materials. Examples of structures under investigation in the mid-1980s are illustrated in Figure 38.10(c).

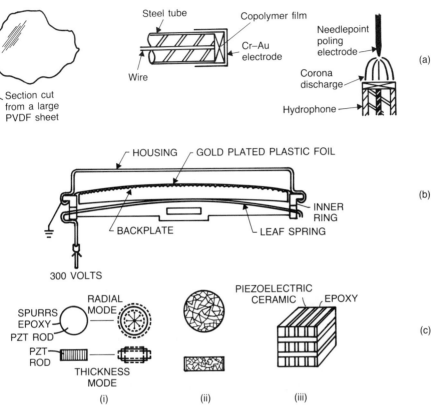

Figure 38.10 Low acoustic impedance transducers: (a) PVDF sheet, available in large sizes; construction and corona discharge poling details, after Chan *et al.* (1989); (b) Polaroid electret; (c) composite transducers: (i) rods in epoxy (after Gururaja *et al.*, 1985 © 1985, IEEE); (ii) fractured ceramic 'repaired' by epoxy (described by Montero de Espinosa *et al.*, 1986, © 1986 IEEE); (iii) stacked layered structure (after Hashimoto and Yamaguchi, 1986, © 1986 IEEE)

38.5.5 Sheet Metal Horns as Local or Remote Flexural Antennas for Ordinary or Extreme Fluid Environments

One approach to partly overcome the inefficient energy transmission normally associated with the five orders of magnitude steel/air impedance mismatch is to radiate energy from an extended aperture or surface whose area greatly exceeds that of the actual transducer. The efficiency of large-area leakage is enhanced by operating the radiator as a lowest-order asymmetrical flexural source at a frequency–thickness product such that the flexural phase velocity c_f only slightly exceeds the speed of sound in the adjacent fluid, namely, air (Sunthankar, 1973; Pierce, 1981). As a numerical example, at $f = 100$ kHz, $c_f = 1000$ m/s in stainless steel that is 1 mm thick. For small fd, c_f is proportional to the square root of fd the frequency–thickness product, one can readily see that a stainless steel sheet

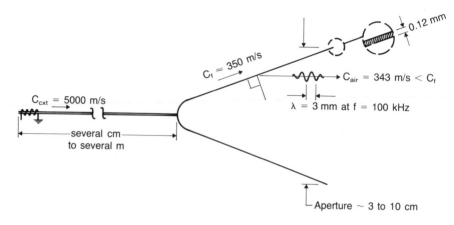

Figure 38.11   Stainless steel sheet metal horn radiating 100 khz waves into air

only 0.1 mm thick and carrying 100 kHz a_0 waves cannot radiate into air at 20°C ($c = 343$ m/s). But a slightly thicker sheet, say where $c_f = 350$ m/s, should be an efficient radiator of 100 kHz waves. This slow wave provides the basis for designing a thin-wall horn or antenna that can be excited into flexure by various means, e.g. by mode conversion from an extensional (compressional) source (Figure 38.11).

38.5.6   Non-contact Techniques

The two most widely used non-contact ultrasonic 'coupling' techniques currently in use are emat (electromagnetic transducer) based on the Lorenz force, and laser. Sometimes they are used in combination, e.g., the laser for transmission, the emat for detection, as in some mid-1980s R&D studies in the USA at NIST (National Institute for Science and Technology; formerly NBS or National Bureau of Standards) and BNWL (Battelle Northwest Laboratories) aimed at deriving the temperature profile from sound speed measurements across different paths in hot steel. Fibre-optic waveguides have been suggested as a means of conveying laser pulses to and from precise points on materials, thereby avoiding losses that might otherwise occur in the air path (Dudderar et al., 1989).

38.6   FLOW MEASUREMENT

38.6.1   Background

When describing fluids in terms of their flow, temperature, density or other parameters that vary in space and/or time, one may need to be careful to distinguish among point, line-average and area-average, and among instantaneous, r.m.s. and long-time averaged values. Because ultrasonic beams can be steered through optically opaque walls, because interactions can be range-gated, and because response occurs at the speed of sound, ultrasonic instrumentation has unique opportunities to disclose fluid dynamic details unavailable from most other disciplines. There is, however, the opportunity to be confused by uncertainties in the interactions between the interrogating beam and the flowing fluid. In many instances it suffices to report only one of the following: area-averaged flow velocity V, volumetric flowrate Q, mass flowrate M, or energy flowrate H. In Sections 38.6.2–5 are briefly treated only three of the dozen ultrasonic flowmetry categories covered in Lynnworth (1989)-contrapropagation transmission, reflection from scatterers, and tag cross-correlation. These three are represented by entries 1, 2, and 4 in Figure 38.12. Of these three, the contrapropagation method has so far achieved the highest accuracy, fastest response (1 ms), operated from lowest to highest velocity V, in conduits from 1 mm ID to river paths of 1 km, in clean liquids and gases and in some not-so-clean fluids, cryogenic to high temperature, with wetted, non-wetted or hybrid transducers.

38.6.2 Transmission Upstream and Downstream—Single Path, Multipath

In Figure 38.13(a), where V_0 is the flow velocity on the axis between the transducers that we imagine to occupy only the on-axis points A and B, the upstream–downstream transit time difference Δt can be shown to be equal to $2LV_0/c^2$ provided $V_0^2 \ll c^2$. This intrusive configuration makes it easy to derive the basic relation between V_0 and Δt, but is usually impractical. A better arrangement is shown in (b), but two holes are required. In (c) the transducers are outside the pipe. As is clear from the end view shown in (d), flow is sampled only along tilted diameter AB. The observed line-averaged V_d generally exceeds V by 5%–33% depending on the Reynolds number Re. $KV_d = V$ where $K = 1/[1.119 - 0.011 \log \text{Re}]$ for fully developed turbulent flow, Re > 4000. However, $K = 0.75$ for laminar flow, Re < 2000. For transitional flow, Re between 2000 and 4000, K is uncertain.

Methods of sampling 100% of the duct cross-section include Figure 38.13(e)— use of an external reflector (Drost, 1980), (f) square conduit, and wetted or unwetted axial flowcells (g,h), the latter due to Lynnworth (1990a). For cross-sections too large to sample everywhere, one may choose mid-radius chords (Figure 38.13(i)) that very nearly yield V for axisymmetric laminar or turbulent flow, or special chords that weight the paths according to the quadrature methods of Gauss, Chebyshev (j) or others (k). In large pipes that may contain swirl or cross-flow, the most general solution for highest accuracy, better than 1%, is crossed multipaths. Depending on conduit material, diameter and wall thickness, and required accuracy, external transducers may be clamped, strapped or snapped onto conduits from 1 mm ID to 10 mm OD. An example of transducers clamped

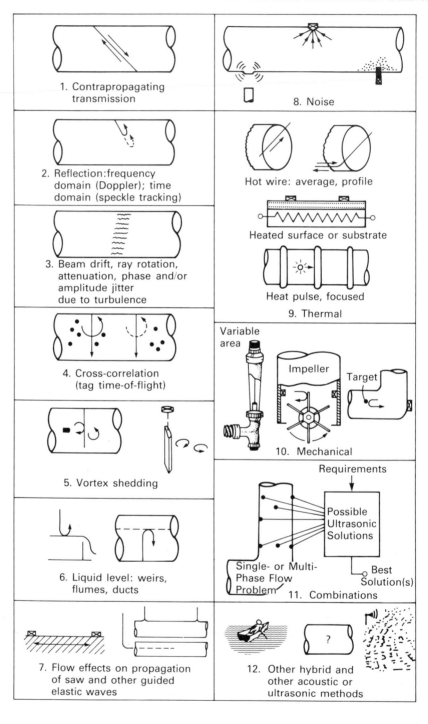

1. Contrapropagating transmission

2. Reflection:frequency domain (Doppler); time domain (speckle tracking)

3. Beam drift, ray rotation, attenuation, phase and/or amplitude jitter due to turbulence

4. Cross-correlation (tag time-of-flight)

5. Vortex shedding

6. Liquid level: weirs, flumes, ducts

7. Flow effects on propagation of saw and other guided elastic waves

8. Noise

Hot wire: average, profile

Heated surface or substrate

Heat pulse, focused

9. Thermal

Variable area

Impeller

Target

10. Mechanical

Requirements

Possible Ultrasonic Solutions

Single- or Multi-Phase Flow Problem

Best Solution(s)

11. Combinations

12. Other hybrid and other acoustic or ultrasonic methods

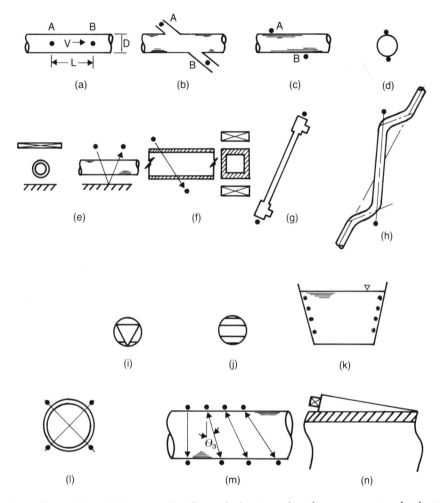

Figure 38.13 Flowcells for measuring flow velocity at a point, along one or more chords or over the full duct area (see text).

at the ends of crossed paths in order to reduce errors due to cross-flow and asymmetry is given in Figure 38.13(l). The four-path arrangement in (m) was derived from a proposal by E.H. Carnevale (private communication, ca. 1988) as a way to avoid loss of a narrow beam despite large changes in refracted angle θ_3 as the liquid's speed of sound changes due to temperature or composition. When this wide-ranging problem occurs in ducts that have a wall thin compared

Figure 38.12 Twelve categories of acoustic or ultrasonic flow measurement principles and methods, with examples. Flow is generally from left to right. (After Lynnworth, 1989, © 1989 Academic Press)

to wavelength, one alternative solution is to generate a wide beam in the liquid by starting with a narrow beam in the wedge, incident on the duct near grazing incidence Figure 38.13(n).

38.6.3 Reflection—Doppler, Double-pulse, Range-gating

When the fluid contains entrained air, particles, or is otherwise not all of one phase or not of uniform impedance, flow can be measured by scattering and/or tag cross-correlation. Scattered waves are Doppler-shifted, and the frequency difference Δf between transmitted and received energy is proportional to the Mach number V/c multiplied by the cosine of the angle between the transmitter and receiver rays. If the interrogating wave is introduced through a wedge, refraction in the fluid occurs according to Snell's law, which compensates for c and results in Δf being proportional to V in the scattering cell. Apart from flow profile and attenuation limits, and differences between scatterer and matrix velocity, it is evident that as V tends to zero so too does Δf. When V falls below 0.1 m/s many Doppler-arrangements fail to provide the sought accuracy.

Another way to measure scatterer velocity is to employ double-pulse ranging. The interval between pulse pairs can be selected automatically to accommodate a wide range of velocity V, including V below 0.1 m/s and in principle down to 1 mm/s or less.

Range-gating, applicable to Doppler as well as to double-pulse methods, yields flow profile information. In cases where the flow profile could be measured accurately enough so that the uncertainty or error in the meter factor K would be reduced below 1%, perhaps down to 0.25 to 0.5%, and if the conduit OD is calipered and its wall thickness measured ultrasonically, the accuracy of a clamp-on flowmeter could closely approach, equal or possibly exceed that of wetted transducers or insertable (non-ultrasonic) calibrated spool pieces.

38.6.4 Transmission and Reflection Combined in One Flowmeter

Even fluids that are normally single phase may become multiphase during upset conditions, e.g. temperature overshoot, temporary introduction of a highly volatile component, cavitation when back pressure falls. Multichannel instruments based on the patent of Jacobson et al. (1988) became available by 1990 that could interrogate fluids in transmission and reflection modes ('transflection'), see Figure 38.9(e), and automatically select the more appropriate mode as fluid conditions changed.

38.6.5 Density and Mass Flowrate

In many pure, well-defined single-phase liquids the density ρ bears a linear relationship to the speed of sound c. In such liquids, the same transducers that

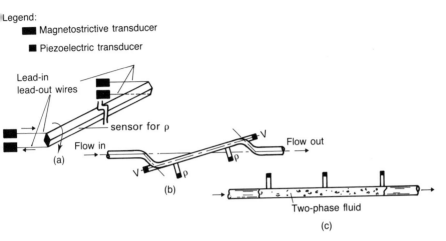

Legend:
- ■ Magnetostrictive transducer
- ■ Piezoelectric transducer

Figure 38.14 Measurement of V, ρ, and M: (a) torsional sensor comprising intrusive waveguide of diamond cross-section; (b) non-invasive streamline M flowcell utilizes flexural ρ sensor comprising the conduit wall, contacted on the outside only; (c) tag cross-correlation method

measure V by contrapropagation can measure c. The ρV product leads to the mass flowrate: $M = K\rho VA$ where A = duct area.

In two-phase fluids, or unknown mixtures of single-phase liquids, c and temperature T are inadequate to determine ρ to high accuracy. Alternative ways to sense ρ ultrasonically include measuring the velocity of certain guided waves such as torsional or flexural waves (Lynnworth, 1989; 1990(a, b)); Jen *et al.*, 1990(a, b). This leads to intrusive and non-intrusive M (mass flowrate) flowcells as shown in Figure 38.14, where ρ and V are measured between the correspondingly designated transducers. Furthermore, if ρ is measured in two axially displaced zones, and if ρ fluctuates enough in these zones in response to instantaneous fluid quality or composition variations, then the cross-correlation method may be applied to these ρ tags, again leading to M.

38.7 TEMPERATURE MEASUREMENT

38.7.1 General Remarks

Ultrasonic thermometers respond to the temperature dependence of the speed of sound, c. In the range −80 to +250°C the quartz thermometer manufactured by Hewlett-Packard offers resolution of 0.1 millidegree and linearity better than platinum resistance sensors. At higher temperatures, approaching 3000°C in some nuclear fuel studies, thin-wire thoriated or K–Al–Si doped W waveguide sensors yielded accuracy, simpler five-zone profiling, faster response and longevity advantages over thermocouples. Media such as gases, plasmas, hot water and hot steel have been used as their own sensors. Air was used in this thermometric manner as early as 1873. In gases, c is proportional to the square root of absolute temperature T; in solids, to the square root of stiffness E or G (E = Young's modulus, G = shear modulus).

Another way to measure temperature is to use the non-linear effect due to the propagation of a higher-frequency (e.g. 2.5 MHz) probing pulse that is superimposed on a lower-frequency (e.g. 300 kHz) pump pulse. So far this method has been used in hyperthermia studies, with accuracies of .2°C reported (Ueno *et al.*, 1990).

38.7.2   Medium as its own Sensor—Hot Gas, Hot Steel

Air, inert gases, gas mixtures and plasmas were used as their own temperature sensors in some cases to 20 000°C in laboratory experiments conducted at Panametrics in the 1963–67 period. As a numerical example of the temperature sensitivity of the speed of sound, in air at 293 K, $c = 343$ m/s, increasing by 1% at 299 K ($\Delta c/c = 0.5\ \Delta T/T$). This means in a path of 300 mm the transit time would decrease from 1000 μs to 990 μs, an easily detected change. By using an instrument that resolves this 10 μs decrement to 0.1 μs, the 6 K increment could be resolved to 0.06 K. In 1970 the author suggested (but did not demonstrate) that by using a number of paths in gaseous or solid media, the T profile could be extracted; furthermore, during heat treating an ingot's asymptotic approach to thermal equilibrium could be monitored.

By 1990 the most important and widespread use of the c vs T relation for gases appeared to be the measurement of boiler and stack gas temperature profiles. By the end of 1990, in installations including chemical recovery boilers, refuse-fired furnaces and in coal-fired power plants on the order of 0.1–1 GW capacity, over thirty tomographic profiling systems were operating successfully. Some were manufactured by Codel in the UK and the rest mostly by SEI in the USA (Kleppe, 1990; private communication). Summaries of the pioneering work in the mid-1980s in this area by Green, (1985) in the UK and Nuspl *et al.*, (1986) in the USA appear in the author's 1989 book and in Kleppe (1989). In these applications the temperature distributions were measured in boilers of maximum dimension 25 m, with $T_{max} = 1500°C$. Transducers were isolated from the high temperature by air-cooled conduit, through which the interrogating waves were transmitted. Strictly speaking, these are not 'ultrasonic thermometers' because the sound waves were reduced to the audio range, 500–3000 Hz, to overcome attenuation. The octave 1000–2000 Hz, approximately, sometimes yields the highest signal-to-noise ratio.

About the same time that hot gas acoustic profiling was proving successful, USA investigations at BNWL and NIST were beginning to show promise that hot steel temperature profiles eventually would be measured ultrasonically. The initial experiments, reported in 1986, used laser excitation, emat reception, and

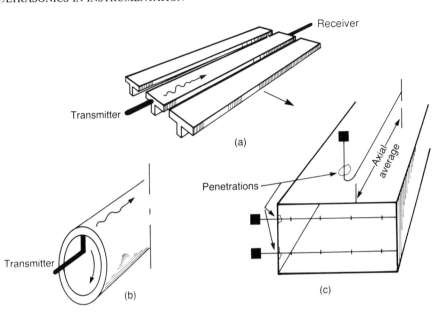

Figure 38.15   Thermometer designs proposed for a bakery: (a) moving hearth plate is its own sensor; (b) rolling drum is its own sensor; (c) multi-zone waveguide sensor

yielded reasonable agreement up to 700°C with thermocouples embedded in an SS304 billet. Later experiments used emats only and extended the temperature to 1000°C. In contrast to the hot gas work, already successful in over twenty large boilers by 1990, the hot steel thermometry work still appears to be at the laboratory stage, in general. (See, however, Section 38.8.4) dealing with accurate ultrasonic thickness measurement on red-hot steel in a steel mill.) On the other hand, at the less than 300°C temperatures encountered in baking, for temperature measurement of moving steel hearth plates or rolling steel drums, each acting as its own sensor, ultrasonic thermometry looks feasible (Figure 38.15 (a), (b)). Ultrasonic paths are in the 1 to 10 m range in some bakeries. Accuracies of about 1% are achievable and appear adequate.

Another way to determine hot steel temperature T_s for a recurring process is to compute T_s from the gas temperature profile, as in some heat-treating operations (Kleppe, 1990, private communication). This may be thought of as a hybrid case where the gas is a 'foreign sensor' surrounding the medium in which temperature must be determined. (Compare with Section 38.7.4.)

38.7.3   Foreign Sensors—Quartz Resonator, Multizone Wire Waveguide

Hewlett-Packard (HP) introduced its first quartz thermometer in 1965. Within its temperature range, −80 to +250°C, according to its manufacturer, the HP 2804A quartz thermometer has advantages over other types of thermometers such as th following: it is an easy-to-use, high-resolution digital thermometer that ap proaches standards-level accuracy; it is more rugged and much easier to use tha a standard platinum resistance thermometer and Mueller bridge, yet costs con siderably less; there is no tedious bridge balancing or calculation of temperatur from measured resistance. The 2804A is reported to outperform industrial plati num, thermistor, and thermocouple thermometers in terms of accuracy, resolu tion, repeatability, stability and probe interchangeability. The quartz temperatur sensor is a precisely cut, polished, and gold-plated quartz disk about the size o a contact lens.

The quartz thermometer has found many applications because of its rathe unusual characteristics relative to other temperature sensors. Among the mos important of the characteristics are portability, stability over extended peri ods, temperature resolution, direct reading linear display, and rapidity of meas urement. Because of these features, many instruments are used a working standards for the calibration and verification of liquid-in-glass thet mometers, thermistors, thermocouples and temperature regulators and control lers.

The multizone wire waveguide thermometer evolved in the late 1960s to mee the needs of researchers interested in nuclear fuel pin centreline profiles u; to the melting point of UO_2,~ 3000°C. Although interest in measuring high tem peratures continues (e.g. Wilkins, 1992), it appears that an important and evol ving use of this technology occurs at temperatures roughly an order of magnitud, lower, e.g., baking ovens. The advantages over thermocouples or resistance wire include easier installation and maintenance; measurement of profile and/or aver age over a long path and lower cost (see Figure 38.15(c)). The intervalomete can be the same as in section 38.6 with a temperature program replacing the velocity program.

38.7.4   Conduit Wall Used as a Sensor

Referring back to Figure 38.14(b), the speed of a flexural wave c_f in thin condui wall can indicate the fluid density ρ when c_f is a known function of ρ. But c_f i: also a function of T. To separate the variables ρ from T one can use another wave e.g. a high-frequency Rayleigh (surface) wave in the wall, say $f = 2, 5$ or 10 MHz By making the Rayleigh wavelength $\lambda_R \ll x$, where x = wall thickness, the Ray leigh wave propagates mainly on the outer skin of the conduit at a speed c_i responsive to T but independent of the liquid inside. Although the T-depend ence of c_R is generally unavailable in the literature, it is calculable from the longitudinal velocity c_L and the shear velocity c_s. For many stainless steels c_R i: about 93% of c_s. The temperature dependence of c_s is known for many materials For example, for SS316 c_s decreases about 10% as T increases from 20°C to 500°C.

38.8 INTERFACE MEASUREMENT

38.8.1 Introduction

The ultrasonic measurement easiest to understand, and probably the most widely used in nature as well as in industry, is ranging, or measuring the distance z to an interface. If c is constant over the path, $z = ct/2$ where t = round trip transit time. Complications occur if c is not constant, if the interface is not well defined, if other targets generate interfering echoes, or if the medium is not transmissive.

38.8.2 Ranging in Air—Robotics

Low-cost housed and ready-to-use transducers that are available from Polaroid, Massa and others are easily adapted to ranging in air. Partly for this reason, many engineers now use ultrasonics for robotic and other air distance measurements. Most air ranging occurs below 200 kHz, but for short ranges or high resolution (e.g. 0.2 μm) frequencies as high as 8 MHz have been used.

According to Schoenwald (1985),

Robotics is an acknowledged critical component of present and future manufacturing methods. The competitive drive for still higher productivity is pressing robotics technology to provide machine intelligence capable of accurate, real-time adaptive behavior. Sensing is considered a major component of machine intelligence. For manufacturing, repair and maintenance, a sense of touch, proximity and depth perception, traditionally considered uniquely human activities, are among the key requirements for any machine to emulate—or surpass. Acoustic and ultrasonic based sensors are demonstrating a capacity to provide perceptual information for range, object recognition, tactile force and shape determination. They also complement machine vision. Numerous systems make use of acoustics/ultrasonics and/or rely on the piezoelectric properties of transduction sensors. The device physics, data acquisition, signal processing and feature extraction capabilities are always interrelated.

Apart from robotics, ranging in air (or water) can also aid human subjects directly. As an example, a multitransducer headset system is reported to enable a quadriplegic person to control the cursor of a computer display by head motion.

38.8.3 Ranging in Liquids—Liquid Level

Apart from liquid level, which is measurable by ranging from above or below, interface applications include determining hydraulic piston position, determining range or proximity of objects as used in mobile robot navigation, robot arm control, micro-distance (e.g. range resolution of 0.2 μm in air at $f = 8.4$ MHz), and prosthetic aids, determining shape of objects, i.e. acoustic imaging, tactile sensing, and underwater applications analogous to those in air, especially searching, manipulating objects, assisting divers; underwater military applications for

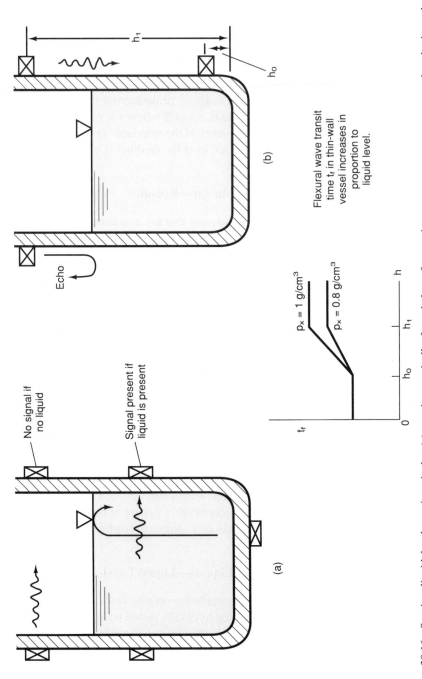

Figure 38.16 Sensing liquid level non-invasively: (a) ranging vertically from below, for continuous measurements, compared to horizontal interrogation for presence; (b) flexural sensing of level in a thin-walled vessel by reflection (left side) and transmission (right side)

ınmanned undersea vehicles including monitoring the vehicle's location, avoid-
ng obstacles, locating targets, detecting threats such as mines, vessels or torpe-
Joes, and imaging.

In some cases it is feasible to determine presence or absence of liquid at a
ıarticular level based on transmission across the vessel, or based on ringdown in
.he vessel wall (Figure 38.16(a)). By using flexural waves, whose phase velocity
.s reduced when liquid is present, one can remotely sense liquid presence and level
ıt least in thin-walled vessels from the flexural transit time (Figure 38.16 (b)).

The simplicity of any non-invasive liquid level sensing method is sometimes
ıffset by complications arising from residue, foam, internal obstructions, access
·estrictions, unsettled surface, unknown speed of sound c or high attenuation
ɔoefficient, α. Note that in Figure 38.16 (a) the diametral interrogation yields c
ıt that level, which may suffice for interpreting the vertical transit time.

38.8.4 Ranging in Solids—Thickness Measurement

Ranging in solids most often yields thickness, with the solid serving as its own
sensor. Solid or hollow waveguides, however, can also be used to sense a liquid
level, as when the waveguide transmits torsional or flexural waves. For simplicity, we limit this discussion to thickness gaging. Because the longitudinal
velocity in steel and aluminum is around $6000 \, \text{ms}^{-1}$ measurement of travel times
to ± 1 ns of ultrasonic pulses in the 2 to 10 MHz range to leads to thickness
resolution of the order of $\pm 3 \, \mu\text{m}$, the same as a machinist's micrometer, but with
access required on one side only. Transducers and electronics are available to
measure metals and non-metals from 25 μm to 1 m or more. Frequency may be
below 1 MHz for high-α materials like fibre-reinforced plastic (e.g. boat hulls),
ɔr above 10 MHz for gauging thin coatings such as plastic films over glass.

High-temperature examples include ablation gauging of re-entry vehicle heat
shields, erosion monitoring of steel pipe used to convey coal conversion (lique-
faction) slurries at 410°C, measuring eccentricity of hot forged tubes at 600–
950°C, and accurate ($\pm 1\%$) gauging of seamless steel tubing at 700–1100°C at a
Timken piercing mill since 1985. This last application involves dozens of alloys,
wall thicknesses ranging from 12.7 to 50 mm and outside diameters from 100 to
300 mm.

38.9 OTHER APPLICATIONS

38.9.1 Sources and Examples

Other ultrasonic applications consisting of either measurement of a measurand
not previously demonstrated, or of a well-known measurand occurring in an

especially harsh environment or in a new situation or measured to higher accuracy than before, are reported in periodicals such as *J. Acoust. Soc. Am.; IEEE*
Trans. UFFC; Rev. Sci. Instr.; Sensors; ISA Trans., and similar journals, and in
proceedings of various technical meetings devoted to sensing, instrumentation,
acoustics or ultrasonics.

This chapter has touched on thickness, which is but one of a number of NDT
measurands successfully addressed worldwide. Microstructure, flaws, bonds, degree of cure, stress, strain, . . . begin to round out the scope of NDT.

Pressure and density sensors exist based on quartz resonators, analogous to the
quartz thermometer, but again not describable within these space restrictions.

One way to summarize the scope of ultrasonic instrumentation is to list a
number of industrial applications, some of which have been included in this
chapter. Readers desiring more details may refer to Lynnworth (1989), or to some
of the thousand or so references cited therein.

38.9.2 Measurands and Industrial Applications Addressable by Ultrasound

The following list demonstrates the range of measurands that have been measured
successfully with ultrasonic methods.

- Flow
- Temperature
- Density
- Porosity
- Pressure
- Dynamic force
- Vibration
- Acceleration
- Viscosity in fluids
- Transport properties
- Level
- Location of low-reflectivity interfaces
- Phase
- Microstructure
- Nodularity

- Thickness
- Position
- Composition
- Anisotropy
- Texture
- Non-destructive testing to determine departures from specifications
- Grain size in metals
- Stress and strain
- Acoustic emission
- Imaging
- Holography
- Microscopy
- Elastic properties

- Bubbles and particles sensing
- Gas leaks

- Interrupted sound beam to determine object presence

38.9.3 Summary of Principal Reasons for Using Ultrasound to Measure and Control Materials or Processes

The following checklist will assist selection of ultrasonic measurements for given applications.

- Non-invasive or minimally invasive measurement
- High accuracy (measurands usually transformed to time or frequency measurements)
- Reliability (no moving parts, in the usual sense)
- Fast response (can be less than 1 ms)
- Remote sensing, sometimes with no physical contact
- Average reading over an extended region
- Profile information (point by point, or small-path average)
- Computer compatibility of time, frequency or, say, 8-bit amplitude data
- Low cost, especially for multiplexed and/or mass-produced sensors
- Small size, small mass
- Avoidance of problem(s) associated with competing (non-ultrasonic) technologies
- Data or results unobtainable any other way.

38.10   SOURCES OF FURTHER ADVICE

The reader is referred to the following textbooks and journals for further information on ultrasonics in instrumentation.

Books

Alippi, A. (ed.) (1989) *Ultrasonic Signal Processing* World Scientific, Singapore.

Christensen, D. A. (1988) *Ultrasonic Bioinstrumentation* John Wiley, New York.

Ristic, V. M. (1983) *Principles of Acoustic Devices* John Wiley, New York.

Journals

IEEE Transactions on Ultrasonics, Ferroelectrics and Frequency Control IEEE, USA.

Journal of the Acoustical Society of America, American Institute of Physics, USA.

Measurement Science and Technology (formerly *Journal of Physics E: Scientific Instruments*), Institute of Physics, UK.

Ultrasonics Butterworths Scientific, UK.

ACKNOWLEDGEMENTS

The author gratefully acknowledges the permission of Academic Press and the IEEE for allowing some of their copyrighted material to be used here. The manuscript was typed by Barbara A. Chiacchio and Tamara M. Stearns.

REFERENCES

Brockelsby, C.F., Palfreeman, J.S. and Gibson, R.W. (1963). *Ultrasonic Delay Lines*, Iliffe, London.

Chan, H.L.W., Ramelan, A.H. Guy, I.L. and Price, D.C. (1989). 'VF2/VF3 copolymer hydrophone for ultrasonic power measurements', *1989 Ultrasonics Symp. Proc.*, IEEE, 617–200.

Drost C.J., (October 14, 1980). *Volume Flow Measurement System,* US Patent No. 4,227, 407.

Dudderar, T.D., Peters, B.R. and Gilbert, J.A. (1989). 'Fiber optic sensor systems for ultrasonic NDE: state-of-the-art and future potential', *1989 Ultrasonics Symp. Proc.*, IEEE, 1181–90.

Green, S.F. (February 1985). 'An acoustic technique for rapid temperature distribution measurement', *J. Acoust. Soc. Am.* **77** (2), 759–63.

Gururaja, T.R., Schulze, W.A., Cross, L.E., Newnham, R.E., Auld, B.A. and Wang, Y.J. (July 1985). 'Piezoelectric composite materials for ultrasonic transducer applications. Part I: Resonant modes of vibration of PZT rod-polymer composites', *IEEE Trans. Sonics and Ultrasonics* **SU-32**(4), 481–498.

Hashimoto, K.Y. and Yamaguchi, M. (1986). 'Elastic, piezoelectric and dielectric properties of composite materials', *1986 Ultrasonics Symp. Proc.*, IEEE, 697–702.

Jacobson, S.A., Lynnworth, L.C. and Korba, J.M. (November 29, 1988). *Differential Correlation Analyzer,* US Patent No. 4,787,252.

Jacobson, S.A., Korba, J.M., Lynnworth, L.C., McGrath, W.F. and Nguyen, T.H. (1992). *Downhole Geothermal Flow Tool*, Final Report, DOE Contract No. DE-AC0288ER80587.

Jen, C.K., Piche L. and Bussiere, J.F. (July 1990a). 'Long isotropic buffer rods', *J. Acoust. Soc. Am.* **88**(1), 23–25.

Jen, C.K., Oliveira, J.E.B., Yu, J.C.H., Dai, J.D. and Bussiere, J.F. (May 28, 1990b). 'Analysis of thin rod flexural acoustic wave gravimetric sensors', *Appl. Phys. Lett.* **56**(22), 2183–5.

Khuri-Yakub, B.T., Kim, J.H., Chou, C-H., Parent, P. and Kino, G.S. (1988). 'A new design for air transducers', *1988 Ultrasonics Symp. Proc.*, IEEE, 503–6.

Kleppe, J.A. (1989). *Engineering Applications of Acoustics*, Artech.

Kleppe, J.A. (May 1990). 'The measurement of combustion gas temperature using acoustic pyrometry', *Proc. 33rd Annual Power Instruments Symp.*, Toronto, ISA.

Kunkel, H.A., Locke, S. and Pikeroen, B. (July 1990). 'Finite-element analysis of vibrational modes in piezoceramic disks', *IEEE Trans. UFFC*, **37**(4), 316–28.

Lynnworth, L.C. (1989). *Ultrasonic Measurements for Process Control*, Academic Press.

Lynnworth, L.C. (June 29, 1990a). *Snap-On Flow Measurement System*, US pat. pending, Serial No. 546, 586.

Lynnworth, L.C. (1990b). 'Flexural wave externally-attached mass flowmeter for two-phase fluids in small-diameter tubing, 1-mm ID to 16-mm ID', *1990 Ultrasonics Symp. Proc.*, IEEE, pp. 1557–62.

Lynnworth, L.C., Fowler, K.A. and Patch, D.R. (October 27, 1981). *Sealed, Matched Piezoelectric Transducer*, US Patent No. 4,297,607; see also, Lynnworth, L.C. Patch, D.R. and Mellish, W.C. (March 1984). 'Impedance-matched metallurgically sealed transducers', *IEEE Trans.* SU **31**(2), 101–4.

Mattiat, O.E. (ed.) (1971). *Ultrasonic Transducer Materials*, Plenum Press, New York, London.

Montero de Espinosa, F.R., Pavia, V., Gallego-Juarez, J.A. and Pappalardo, M. (1986). 'Fractured piezoelectric ceramics for broadband ultrasonic composite transducers', *1986 Ultrasonics Symp. Proc.*, IEEE, 691–6.

Nuspl, S.P., Szmania, E.P., Kleppe, J.A. and Norton, P.R. (October 19–23, 1986). 'Acoustic pyrometry applied to utility boilers', Paper 86-JPGC-PTC-2, *Joint ASME-IEEE Power Generation Conf.*, Portland, Oregon.

Pierce, A.D. (1981). *Acoustics—An Introduction to Its Physical Principles and Applications*, McGraw-Hill, N. Y.

Sachse, W. and Hsu, N.N. (1979). 'Ultrasonic transducers', *Physical Acoustics 14*, W.P. Mason and R.N. Thurston (eds.), Academic Press, 277–406.

Schoenwald, J.S. (1985). 'Strategies for robotic sensing acoustics', *1985 Ultrasonics Symp. Proc.*, IEEE, 472–82.

Silk, M.G. (1984). *Ultrasonic Transducers for Nondestructive Testing*, Adam Hilger, Bristol, UK.

Straty, G.C. and Younglove, B.A. (1973). 'Velocity of sound in saturated and compressed fluid oxygen', *J. Chem. Thermodynamics* **5**, 305–12.

Sunthankar, Y. (July 1973). 'A novel ultrasonic radiator', *IEEE Trans. Sonics and Ultrasonics*, **SU-20**(3), 274–8.

Ueno, S., Hashimoto, M., Fukukita, H. and Yano, T. (1990). 'Ultrasound thermometry in hyperthermia', *1990 Ultrasonics Symp. Proc.*, IEEE, 1645–52.

Wilkins, S.C. (1992). 'A single-crystal tungsten ultrasonic thermometer' in: *Temperature – Its measurement and control in Science and Industry*, ISA.

Chapter

39 RC SPOONCER

Fibre Optics in Instrumentation

Editorial introduction

Optical fibre sensors have clearly become an important class of the sensor family. Whilst not having made the impact some expected they would, they certainly have provided solutions to sensing needs that far exceed the performance of alternatives. Development of the suitable electro-optical interfaces, connection methods and practice and understanding of their special capabilities are each mature enough for these sensors to now be used in routine ways. It is to be expected that they will continue to find increasing application as this knowledge finds its way to the wider community.

39.1 INTRODUCTION

During the last fifty years two major advances, the development of the laser and the invention of the transistor, have revolutionized instrumentation technology. Industry is now in the midst of a third major development, the use of optical fibres for communication. Will the spin-off from this advance promote another quantum leap ahead in measurement methods similar to those inspired by the laser and the transistor? Many organizations, academic and industrial, think so. In 1987 more than 400 primary papers on fibre optic sensors (FOS) were listed in the Institution of Electrical Engineering database (Inspec) and, world-wide, over 200 FOS suppliers have been identified, albeit mainly on a small scale (McGeehin, 1989), whilst the annual world market is of the order of $50m (Reed, 1989) with a growth rate forecast (Frost and Sullivan, 1989) approaching 20% per annum.

The potential advantages of fibre optic sensors compared with conventional sensors have been well documented (Giallorenzi *et al.* 1982; Jones, 1985) the most important of which are:

Handbook of Measurement Science, Volume 3
Edited by P. H. Sydenham and R. Thorn
© 1992 John Wiley & Sons Ltd

- freedom from electromagnetic interference (can be used in electrically noisy environments);

- electrically passive (intrinsically safe in certain hazardous environments);

- Lightweight and flexible (valuable in aircraft, medical, offshore and other industries);

- higher sensitivity than conventional sensors in some applications.

In addition the extensive bandwidth of optical fibres offer opportunities for multiplexing FOS on a scale not possible with electrical telemetry.

39.2 PROPAGATION OF LIGHT THROUGH OPTICAL FIBRES

39.2.1 Basic Principles

The propagation of light down optical fibres is governed by Maxwell's equations, solution of which is beyond the scope of this chapter, but a simpler explanation using ray optics can be developed which has the advantage that the physical picture is clear and comprehensible.

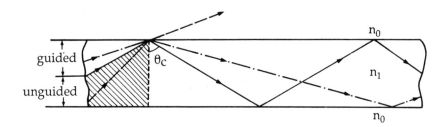

Figure 39.1 Guidance in a slab waveguide

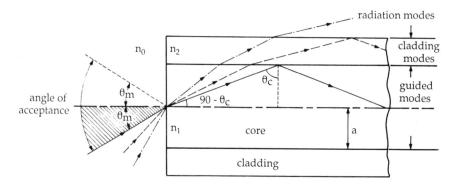

Figure 39.2 Cross-section of a circular step-index fibre

Light is guided down optical waveguides by total internal reflection (TIR). If a ray of light propagates along a parallel-sided transparent dielectric, a so-called slab waveguide, having a refractive index, n_1, higher than that the surrounding medium, n_0 it will be totally internally reflected if the angle of incidence at the interface between the two media is greater than the critical angle, θ_c, and bounce down the waveguide in successive TIRs (Figure 39.1). Such a ray is said to be 'guided'. Rays with an angle of incidence less than the critical angle will escape from the waveguide and be 'unguided'.

Slab waveguides only give guidance at the top and bottom but light can still escape from the edges. All-round guidance is, however, possible in a cylindrical waveguide. TIR may, however, be frustrated and light escape if surface damage or contamination is present. To limit the loss it is customary to protect the surface by a further layer of dielectric of lower refractive index, n_2, known as the cladding (Heel, 1954), Figure 39.2. If the refractive index is constant throughout the core this is known as a step-index fibre.

39.2.2 Numerical Aperture

Figure 39.2 shows a cross-section of a circular, step-index fibre of refractive indices of n_1, n_2 working in a medium n_0 in which a ray, passing through the axis of the fibre and known as a meridional ray, is incident on the end face of the fibre at an angle of θ_m, corresponding to a ray incident at the critical angle, θ_c, on the core/cladding interface. Rays which have an angle of incidence at the interface of more than θ_c will have an angle of incidence on the end face of less than θ_m and will be guided down the fibre. Rays with an angle of incidence on the end face of greater than θ_m will escape from the core. It is important, therefore, to know the value of θ_m defined by the numerical aperture (NA),

$$NA = n_0 \sin \theta_m \tag{39.1}$$

The critical angle is given by

$$\sin \theta_c = n_2/n_1 \tag{39.2}$$

from which it can easily be shown that

$$n_0 \sin \theta_m = (n_1^2 - n_2^2)^{1/2} \tag{39.3}$$

If the outside medium is air ($n_0 = 1$) the NA reduces to $\sin \theta_m$.

If the difference between n_1 and n_2 is small (39.3) may be written

$$\sin \theta_m = n_1(2\Delta)^{1/2} \tag{39.4}$$

where $\Delta = (n_1 - n_2)/n_1$

Putting this into physical terms, NA is a measure of the light-gathering capability of the fibre and is equal to the sine of the half angle of the maximum cone of light which can be guided down the fibre; it is dependent on the refractive index difference between core and cladding. This simple explanation is applicable only to meridional rays.

From the argument so far, it might be inferred that any ray incident on the end face at less than θ_m will be guided down the fibre, but this is not so. Maxwell's equations, when solved for a given set of boundary conditions, give solutions corresponding to specific quantized angles of incidence at the core–cladding interface; the effect may be illustrated by referring back to the slab waveguide.

Light may be thought of as travelling along the waveguide as a series of plane waves having an angle of incidence at the interface, θ. The waves propagate with a phase change per unit length, $n_1 k$, known as the *propagation constant*, in a direction normal to the wavefront ($k = 2\pi/\lambda_0$; λ_0 is the free space wavelength). The propagation constant may be resolved in two orthogonal directions (Figure 39.3), one of which, in the z direction, $\beta = n_1 k \sin \theta$ represents a travelling wave transporting energy along the fibre.

Transversely, however, rays are reflected to and fro across the fibre between the guide boundaries and interference occurs which can set up a standing wave pattern. This can only occur when the interference is constructive; the condition for this is that when the light has been reflected backwards and forwards across the waveguide and returned to its starting point, measured in the y-direction, the round-trip phase shift must be an integral multiple, m, of 2π. Therefore,

$$2W n_1 k \cos \theta + 2\varphi_{12} = m2\pi \qquad (39.5)$$

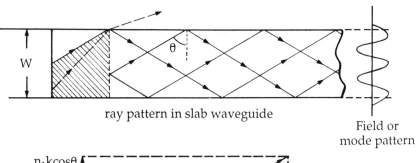

ray pattern in slab waveguide

Field or mode pattern

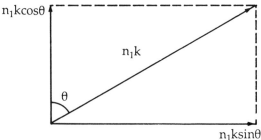

Resolution of propagation constant
axially and transversely across the waveguide

Figure 39.3   Formation of modes in a slab waveguide

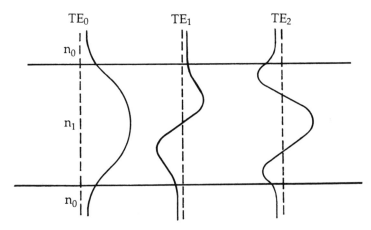

Figure 39.4   Some lower-order modes in a slab waveguide

where $W =$ the width of the waveguide

$\varphi_{12} =$ the phase change on reflection at the waveguide boundary.

Solving equation (39.5) for θ gives

$$\theta = \cos^{-1}\{(m\pi - \varphi_{12})/Wn_1k\} \qquad (39.6)$$

For each value of θ, determined by m, which satisfies this equation there is an allowable ray pattern corresponding to an unique interference system known as a mode. In addition θ must be greater than the critical angle which results in a finite number of discrete solutions. Figure 39.4 shows, diagrammatically, the modal field patterns for the first three modes in a slab waveguide. The electric field in the x-direction does not change as the wave propagates in the z-direction. For a cylindrical fibre the analysis is more complicated but modal patterns exist with both radial and circumferential dependencies (Gloge, 1971).

Most fibres support hundreds or even thousands of modes. Such fibres are called multimode, but as the fibre diameter and the NA are reduced so too is the number of modes until at a fibre diameter of about $10\,\mu$m only one mode can be supported. Such a fibre is known as a *single-mode fibre*.

39.2.3   Fibre Properties

Bandwidth and attenuation are the two most important properties of optical fibres of which attenuation, normally measured in dB km^{-1}, is the more relevant to FOS. Fibres have two intrinsic loss mechanisms, absorption and scattering.

All dielectrics absorb strongly in the ultraviolet and the infrared regions. Tails of these absorption bands extend into the visible part of the spectrum but at visible wavelengths they are weak and pure fused silica has negligible absorption in the near-infrared part of the spectrum (Figure 39.5). It is difficult, however, to make impurity-free silica and in practice various absorption peaks are super-imposed on the absorption tails. Transition metals such as copper and iron have absorption lines near the visible part of the spectrum and transition metals must be held below 1 ppb to keep the loss below 1 dB/km. Water is another undesirable contaminant. The hydroxyl OH⁻ radical of the water molecule vibrates at 2.7 μm and has harmonics at 0.7, 0.95 and 1.4 μm, the effect of which can be seen in Figure 39.5. These peaks fall in the preferred communication waveband so communication fibre must be kept free from water contamination.

Scattering is the second mechanism which causes light to be lost from a fibre. The most fundamental type is Rayleigh scattering which is inversely proportional to the fourth power of the wavelength. Even if all absorption losses could be avoided Rayleigh scattering would still cause power loss; its contribution can be seen in Figure 39.5. Longer wavelengths are, therefore, preferred for optical fibres provided that increasing losses from the infrared absorption tail do not outweigh the reduced scattering loss. At the same time wavelengths corresponding to dis-crete absorption peaks must be avoided. The low-loss window at 825 nm, for which cheap sources and detectors are available, is commonly used for FOS.

Bending losses also occur in fibres. Any minor bend, known as a *microbend*, in a fibre may distort the ray path and cause some energy to be scattered from

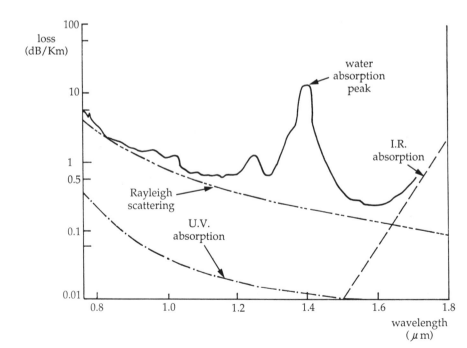

Figure 39.5   Spectral attenuation in a typical optical fibre

guided modes into the cladding, but the loss is small unless the radius of curvature is of the order of centimetres or even millimetres. More important are microbending losses which are caused by irregular changes in the fibre axis and exaggerated by scattering centres at the core–cladding interface. Changes in bend radii caused by temperature or vibration can induce considerable variation in the magnitude of bending losses and cause errors in intensity–modulated forms of sensor.

39.3   CLASSIFICATION OF FOS

Passive FOS may be classified as *intrinsic* or *extrinsic*. Intrinsic sensors use the fibre itself as the sensor; in extrinsic sensors the fibre acts merely as a light guide to and from a transducer which uses optical means to modulate some quality of the light coupled between input and output fibres (Figure 39.6).

Intrinsic sensors have the advantage that light is retained within the fibre; they are generally very sensitive but tend to suffer from sensitivity to environmentally induced effects. They may also act as distributed sensors to monitor the spatial variation of a measurand along the length of a fibre. Extrinsic sensors are simple

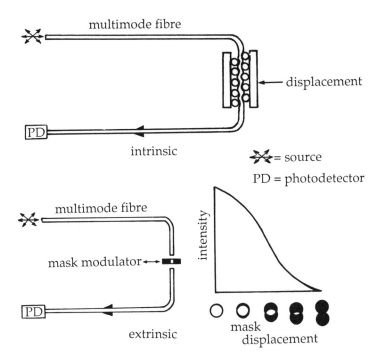

Figure 39.6   Intrinsic and extrinsic fibre-optic sensors

and flexible but light is lost between input and output fibres no matter how good the collimation.

In both types of sensors light may be modulated by:

● phase;

● polarization;

● intensity;

● wavelength;

● rate.

In a multimode system, intensity, wavelength and rate can be measured as meaningful bulk properties using incoherent sources but phase and polarization of a multimode system have limited meaning and are more readily measured in a single-mode system illuminated by a coherent source. Figure 39.7 shows the modulation methods most commonly used with each sensor class.

Although all electromagnetic wavelengths may be represented by Maxwell's model, the resolving power of an electric field as a measuring tool tends to increase as the wavelength decreases. Hence, a number of effects can be monitored using optical methods which cannot be measured by electrical means. Examples are: variation of refractive index with temperature or pressure, electric or magnetic

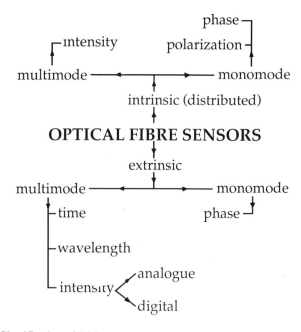

Figure 39.7   Classification of FOS

fields using the Pockels or Kerr effect and inelastic scattering processes such as fluorescence or Raman scattering. This property extends the range and accuracy of FOS as measuring systems beyond that achievable with conventional sensing methods. Table 39.1 shows some of the reported effects measurable by FOS.

Most physical measurands can, however, be transduced into displacement or force which may be used as the input to a FOS. This offers a simple and attractive way of standardizing and making progress, without too much innovation in one step, towards a generic multiparametric range of sensors.

Table 39.1   Classification of Multimode Optical Fibre Sensors

Intensity		Wavelength	Time
Digital	Analog		
On–off	Mask or shutter	Dispersing devices (grating prisms)	Mechanical resonators
Multiple position	Fibre displacement	Filters (colour, interference)	Electrical resonators
Encoders	Variable attenuators	Fluorescence	Fluidic resonators
Moiré fringes	Reflective	Thermal radiation (bolometers)	Decay-time
	FTIR	Birefringence	Echo-sensor
	Refractive index	Band-edge	Doppler effect
	Defocusing		
	Deformable components		
	Evanescent coupling		
	Microbending		
	Polarization		
	Scattering		

39.4   SINGLE-MODE FIBRE OPTIC SENSORS

39.4.1   Background

Single-mode FOS may use any of the light modulation methods listed in Section 39.3 but phase modulation, detected interferometrically, exploits most effectively the specific properties of single-mode fibres. Optical interferometry used in a bulk optic configuration, as in a Michelson interferometer, shows an intrinsically high sensitivity. The use of interferometry as a measuring tool has grown rapidly since the development of the laser which enhances the dynamic range as a result of the increased source coherence length. The advent of optical fibres prompted the development of the all-fibre interferometer, bringing the additional advantages of avoidance of the use of bulk optic components such as beam splitters, lenses and mirrors, simplification of alignment, increased sensitivity (a function of the length of fibre exposed to the measurand) and the potential for use as a distributed sensor to monitor an environmental parameter along the length of a fibre. Simplistically the measurand is used to vary the optical path length in one arm of an interferometer compared with that of a reference fibre not exposed to the measurand field.

In an all-fibre Mach–Zehnder interferometer (Figure 39.8) coherent light is launched into a monomode fibre and split by a directional coupler into a signal path exposed to the measurand field and a reference path shielded from the environment. The two paths recombine and produce two outputs in antiphase which represent the difference in path lengths between the arms. Combining these in a differential amplifier eliminates extraneous noise and gives an output i which is proportional to the cosine of the phase difference between signal and reference signals.

$$i = k\, I_0\, V \cos \varphi_D \tag{39.7}$$

where i = amplifier output

    I_0 = light intensity launched into the interferometer from the source

    V = fringe visibility

    k = constant

    φ_D = the phase difference between the outputs from the signal and the reference arms of the interferometer.

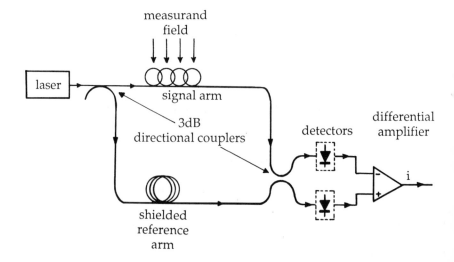

Figure 39.8 All-fibre Mach–Zehnder interferometer

Sensitivity is at a maximum when the phase difference is an odd-numbered multiple of $\pi/2$, $\sin \varphi_D = 0$, called the *quadrature condition*. Unfortunately φ_D is the sum of the measurand and drift induced phase shift, and environmental changes can cause the optical path length to vary randomly causing rapid fading. Several methods have deen devised to compensate for this.

39.4.2   Active Homodyne Processing

A fibre stretcher may be incorporated in the reference arm and used as part of a servo feedback loop in order to maintain the interferometer at quadrature (Figure 39.9). The system has the inherent disadvantage that the tracking range is limited to 2π (Jackson *et al.*, 1980).

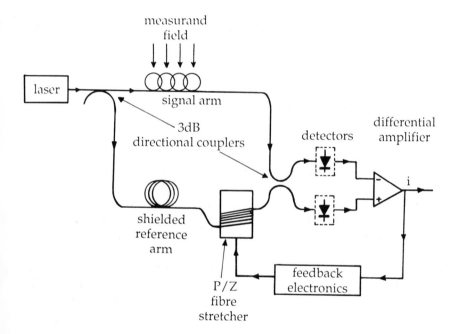

Figure 39.9   Active phase tracking in a Mach–Zehnder interferometer

39.4.3   Heterodyne Detection

Heterodyne processing in interferometers is an attractive technique giving linear response with infinite tracking range. The optical frequencies in the two arms are made unequal, usually achieved by using an acousto-optic modulator such as a Bragg cell as a frequency shifter, Figure 39.10. The output thus becomes a phase-modulated heterodyne carrier. The system has the disadvantage of lack of

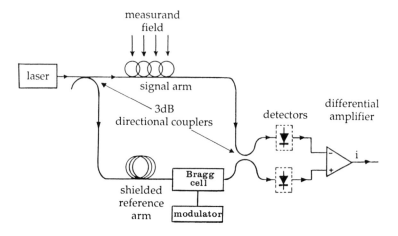

Figure 39.10 Heterodyne detection in a Mach–Zehnder interferometer

compatibility between a bulk-optic component, the Bragg cell, and the fibre interferometer.

39.4.4 Pseudo-heterodyne Modulation

This technique eliminates the use of bulk-optic components. The laser source in an unbalanced interferometer may be frequency-ramped by varying the laser injection current to give a shift in frequency during the rising part of the ramp, thus producing the required heterodyne carrier (Jackson *et al.*, 1982). A bandpass filter is inserted in the photodetector output to eliminate the harmonics introduced by the ramp flyback. Optical phase shift induced in the interferometer phase-modulates the carrier; the signal may be recovered by FM demodulation techniques (Figure 39.11).

39.4.5 Analysis of Single-mode Interferometers

The advantages and disadvantages of interferometric uses of OFS use are

Advantages

- Extremely high sensitivity to strain, refractive index and pressure;
- very versatile—most measurands can be expressed as force;

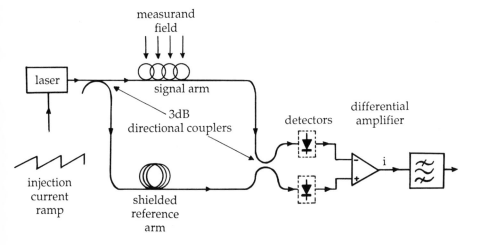

Figure 39.11   Pseudo-heterodyne detection in a Mach–Zehnder interferometer

- ideal for a.c. fields;

- sensitivity a function of fibre length;

- can be used as distributed sensors.

Disadvantages

- High cross-sensitivity;

- high down-lead sensitivity;

- incremental—difficult to determine slow-moving measurands;

- signal-processing complex

39.5   INTENSITY-MODULATED SENSORS (IMS)

39.5.1   General Overview

Intensity-modulated sensors, which depend on varying the intensity of light coupled between input and output fibres, are simple, reliable and cheap in that they can use broadband or narrow band incoherent sources, multimode fibres, PIN photodetectors and uncomplicated signal processing. As a result no less than 50% of FOS patented in the USA adopt some form of intensity modulation, either

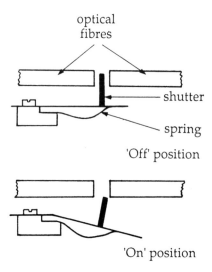

Figure 39.12   Displacement actuated on–off optical switches

reflective or transmissive (Zilber, 1984). In their simplest form IMS operate in the on-off mode as optical switches (Figure 39.12) triggered by displacement (Place, 1984). This can readily be extended to a digital system as in the absolute linear encoder shown in Figure 39.13 which uses a sensor head in transmission

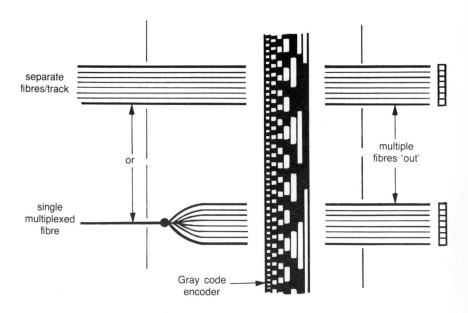

Figure 39.13   Digital optical displacement encoder

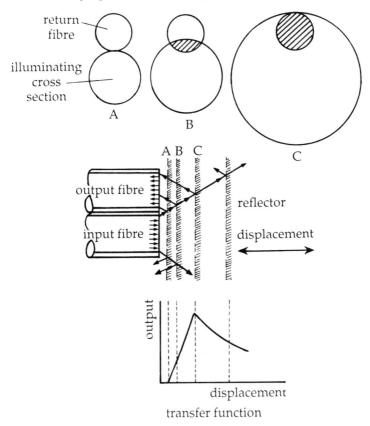

Figure 39.14   Fibre-optic reflective displacement sensor

mode but requires a fibre for each track (Gardiner and Edwards, 1987). Resolution can only be improved at the expense of additional fibres unless some form of multiplexing is adopted (Dakin and Liddicoat, 1982).

Analog modulation is more versatile than digital and a number of possibilities have been reported dating back to 1967 when Kissinger (1967) patented a reflective displacement sensor (Figure 39.14). Other methods use displacement to interpose various masks or gratings between fibres, (Jones and Spooncer, 1982), or to alter the alignment between fibres (Spillman and Gravel, 1980). Table 39.1, given earlier, lists other examples. Of especial interest is a microbending sensor which is used as a safety mat (Oscroft, 1987).

Analog intensity-modulated sensors can have a dynamic range of better than 1 in 10^6 using lock-in amplification, in the ultimate being limited by detector shot noise. They have the disadvantage that adventitious changes in light level due to variations in source emission, link attenuation or detector responsivity may affect the output, and to ensure good repeatability, some form of intensity referencing is necessary to compensate for changes in light level not connected with the measurand.

Two main methods have emerged:

- provision of a separate reference path which may use a second fibre or preferably be wavelength multiplexed on a single fibre;
- use of an optical bridge to balance out intensity changes.

Several versions of a two-wavelength intensity-referenced system have been reported in which one of the wavelengths is modulated by the sensor and carries the signal data whilst the other passes through the sensor unchanged and is used to normalize the signal intensity.

A grating displacement sensor (Figure 39.15) can show high resolution (better than 0.01% of span) but intensity-referencing is needed to obtain adequate repeatability. This can be achieved by forming the bars of the grating as a volume

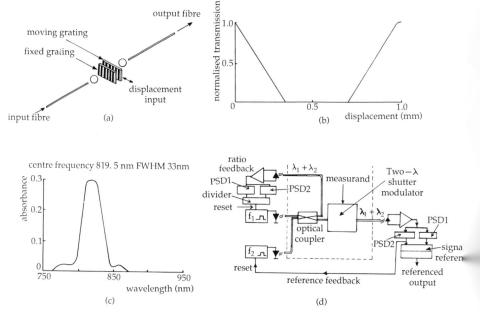

Figure 39.15   Two-wavelength intensity-referenced grating displacement sensor

hologram which acts as a narrow-band notch filter reflecting wavelengths specific to the notch (Fig. 39.15(c)); wavelengths outside the notch cannot be seen by the grating and pass through unmodulated. The modulated wavelength intensity is normalized against the reference wavelength intensity. A holographic grating linked to a diaphragm to form an optical pressure sensor has a repeatability of better than ±2% of span over a light intensity range of 10 dB (Jones and Spooncer, 1984)

Accurate, two-wavelength, referencing requires the source outputs, fibre attenuations and detector sensitivities at the two wavelengths to remain constant in relationship to each other. A balanced bridge, twin-fibre system avoids differential wavelength effects but requires two sources and two detectors (Figure 39.16). Twin sources modulated at different frequencies are launched into separate fibres, led to the sensor head and injected into opposite ends of the bridge structure, one leg of which incorporates the intensity-modulated sensor. The outputs from the opposite diagonal of the bridge are detected and the contribution from each source determined by phase-sensitive detection. The system output (P) is given by

$$P = (M1 \cdot R2)/(M2 \cdot R1) \qquad (39.8)$$

where P involves only quantities determined by the bridge. Effects external to the bridge are nulled or compensated. A differential change in link attenuation of 20 dB in one fibre caused a change in P of only 1%. Drift may still occur if variation in the coupler splitting ratio occurs, (Giles *et al.* 1985).

39.5.2   Analysis of Intensity-modulated Sensors

Advantages and disadvantages of using intensity-modulated sensors are now given.

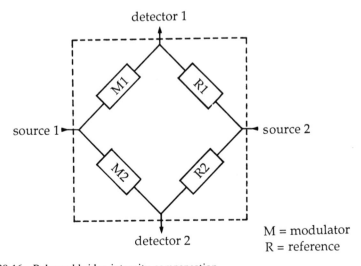

Figure 39.16   Balanced bridge intensity compensation

Advantages

- Simple;
- wide dynamic range;
- versatile;
- excellent for on–off and digital applications.

Disadvantages

- High environmental sensitivity;
- intensity-referencing essential for good repeatability.

39.6   WAVELENGTH-MODULATED SENSORS

39.6.1   General Overview

The referencing problems of analog intensity-modulated sensors may be over-come if a digitally modulated system is used. Pseudo-digital modulation is possible if the return spectrum from the sensor is coded to represent the measurand. The method may be regarded as an extended version of a two-wavelength referencing system. It requires a :

- broad-band source;
- fibre with low attenuation over the source bandwidth;
- wavelength-measuring system with a resolution of 0.1%.

Tungsten-halogen lamps are often used as broad-band sources but these have a limited life, of about 2000 hours, and the power spectral density varies with temperature both in shape and magnitude. Much of the power available from an incandescent source is wasted because the spectral power distribution spreads beyond the low attenuation window of most fibres.

A broad-band LED can, over a narrow bandwidth, launch greater power into a fibre than an incandescent source but this limits the useful bandwidth available and, hence, for a given spectrometer resolution, reduces the dynamic range. This creates a dichotomy. Any attempt to increase the dynamic range by improving the spectrometer resolution will reduce the signal strength and hence, the signal-to-noise ratio, imposing a limit on the sensor resolution.

A resolution of 0.01 nm can readily be obtained with a commercial spectrometer but these are expensive and insufficiently reliable for use in an industrial

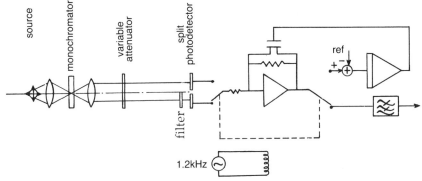

block diagram of wavelength discriminator

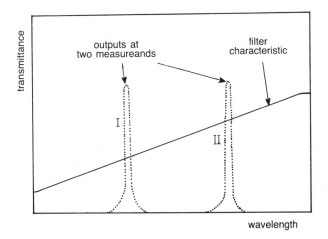

Ideal filter characteristic for
two wavelength discriminator

Figure 39.17   Two-wavelength spectral decoder

environment. A cheap solid-state spectrometer may, however, be built using a
pair of photodetectors of different spectral sensitivities stacked one above the
other (such as Sharp PD 150 series). Provided that the detectors are illuminated
by a waveband which contains no secondary peaks the ratio of the outputs from
the two detectors represents the wavelength uniquely. Figure 39.17 illustrates the
principle.

Simple wavelength-encoded sensors have been made using displacement-
linked dispersing elements such as diffraction gratings, prisms and zone plates
to modulate the transmitted spectrum (Hutley, 1984); direct modulation using
birefringent filters (Jones and Spooncer, 1983), fluorescence (Wickersheim and

Alves, 1979), thermal radiation (Diles, 1983) and thermochromic effects (Snitzer *et al.*, 1983) have also been reported. Several of these devices have been commercialized.

39.6.2 Analysis of Wavelength Modulation

Advantages and disadvantages of modulating the wavelength in optical fibre sensors are now given.
Advantages

● Pseudo-digital—may not need referencing;

● can be used to monitor, directly, effects involving wavelength changes, e.g., fluorescent sensors;

● can monitor chemical change involving colour, e.g. pH;

● two-wavelength referenced systems can use simple filters.

Disadvantages

● Instability of source colour temperature;

● temperature dependency of spectral transmission of fibres, splitting ratios of couplers and spectral responsivity of photodetectors;

● resolution depends on

— wide spectral bandwidth which implies increased fibre attenuation losses;
— narrow-band spectrometer resolution which reduces received signal power.

Both result in lower signal-to-noise ratio.

39.7 HYBRID SENSORS

39.7.1 Introductory Remarks

The process industries, whilst recognizing the benefits of electromagnetic interference free and intrinsically safe telemetry, have been reluctant to exchange established measuring methods for as yet untried optical methods. As a compromise the reliability and performance of traditional sensors may be combined with the advantages of optical fibre telemetry in what is known as a *hybrid* sensor. This consists essentially of an established transduction unit which produces an optical output. A hybrid sensor may be powered in its normal manner, from batteries or from an optical power line (McGeehin, 1987).

39.7.2 Battery-powered Hybrids

Figures 39.18 and 39.19 show a fibre optic temperature sensor, now available commercially, in which an industry standard platinum resistance thermometer is housed in a thermowell, the head of which contains optical pulse generation equipment powered by a long-life lithium battery. Temperature is determined by measuring the time constant of a capacitor shunted by the resistance thermometer. These data are transmitted as a series of pulses through a multimode fibre to a photodetector housed in a 'safe' control area. Temperature is represented by the time between pulses; accuracy is within ±0.1% of span. The transmitter has been approved for use in hazardous areas (Philp, 1988).

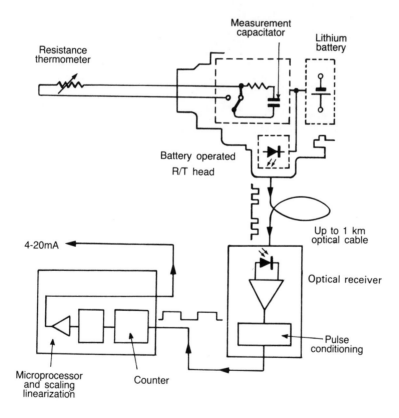

Figure 39.18   Battery-powered hybrid optical temperature transmitter

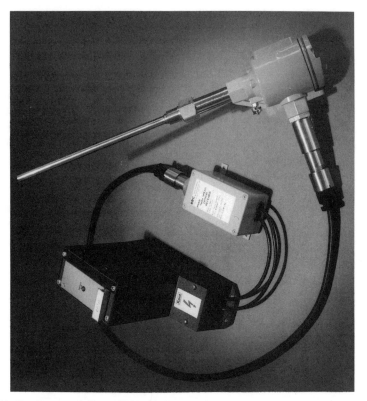

Figure 39.19   Commercially available platinum resistance temperature transmitter

39.7.3   Optically Powered Resonant Hybrids

From a maintenance/safety point of view it may be preferable to power the transmitter optically. This has led to the development of light-powered hybrids in which optical power, transmitted over a fibre, is photovoltaically converted into electrical energy at the sensor head to provide power to local electronics and to drive the output light source. The efficiency of such units is low, restricting the power available and limiting the intelligence which can be incorporated.

Most of the configurations reported have, therefore, used resonant sensing elements of high quality factor (Q) to reduce the electrical demand on the system. Resonant systems also have the advantages of being intensity independent and free from corruption compared with analog systems. Analog signals can be converted electronically into a digital output but there always remains some analog element which is subject to drift and distortion. It is preferable, therefore, to use primary sensors which are intrinsically oscillatory (Gast, 1985). If the sensor forms part of a feedback loop, controlling the optical source, a *sing-around* oscillator results, the frequency of which represents the value of the measurand.

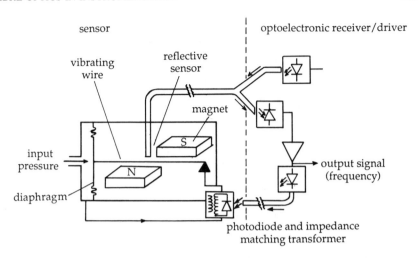

Figure 39.20   Optically powered vibrating wire hybrid sensor for gas density measurement

Not only is the stability, sensitivity and selectivity improved by an increase in Q but the power demand of the sensor is reduced.

Optically powered hybrid resonant sensors have been reported which use as the resonant element either a quartz crystal (McGlade and Jones, 1984; Grattan *et al.*, 1987; Spooncer *et al.*, 1987) or a vibrating mechanical structure (Halliwell and Jones, 1987; Bois *et al.*, 1989; Jones and Philp, 1983). The last has been used to measure gas density in industrial switchgear (Figure 39.20).

Figure 39.21 shows a commercially produced vibrating wire optical gas density transmitter.

39.7.4   Hybrid Actuation

The advantages of optical telemetry cannot be fully realized if electrical power is still needed to control actuators. This philosophy has led to the development of a number of optically controlled pneumatically powered actuators (Collier *et al.*, 1984; Liu and Jones, 1988). The latter is illustrated in Figure 39.22.

39.7.5   Analysis of Resonant Hybrid Sensors

A comparison of resonant hybrid sensors is now given.

Advantages

●   Combines the reliability and performance of traditional sensors with the benefits of optical fibre telemetry;

- avoids the use of immature technology;

- can incorporate intelligence in the sensor head;

- good stability and high selectivity;

- digital output is intensity-independent;

- low power consumption.

Disadvantages

- Not fully electrically passive;

- rejects the use of optical measuring methods.

39.8 OPTICAL MULTIPLEXING

39.8.1 Justification for Optical Multiplexing

Potentially the unit cost of FOS may be reduced by multiplexing a number of sensors on a single-fibre bus. Several feasible schemes have been reported (Dakin, 1987) but no generalization can be made on their financial benefits although it is self-evident that as the fibre link gets longer the cabling cost becomes an increasing and often dominating component of the cost. If in addition the sensor requires an expensive source or signal processing which may be shared the argument for multiplexing is strengthened.

In a shared network sensors may be identified by time, wavelength or frequency in what are called time-division, wavelength-division or frequency-division multiplexing.

39.8.2 Time-division Multiplexing (TDM)

TDM commonly uses optical time-domain reflectometry techniques (Barnoski and Jensen, 1976) for which commercial transceivers are available to identify and separate sensor signals. A schematic of a system is shown below (Figure 39.23). A probe signal, normally a laser diode pulse, is launched into a fibre bus via a directional coupler. Part of the pulse is tapped off at each of a number of sensor stations, spatially separated. The sensors may be reflective or transmissive. Return signals arrive at the detector separated in time according to the spatial separation of the sensors. Up to ten sensors have been multiplexed in a laboratory simulation. Intensity modulation is normally adopted.

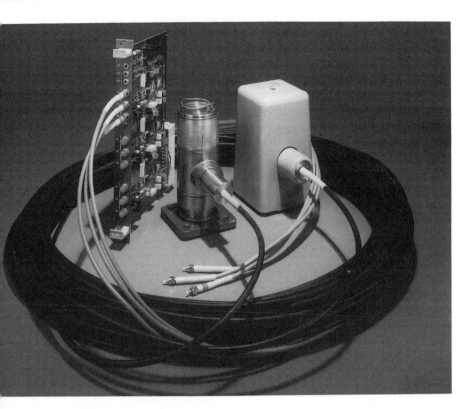

Figure 39.21   Optical gas density transmitter

39.8.3   Wavelength-division Multiplexing (WDM)

WDM uses a broadband source, e.g. a white light source, the output of which is split into several wavebands, one of which is allocated to each sensor. At each sensor the appropiate waveband is filtered off the optical highway, modulated, usually by intensity, and reinjected into the highway. The return signals are separated by a wavelength dispersive device, e.g. a prism or a grating and individually detected (Figure 39.24).

39.8.4   Frequency-division Multiplexing (FDM)

FDM is similar to WDM but each sensor is now allocated a frequency band which may be rate-modulated by the measurand. Each sensor contains a resonant element, either mechanical or electrical, which acts as a channel filter analogous to the wavelength filter in the WDM system. Light from a laser diode or an LED, modulated at several frequencies, is launched into a fibre bus exciting the corre-

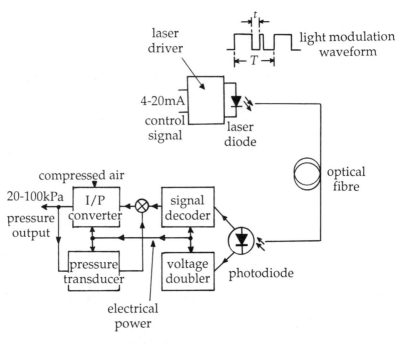

Figure 39.22   Optically controlled pressure actuator

sponding resonant elements the natural frequencies of which are functions of the measurands. The pulsed optical output from the sensor corresponds to the natural frequency of the resonant element. This is fed back to control the source modulation rate which in turn excites the the oscillatory element in synchronism.

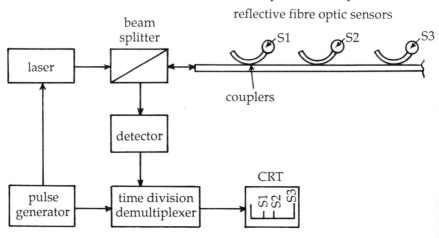

Figure 39.23   Time-division multiplexing of fibre-optic sensors

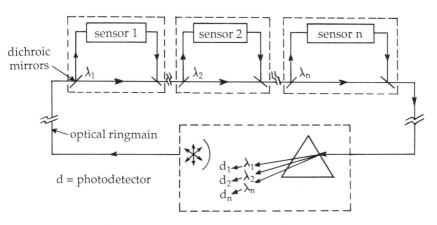

Figure 39.24   Wavelength division multiplexing of fibre- optic sensors

A frequency-multiplexed system may be seen as a number of self-resonant loops in parallel, the frequencies of which represent the measurands. The frequency excursions of each sensor must remain within the allocated waveband and adequate separation must be provided between channels to avoid cross-talk.

39.9   CONCLUDING REMARKS

As a result of the massive investment in optical fibres for telecommunications the technology of fibre optics is proceeding rapidly on all fronts and most of these developments will influence the industrial implementation of fibre optic sensors. FOS have already been adopted for specialized applications which cannot easily be satisfied by conventional sensors; such opportunities have already been identified in the aerospace, automotive, medical and mining industries and for remote on-line chemical analysis. The more general replacement of conventional sensors by FOS is less imminent but hybrid sensors are an intermediate development which may bridge the gap until all-fibre versions, matching conventional sensors in price and performance, become available.

39.10   SOURCES OF FURTHER ADVICE

The reader is referred to the following textbooks and journals for further information on fibre optics in instrumentation.

Books

Corke, M. (1991) *Fibre Optic Sensors: Design and Implementation* McGraw-Hill, New York.

Dakin, J. P. and Culshaw, B. (eds) (1988)—*Optical Fibre Sensors: Volume 1—Principles and Devices, Volume 2—Systems and Applications* Artech House, London.

Krohn, D. A. (1988) *Fibre Optic Sensors—Fundamentals and Applications* Instrument Society of America, Research Triangle Park, NC.

Ottos, W. (1991) *Fibre Optic Chemical Sensors and Biosensors* CRC Press, Boca Raton, FL.

Udd, E. (ed.) (1991) *Fibre Optic Sensors: An Introduction for Engineers and Scientists* John Wiley, Chichester.

Journals

Applied Optics Optical Society of America, USA.

Electronics Letters IEE, UK.

IEE Proceedings J. Optoelectronics IEE, UK.

Journal of Lightwave Technology. IEEE, USA.

Optics Letters Optical Society of America, USA.

REFERENCES

Barnoski, M.K. and Jensen, S.M. (1976). 'Fiber waveguides: a novel technique for investigating attenuation characteristics', *Appl Optics*, **5**, 2112–15.

Bois, E., Spooncer, R.C. and Jones, B.E. (1989). 'A hybrid resonant differential pressure transmitter with wavelength multiplexed power and data channels', *Proc. 6th Intl, Conf. OFS '89*, Paris, France, Sept. 18–20, *Proceedings in Physics*, (Springer Verlag Berlin), 478–83.

Collier, M.J., McGlade, S.K. and Stephens, P.E. (1984). 'The optical actuation of a process control valve' *GEC J. Res.* **2**, 125–8.

Dakin, J.P. (1987). 'Multiplexed and distributed optical fibre sensor systems', *J. Phys. E. Sci. Instrum.*, **20**, 954–67.

Dakin, J.P. and Liddicoat, T.J. (1982) 'A wavelength multiplexed shaft encoder', *J. Meas. Control*, **15**, 176–7.

Diles, R.R. (1983). 'High temperature optical fibre thermometer', *J. Appl. Phys.*, **54**, 1198–1201.

Frost and Sullivan, Inc. (1989) *European Market in Fibre Optic Sensors*, Frost and Sullivan, New York, USA.

Gardiner, P.T. and Edwards, R.A. (1987). 'Fibre optic position sensor for aircraft flight control systems', *IOP Short Meetings Series* **7** (London: IOP), 115–127.

Gast, T. (1985). 'Sensors with oscillating elements'. *J. Phys. E.:Sci. Instrum.*, **18**, 783–9.

Giallorenzi, T.G. Bucaro, J.A., Dandridge, A., Sigel, G.H., Rashleigh, S.C. and Priest, R.G. (1982). 'Optical fibre sensor technology', *IEEE J. Quant. Electron.*, **QE-18**, 626.

Giles, I.P., McNeill, S. and Culshaw, B. (1985). 'A stable remote intensity based optical fibre sensor', *J. Phys, E.:Sci. Instrum.*, **18** 502–4.

Gloge, T. (1971). 'Weakly guiding fibres', *Appl. Optics*, **10**, 2252.

Grattan, K.T.V., Palmer, A.W. and Saini, D.P.S. (1987). 'Optical vibrating quartz crystal pressure sensor using frustrated-total-internal-reflection readout'. *IEEE J. Lightwave Techn.*, **LT-5**, 972–9.

Halliwell, M.J. and Jones, B.E. (1987). 'Force transducer using a double-ended tuning fork with optical fibre links', *Proc. Eurosensors, 3rd Conf. on Sensors and their Applications*, Cambridge, (IOP, Bristol), 145–147.

Heel, A.C.S. van (1954). 'A new method of transporting optical images without aberrations', *Nature*, **173**, 39.

Hutley, M.C. (1984). 'Wavelength encoded Optical Fibre Sensors', *2nd Intl. Conf. on Optical Fibre Sensors*, Stuttgart, (*Proc. SPIE*, **514**), 111–16.

Jackson, D.A., Priest, R., Dandridge, A. and Tveton, A.B. (1980). 'Elimination of drift in a single-mode optical fibre interferometer using a piezo-electrically stretched coiled fibre', *Appl. Optics*, **19**, 2926.

Jackson, D.A., Kersey, A.D., Corke, M. and Jones, J.D.C. (1982). 'Pseudo-heterodyne detection scheme for optical interferometers', *Electron Lett.*, **18**, 1081.

Jones, B.E. (1985). 'Optical fibre sensors and systems for industry', *J. Phys. E.: Sci. Instrum.*, **18**, 770–82.

Jones, B.E. and Philp, G.S. (1983). 'A fibre optic pressure sensor using reflective techniques' *Proc. International Conference on Optical Techniques in Process Control*, The Hague, June (BHRA Fluid Engineering Cranfield), 11–25.

Jones, B.E. and Spooncer, R.C. (1982). 'Simple analogue and digital optical transducers with optical fibre links', *IEE Coll. Digest No. 1982/60* (IEE, London), 7/1–7.

Jones, B.E. and Spooncer, R.C. (1983). *Photoelastic Pressure Sensor with Optical Fibre Links Using Wavelength Discrimination*, IEE Conference Publication No. 221 (IEE, London), pp. 173–7.

Jones, B.E. and Spooncer R.C. (1984). 'An optical fibre pressure sensor using a holographic shutter modulator with two-wavelength intensity referencing' *Proc. 2nd Intl. Conf on Optical Fibre Sensors*, Stuttgart, (*Proc. SPIE*, **514**, 223–6.

Kissinger, C.D. (1967). *Fiber Optic Proximity Probe*, US Patent 3, 327, 584.

Liu, K. and Jones, B.E. (1988). 'Optoelectronic control of pneumatic pressure using a local micropower feedback sensor', *Conf. Proc. Sensor '88*, Nurnberg, (ACS organizations, Wunstorf), Part B, 297–313.

McGeehin, P. (1987). *Hybrid Sensors with Optical Fibre Links'*, OSCA Document No. 87/1989–18th October (SIRA, Chislehurst, Kent, UK).

McGeehin, P. (1989). *Optical Fibre Sensors*, Vol. 2, Culshaw, B. and Dakin, J. (eds.), Artech House, USA.

McGlade, S.M. and Jones, G.R. (1984). 'An optically powered quartz force sensor', *GEC J. Res.*, **2**, 135–8.

Oscroft, G. (1987). 'Intrinsic fibre optic sensors', *Fifth Intl. Conf. Fibre Optics and Opto-electronics*, London, (*Proc. SPIE* **734**), 207–13.

Philp, G.S. (1988). 'Fibre optics in field transmitters', *Contr. Instr.*, 79–80.

Place, J.D. (1984) 'Optical control and alarm switches for the process industries', *Proc. Conf, PROMECON* (IMC, London), 157–63.

Reed, G.T. (ed.) (1989). *Fibre Optic Sensors Developments in the USA*, OSTEM Report (DTI/ERA, London).

Snitzer, E., Morey, W. W. and Glenn, W. H. (1983). 'Fiber optic rare earth sensors', *IEE Conf. Pubn. No. 221* (IEE, London), 79–82.

Spillman, W. B. Jr. and Gravel, R. L. (1980). 'Moving fiber-optic hydrophone' *Opt. Lett.*, **5** 30–1.

Spooncer, R.C., Jones, B.E., and Ohba, R. (1987). 'A pulse modulated optical fibre quartz temperature sensor', *Fiber Optic Sensors II*, A. M. Scheggi (ed.), *Proc. SPIE*, **798**, 137–141.

Wickersheim, K.A. and Alves, R.B. (1979) 'Recent advances in optical temperature measurement', *Industrial Research Development*, December, 82–9.

Zilber, J. (1984). 'Fiber optic sensor market development', *Proc. 2nd Intl. Conf. on Optical Fibre Sensors*, Stuttgart, *Proc. SPIE*, **514**, 177–83.

Chapter

40 I. KARUBE

Biosensors

Editorial introduction

The newest class of sensors are those using and measuring biological properties at their physical (some would say chemical) world interface. Sensing of chemical properties is the most difficult sensing interface to engineer. The main difficulty is that, as well as involving energy transfer in their operation—which can be a long-life virtually non-destructive process if designed well—they also involve transfer of mass that has very poor life characteristics. This account explains the difficulties to be overcome before biosensors will become as routine in use as sensors for physical measurands. For comparison with routine chemical analytical measurements the reader is referred to Chapter 27 in Volume 2 of this Handbook.

40.1 INTRODUCTION

Updike and Hicks (1967) presented a biosensor which was a combination of an immobilized enzyme element and an oxygen electrode. Since their first introduction enzyme electrode biosensors have become a powerful methodology for analysing biological substances and organic compounds.

Biosensors enable detection of complex organic compounds which were previously difficult to determine by other methods. This is especially so in the case where the analytes are biological compounds. Biosensors have provided high sensitivity and exclusive selectivity for optical isomers. Biosensors provide rapid measurement and have only a small sample requirement. These features are very important in clinical analysis.

Conventional biosensors employ an electrochemical electrode as the transducer, while other transducers such as optical fibre sensors, imaging sensors,

Handbook of Measurement Science, Volume 3
Edited by P. H. Sydenham and R. Thorn
© 1992 John Wiley & Sons Ltd

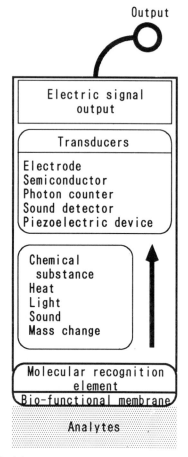

Figure 40.1   Biosensor principles

piezoelectric devices and SAW devices are currently used for biosensors. Novel
biosensors utilizing these new types of biosensor are described in later sections
of this chapter.

   In this section, fundamentals and applications of several kinds of biosensor are
introduced.

40.2   OVERVIEW OF BIOSENSORS

Biosensors, composed of a molecular recognition element and a transducer, have
been developed and applied to chemical, clinical of food analysis and monitoring
and to environmental measurement (Cass *et al*, 1987; Turner *et al*., 1987).
Enzymes, micro-organisms and antibodies may be used as the molecular recognition element (Figure 40.1).

Electrochemical devices have often been used as transducers which convert chemical or physico-chemical signals to electrical signals. Various enzymes and enzymatic reactions have been used as molecular recognition elements. An *enzyme electrode* is composed of an immobilized enzyme membrane and an electrode.

The electrode detects an electroactive species produced or consumed by the enzyme reaction. Both the rate of generation or degeneration and steady-state value of electric current can be used as an indicator of the analyte's concentration in a sample. For example, a conventional glucose sensor is composed of immobilized glucose oxidase (GOD) and an electrode. GOD oxidizes glucose, consuming oxygen, to produce gluconolactone and hydrogen peroxide. The consumption of oxygen is measured with an oxygen electrode or the production of hydrogen peroxide with a hydrogen peroxide electrode. The concentration of glucose can be determined from the steady-state value of the oxygen or hydrogen peroxide concentration. In the steady state the rate of diffusion of glucose toward the sensor and its rate of consumption by GOD are balanced. This type of glucose sensor has been commercialized for diagnosis of diabetes. In principle, any oxygen oxidoreductase can be replaced by GOD and its substrate measured in this way. Many kinds of biosensors using the same principle and devices have been developed for use in the fields of clinical analysis and analysis of foodstuffs.

Micro-organisms have also been utilized as molecular recognition elements. A *microbial* sensor consists of an immobilized micro-organism membrane and an electrode. Many kinds of microbial sensors have been developed and applied to the measurement of biological compounds. Sensor function is based on either the change of respiration or the amount of produced metabolite as a result of assimilation of substrates by the micro-organism. Furthermore, auxotrophic mutants can be used to selectively determine many kinds of substances. For example, the vitamin B_{12} sensor was constructed by using immobilized *Escherichia coli* 215 (Karube *et al.*, 1987): *Escherichia coli* 215 strain requires vitamin B_{12} for the growth. The linear relationship was obtained in the range between 5×10^{-9} and 25×10^{-9} gl^{-1}. Within 25 days the decrease in the response was approximately only 8%.

Microbial sensors using thermophilic bacteria have been developed. The use of thermophilic bacteria and hence high temperatures can reduce contamination by other micro-organisms and yield greater long-term stability. For example, a biological oxygen demand (BOD) and a carbon dioxide sensor have been constructed by using thermophilic bacteria isolated from a hot natural water spring. Good linear correlation was observed between the BOD sensor response and BOD value in the range 1–10 mgl^{-1} BOD (the Japanese Industrial Standard, JIS) at 50°C (Suzuki *et al.*, 1988a). The sensor signal was stable and reproducible for more than 40 days. For the carbon dioxide sensor, a linear relationship was obtained for $NaHCO_3$ concentrations between 1 and 8 mM at 50°C and the response time was 5–10 min (Karube *et al.*, 1989). A linear relationship was also observed in the CO_2 concentration range 3–8%.

Although few biosensors actually have been fully commercialized, attention is currently focused on miniaturization and integration of biosensors. Micro-biosensors have many advantages such as the possibility of implantation in the human body and are suitable for *in vivo* measurement. In addition, micro-biosensors can be integrated on one chip and are useful for measuring simultaneously, various substrates in a small amount of sample solution. Application of semiconductor fabrication technology to micro-biosensors makes it possible to develop cheap, mass-produced disposable transducers for biosensing. Furthermore, by combining micro machining technology with biosensors, sample collection and analysis systems will be built on a single chip will be feasible in the future. Chapter 37 addresses the production technology.

Novel biosensors using piezoelectric devices, surface acoustic wave (SAW) devices, imaging sensors and optical fibres are now described in detail.

40.3 MICRO-BIOSENSORS

40.3.1 Micro-biosensors Based on the ISFET

An ion-sensitive field effect transistor (ISFET) was first reported by Bergveld (1970). It was improved using silicon nitride as the gate insulator to construct a micro pH sensitive device (Matsuo and Wise, 1974). The structure of the ISFET is shown in Figure 40.2.

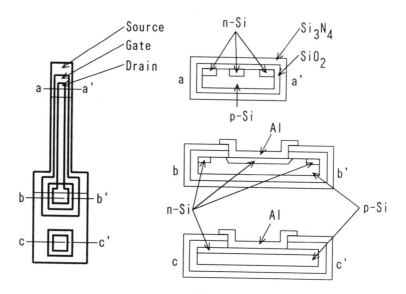

Figure 40.2 Structure of an ISFET

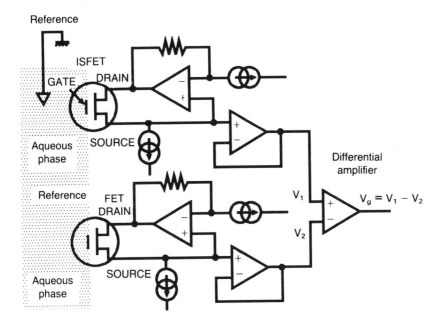

Figure 40.3   Circuit diagram of a differential ISFET driver

Merits of the ISFET are its rapid response, low power consumption, low noise and avoidance of need for a high-impedance interrogation amplifier. A basic circuit diagram for measuring the gate output voltage is shown in Figure 40.3.

In this circuit, both the voltage and current between the source and drain are maintained constant. The Ag/AgCl reference electrode is immersed in the same solution. The surface potential on silicon nitride (Si_3N_4) of the ISFET (Figure 40.2) is affected by the pH of the solution, with resultant change in the gate voltage being proportional to the change in surface potential. Because both drain current and drain–source voltage are fixed, the source potential shifts so that the surface potential change of the gate is cancelled. Therefore, the surface potential change in the ISFET, caused by the change of pH, can be measured as the change in the source (or drain) output voltage.

ISFETs are fabricated using semiconductor technology. Hence it is easy to miniaturize and integrate ISFETs on one chip. The ISFET is used as a potentiometric transducer. Therefore, enzymes which catalyse reactions involving a pH change, such as urease or glucose oxidase, can be used. Since the penicillin-sensitive immobilized pencillinase ISFET was first reported by Caras and Janata (1980), many experimental reports about enzyme-modified ISFETs have been published and ISFETs have been used as the basis of many micro-biosensors.

ISFET biosensors for the measurement of urea and alcohol are now described. An amorphous silicon ISFET, and its application to the biosensors, is also introduced.

40.3.2   Urea Sensor

A urea sensor (Karube *et al.*, 1986) consisted of an immobilized urease membrane and a pH electrode. When an ISFET was used as the pH electrode, pH change caused by the urease-catalysed reaction was converted to a voltage output of the ISFET interrogating circuit. The fabrication process of a micro-urease sensor was as follows. An ISFET was placed inside a vacuum chamber and 3-aminopro-pyltriethoxysilane (3-APTES) was vaporized at 80°C and 0.5 torr for 30 min. This was followed by glutaraldehyde treatment under the same condition. The, new chemically modified, ISFET was covered with a cellulose acetate membrane containing 1,8-diamino-4-aminomethyloctane and glutaraldehyde. The immobilized scheme is shown in Figure 40.4.

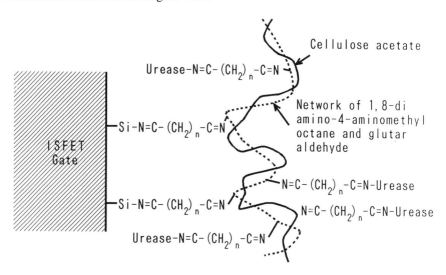

Figure 40.4   The immobilization scheme of an ISFET urea sensor

The ISFET was immersed in urease solution. The urea sensor gave the linear relationship between the initial rate of change of the gate output voltage and the logarithm of urea concentration in the range 16.7–167 mM. It could be used for 20 days with only slight degradation of the enzymatic activity.

40.3.3   Alcohol Sensor

The study of an alcohol sensitive micro-biosensor (Tamiya *et al.*, 1988a) based on an ISFET and the enzyme system existing in the cell membrane has been reported. The cell membrane of acetic acid producing bacteria has a complex enzyme system oxidizing ethanol to acetic acid via acetaldehyde. This system consists of membrane-bound alcohol dehydrogenase (ADH), aldehyde dehydrogenase (ALDH) and an electron transfer system. This complex enzyme system can be used for application with an ISFET.

Whole cells of *Gluconobacter suboxydans* are made to pass through a French press and centrifuged to remove intact cells. The supernatant liquid was then ultra-centrifuged at 100 000g and cell debris was used as the cell membrane fraction. Immobilization of the cell membrane, using calcium alginate gel coated with nitrocellulose, was carried out as follows. The mixture of cell membrane and sodium alginate and pyrroloquinolinequinone (PQQ) was spread on the gate surface of the ISFET, which was then dipped in CaCl$_2$ solution in order to form the calcium alginate gel layer containing the cell membrane. The ISFET was dipped in the acetone solution of nitrocellulose and immediately dried in air, resulting in formation of a nitrocellulose-coated layer.

A membrane coated ISFET without the enzyme was used as a reference ISFET and the output of the differential mode circuit (Figure 40.3) was connected to a chart recorder. The gate potential changes immediately after the injection of the sample and reaches a steady state after approximately 10 minutes. A linear relationship was observed between the differential gate output and ethanol concentration up to 20 mgl^{-1}.

40.4 BIOSENSORS USING THE AMORPHOUS SILICON ISFET

40.4.1 Introduction

The ISFET device described in Section 40.3.3 can only be manufactured by using a silicon wafer for the substrate. In recent years, devices made from amorphous silicon have received widespread attention because of their great potential in applications. Various substrates such as glass and plastics can be used for preparing amorphous silicon, and transistors can be fabricated with a number of different structures such as the needle of a syringe.

40.4.2 Construction of an a-Si ISFET Device

Figure 40.5 shows the structure of the a-Si ISFET (Gotoh *et al.*, 1989). These devices are mainly fabricated by RF plasma discharge.

Using a glow-discharge deposition system, a 0.05 μm of n$^+$ (3000 ppm PH$_3$ in silane) was deposited over an evaporated aluminium layer on a glass (Corning 7059) substrate. Ohmic contact is thus ensured between the a-Si and the aluminium. After etching, amorphous silicon and amorphous silicon nitride layers were deposited successively with the same capacitively coupled glow discharge deposition system operating at 13.56 MHz. The amorphous silicon layer was grown from a mixture of silane and hydrogen and the amorphous silicon nitride layer was formed from silane and ammonia. All three layers were deposited at 300°C. Finally, a silicon oxide layer was evaporated over the amorphous silicon nitride layer.

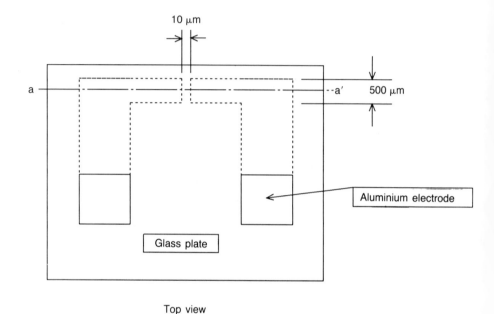

Top view

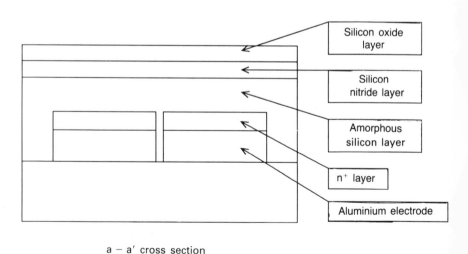

a − a′ cross section

Figure 40.5   Structure of an amorphous silicon ISFET (from Gotoh *et al.*, (1989).

The surface potential on the silicon oxide insulator of the a-Si ISFET is affected by the pH, as described previously for the crystalline ISFET. The a-Si ISFET and the Ag/AgCl reference electrode were immersed in a 10 mM Tris-HCl buffer solution at 18°C for 10–20 min. For comparison of the theoretical surface potential curve obtained on the basis of site-dissociation theory and the pH characteristics of the a-Si ISFET, it is assumed that the logarithm of the dissociation constant (pKa) is 5 and the C_H (the capacity of the electric double layer per unit area) is 20 $\mu F\,cm^{-2}$. The pH dependence of the linear gate voltage of the a-Si ISFET was obtained over the pH range 5–10. The pH sensitivity was approximately 46 mV/pH at 18°C. The pH response of the a-Si ISFET was rapid, less than 30 s being necessary to reach a steady-state value.

This device is suited to mass production. It can be made more cheaply than a crystalline Si FET. For this reason the a-Si ISFET is preferable for disposable sensors. Food analysis is a field where disposable sensors are needed in order to prevent cross-contamination between spoiled and good food during routine testing. Two applications of the a-Si ISFET biosensor are now described.

40.4.3 Hypoxanthine Sensor

To maintain quality, evaluation of freshness is important in the fish industry (Tamiya et al., 1988b). When the fish dies, adenosine-5'-triphosphate (ATP) decomposition occurs in its flesh and adenosine-5'-diphosphate and adenosine-5'-monophosphate and related compounds including hypoxanthine (Hx) are

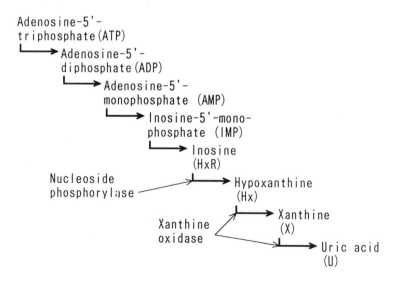

Figure 40.6   Generation of urea from ATP of fish

generated (Figure 40.6). Consequently, Hx accumulation with increase in storage time can be used as an indicator of fish meat freshness. Simple and rapid methods for the determination of Hx are, therefore, required in the industry.

Hypoxanthine is measured on the basis of a reaction catalysed by xanthine oxidase (XO). The pH change caused by the production of uric acid is detected by using an a-Si ISFET. XO is immobilized as follows. Polyvinylbutyral and 1,8-diamino-4-aminomethyloctane are dissolved in dichloromethane. This polymer solution is dropped onto the gate of the a-Si ISFET. The a-Si ISFET is then immersed in glutaraldehyde solution to promote a cross-linking reaction; xanthine oxidase (XO) is thus immobilized on the membrane.

The immobilized XO a-Si FET and the reference electrode were allowed to equilibrate in a phosphate buffer. After the gate output voltage has reached a steady value, hypoxanthine solution was injected into the buffer and the change in the gate voltage recorded. A linear relationship between the logarithm of the hypoxanthine concentration and the rate of change of the gate voltage per minute was obtained in the range 0.02–0.1 mM of hypoxanthine concentration.

40.4.4 Inosine Sensor

The inosine sensor (Gotoh et al., 1988) was fabricated similarly to the hypoxanthine sensor by using nucleoside phosphorylase and XO co-immobilized on an a-Si ISFET. After 90s from injection of inosine solution, the gate voltage gradually increased and reached steady-state in approximately 7 min. Xanthine formed by the decomposition of inosine catalysed by nucleoside phosphorylase was subsequently oxidized to uric acid by XO. The initial rate of the gate voltage change was plotted against the logarithm of inosine concentration, yielding a linear relationship in the range 0.02–0.1 mM. Oxidation of hypoxanthine to uric acid, by xanthine oxidase, is initiated immediately after injection. The response to inosine, however, has a time lag of 90s after injection. This phenomenon is attributed to the three-step reaction. On the basis of this time lag, this sensor can simultaneously determine inosine and hypoxanthine.

40.5 SILICON-BASED MICRO-OXYGEN ELECTRODE BIOSENSOR

40.5.1 Introduction

Clark-type oxygen electrodes have been applied to various biosensors, immobilizing either enzymes or micro-organisms which catalyse oxidation of biochemical organic compounds. Recently, miniaturized and integrated biosensors have found application in clinical analysis. The development of a disposable oxygen electrode, based on conventional semiconductor fabrication technology, and its application to biosensors, is now described.

Several oxygen electrodes, based on conventional semiconductor technology, have been fabricated by several groups, but they have not yet been adapted for mass production. Because the oxygen electrode contains electrolytes in solution, it is difficult to fix the gas-permeable membrane to the substrate: this problem has been the bottleneck preventing mass production of such devices.

Two key improvements are just to use a porous material (e.g. agarose gel) to support the electrolyte solution, and second, to use an hydrophobic polymer (e.g. negative photoresist) as the gas-permeable membrane and to cast it directly over the porous material.

40.5.2 Fabrication Process of a Micro-oxygen Electrode

The oxygen electrode is fabricated (Suzuki *et al.*, 1988b) by the following process, shown in Figure 40.7(a). The basic procedure is similar to a conventional semiconductor fabrication process except in the filling of the U-shaped groove with agarose gel.

(1) The silicon wafers (thickness 350 μm, diameter 2 inches) are washed in hydrogen peroxide and boiling ammonia solution.

(2) The wafer is subjected to thermal oxidation at 1000°C. The thickness of the SiO_2 layer is 1 μm.

(3) The groove pattern is formed with negative photoresist (Tokyo Oka, OMR-83), after which the other side is coated with the same photoresist; 50–60 s of light exposure is suitable when the MA-10 mask aligner is used.

(4) The SiO_2 layer is etched with mixed solution consisting of 50% hydrogen fluoride and 50% ammonium fluoride (1:6). The remaining SiO_2 layer becomes the mask for the anisotropic silicon etching.

(5) The resist is removed in sulphuric acid and hydrogen peroxide (1:2) solution at room temperature.

(6) The silicon is submitted to anisotropic etching in 35% potassium hydroxide solution. The temperature is maintained at 80°C.

(7) The SiO_2 layer is removed with the same solution as that in step (4).

(8) The silicon wafers are then washed with the same solution as that in step (1).

(9) The wafers are submitted to thermal oxidation. The oxidation temperature is 1000°C. The SiO_2 layer is 500 nm thick.

(10) The resist pattern is formed for the gold electrodes by using the same photoresist as that in step (3).

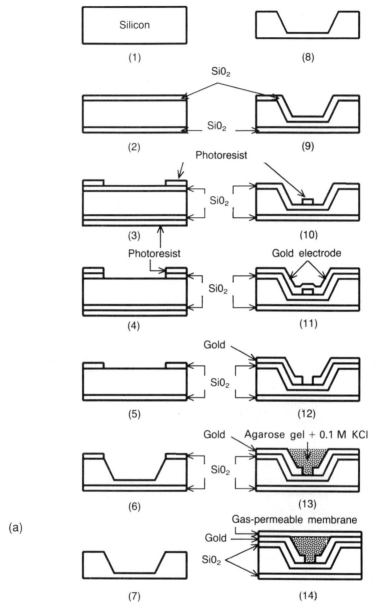

(11) A 50 nm chromium layer is deposited, followed by 1 μm of gold.

(12) The photoresist is then removed in warm sulphuric acid.

(13) The U-shaped groove is filled with heated agarose gel containing 0.1M
      KCl solution using a micro syringe. Then the gel is cooled. The agarose
      gel concentration is 1%.

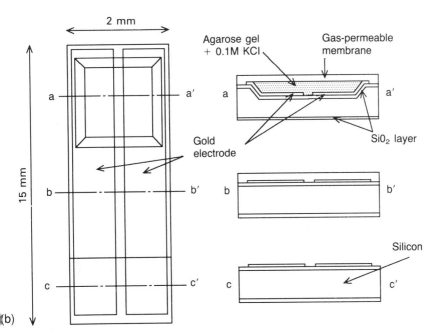

Figure 40.7(a)   Fabrication process of a micro-oxygen electrode; (b) structure of a micro-oxygen electrode ((a) from Suzuki *et al.* 1988b. Reproduced by permission, © 1988 American Chemical Society.)

(14)   A hydrophobic polymer is applied, followed by the photochemical reaction. In this case, the same photoresist (OMR-83) is used as the gas-permeable membrane and for insulation. The photo-resist is spin-coated at 500 rpm for 5 s and 1500 rpm for 20 s, followed by exposure to ultraviolet light.

The structure of this electrode is shown in Figure 40.7 (b). A linear relationship is obtained for the range 1–7.9 ppm (saturated) by using a 2mm wide electrode with a terminal voltage between the two gold electrodes of 0.8 V. Similar calibration curves are obtained by for other oxygen electrodes.

40.5.3   Glucose and CO_2 Sensors

The glucose sensor is fabricated by immobilizing GOD on the sensitive part of an oxygen electrode by cross-linking with bovine serum albumin (BSA) and glutaraldehyde. The immobilized enzyme membrane is formed by dipping the sensitive part into the mixture containing 2 mg of GOD, 20 μl of 10% BSA solution and 10 μl of 25% glutaraldehyde solution. The glucose sensor responded as soon as the glucose solution was injected into the buffer and reached steady-state in 5–10 min. The sensor responded almost linearly for glucose concentration between 0.2 and 2 mM, which is comparable with conventional glucose sensors.

A microbial CO_2 sensor using this oxygen electrode has also been constructed (Suzuki *et al.*, 1989). An autotrophic bacterium identified as S-17, which can grow with only carbonate as the carbon source, was obtained from The Fermentation Research Institute, Japan. Bacterial whole cells were immobilized on the micro-oxygen electrode. The sensitive area of the oxygen electrode was immersed in a 0.2% sodium alginate solution containing S-17 whole cells, then removed and immediately immersed in 5% $CaCl_2$ solution to form a calcium alginate gel containing the bacteria. The negative photoresist forming the gas-permeable membrane is then formed over the bacteria-immobilized gel. The photoresist is only exposed to ultraviolet light for a few minutes. The electrode response time is 2–3 min. Carbon dioxide was supplied by acidification of HCO_3 and the $NaHCO_3$ concentration can be related to CO_2 concentrations. A linear relationship is obtained between the current decrease and $NaHCO_3$ concentration in the range 0.5–3.5 mM. The detection limit was 0.5 mM $NaHCO_3$ within the margin of the noise amplitude. Above 3.5 mM, no significant increase in response was observed.

40.6 GLUCOSE SENSOR WITH ELECTRON TRANSFER MEDIATOR

40.6.1 Introduction

Enzyme-based amperometric biosensors, described in the preceding section, employ an oxygen oxidoreductase such as GOD. This type of enzyme catalyses the oxidation of substrate (glucose, alcohol) and the enzyme itself is changed to the reduced form. Its redox state is restored by consumption of oxygen. Biosensors usually determine the concentration of the substrate through the local change of oxygen or hydrogen peroxide concentration. When, however, the concentration of the substrate is high in this arrangement, oxidation of the enzyme becomes the rate-determining step. In this condition, the sensor response no longer correlates with substrate concentration. Furthermore, the electrochemical reduction of oxygen and the oxidation of hydrogen peroxide require a rather high potential. If electrochemically active species exist in the analyte, keeping the electrode at such high potential also oxidizes or reduces these compounds. For example, in a blood sample ascorbic acid and uric acid are present and are oxidized at a lower potential than hydrogen peroxide at the gold electrode. An undiluted blood sample may contain a high concentration of glucose and too low a concentration of oxygen to oxidize the reduced enzymes.

If an enzyme redox reaction can take place directly on the electrode, the enzyme electrode would be unaffected by the dissolved oxygen concentration and a wide dynamic sensing range would be expected. However, in general, few redox enzymes show direct electron transfer on an electrode. To solve this problem, an transfer mediator is used to oxidize the redox enzymes. An amperometric

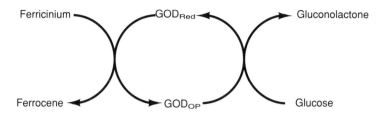

Figure 40.8 Reaction scheme of a mediated glucose sensor

glucose sensor using ferrocene derivatives as the electron transfer mediator has been fabricated and characterized (Cass *et al*, 1987). This glucose sensor exhibited only a small oxygen effect. The reaction scheme is shown in Figure 40.8.

In other work (Foulds and Lowe, 1988) GOD immobilization in ferrocene-modified polypyrrole on the platinum electrode was reported. Polypyrrole (pPy) modified electrodes are prepared by electrochemical oxidation of pyrrole. Ferrocene-modified polypyrrole is deposited electrochemically using a potential-cycle method between 0 and +1.0 V in aqueous perchlorate solution containing pyrrole and [(Ferrocenyl) amidopropyl] pyrrole (FAPP). Entrapment of GOD in the redox copolymer (FAPP/pyrrole) is done by electrochemical deposition, where the electrode potential was cycled between 0 and +0.8 V. The dynamic range for glucose was extended up to 100 mM.

Also reported (Ikeda *et al.*, 1986) is a mediator-based immobilized enzyme biosensor composed of GOD and a carbon paste electrode with p-benzoquinone as a mediator and covered with a nitrocellulose membrane. It was able to detect glucose concentration in the range 10–150 mM. The responses of these sensors, however, are somewhat dependent on the dissolved oxygen concentration.

To overcome this problem, a glucose sensor using $NAD(P)^+$-independent glucose dehydrogenase (GDH) was developed. D'Costa *et al.* (1986) report an amperometric glucose sensor using quinoprotein GDH, purified from *Acinetobacter calcoaceticus*. GDH was immobilized on a 1,1'-dimethylferrocene-modified graphite foil electrode. The current density of the GDH electrode produced was more than twice that for the graphite based GOD electrode at 4 mM glucose concentration. The calibration curve showed that this sensor can detect glucose concentration in the range 0.5–4 mM. When treated with an increased amount of GDH in the presence of glutaraldehyde, response was increased up to 15 mM. However, this treatment reduced the current output. The response of this latter electrode was unaffected by background changes under either anaerobic or aerobic conditions.

40.6.2 Micro-gold Electrode with Mediator

Tamiya *et al.* (1989) have developed a micro-gold electrode using a mediator as described above and reported elsewhere (Umana and Waller, 1986), electropolymerization of pyrrole can be employed to entrap and immobilize GOD on an electrode surface. This method, has the disadvantage in clinical applications that a long time is required to obtain a steady current in an amperometric determination of glucose. Moreover, electrode sensitivity decreases after a week or so. The response becomes saturated at glucose concentration of 15 mM (37°C, pH 7.0). In addition, the electrode is unstable above +0.6 V (versus saturated calomed electrode, SCE). This is due to the degradation of the PPy membrane. The use of an electron mediator was investigated in order to reduce the applied potential and to remove H_2O_2 directly, which leads to stabilization and a rapid response to glucose. Taking into account its lowest redox potential (+0.1 V versus SCE), sufficient reaction rate for the GOD (red) oxidation and insolubility in water (advantageous to the immobilization in a PPy membrane) among ferrocene derivatives, 1, 1'-dimethylferrocene (DMFe) was found to be the best electron transfer mediator between the reduced form of GOD and the PPy-modified electrode.

The micro-gold electrode utilized has two working electrodes and a counter electrode on the device substrate of 1.6 mm width. This electrode is fabricated as follows. First, titanium is sputtered onto a sapphire substrate to improve gold adhesiveness. Then, a gold layer ($ca.$ 1 μm thick) is deposited onto the titanium layer by sputtering. The gold layer is patterned by etching. The gold electrodes are covered with a negative photoresist for insulation, except for the sensitive area and pad parts. Polypyrrole (PPy)-modified electrodes are prepared by electropolymerization of pyrrole in an aqueous solution. GOD immobilization is performed by adsorption of GOD onto the PPy-modified electrode. The PPy-modified electrode is then dipped in the GOD solution (50 mg/ml) overnight at 4°C and afterwards rinsed and dried. With GOD adsorbed, the electrode is dipped into dichloromethane solution containing 1% DMFe and 2% polyvinylbutyral (PVB) for 10 s and dried. DMFe is then trapped in the PVB matrix. The DMFe immobilized GOD/PPy electrode is characterized at the electrode potential of +0.1 V in a nitrogen atmosphere. This glucose sensor has a wide dynamic range and response saturation is not observed below the glucose concentration of 30 mM. This limit is high enough for the measurement of glucose in a whole blood sample.

40.7 MULTI-BIOSENSORS

40.7.1 Introduction

There exists high demand for rapid, simultaneous, multi analyte and small sample volume measurements in clinical analysis. To meet these demands it is necessary to develop micro-multi-biosensors; special technology is necessary. Immobilization of enzymes in a small area and determination of different substrates by closely aligned small electrodes are the required technology.

Sibbald et al. (1984) reported the simultaneous on-line measurement of potassium, sodium, calcium ion and pH in blood using a quadruple function ChemFET (chemical sensitive field effect transistor) attached to the flow cell.

Several methods for preparing the enzyme membranes of multi- biosensors have been developed. For example, a water-soluble photo crosslinkable polymer solution containing an enzyme is spin-coated on the ISFET gate surface and the enzyme is immobilized by ultraviolet light irradiation polymerization (Hanazato *et al.*, 1986). This method is advantageous when the immobilized enzyme membrane is to be patterned by photolithography because different kinds of enzyme membrane can be immobilized in a small area by this method.

In this section, two multi-biosensors based on use of a mediator and four integrated multi-biosensor fabrication technologies are described.

40.7.2   Simultaneous Determination of Glucose and Galactose

Simultaneous measurement of several kinds of biological substrates is important. In this section, an integrated multi-biosensor for the simultaneous measurement of glucose and galactose concentrations based on GDH and galactose oxidase (GAO) is described (Yokoyama *et al.*, 1989a).

GDH, with pyrroloquinolinequinone (PQQ) as a prosthetic group, catalyses glucose oxidation. The reduced form of the mediator (Med (red)) is detected on the gold electrode as shown in Figure 40.9.

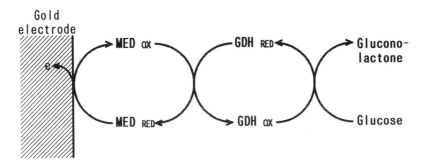

Figure 40.9   Reaction scheme of a mediated glucose sensor with GDH (from Yokoyama *et al.*, 1989a).

This sensor is unaffected by the concentration of dissolved oxygen because no electrons can transfer from PQQ to oxygen. On the other hand, GAO catalyses galactose oxidation as shown in Figure 40.10.

The galactose concentration was determined by measuring oxygen consumption. Thus, it is possible to measure the glucose and galactose concentration in the same solution even if GDH and GAO are immobilized on two electrodes on the same substrate. Glucose can be detected by current increase during oxidation

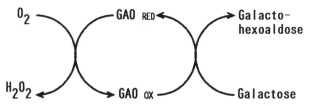

Figure 40.10   Reaction scheme of a galactose sensor

of the mediator. Galactose can be measured by current decrease during oxygen reduction.

Four gold working electrodes and a counter electrode were formed on a glass substrate (Corning 7059) by vapour deposition (Figure 40.11). The area of the working electrode is 0.2 mm^2. Two of the four electrodes were used as the biosensor, one for glucose determination, the other for galactose. The immobilization procedure is described as follows. GDH (5 mg), GAO (4 mg) and BSA (5 mg) were dissolved in 100 μl of HEPES buffer (pH 7.9) containing 10 mM MgSO$_4$ and 3 mM PQQ. The solution was spread on the glass substrate, followed by exposure to glutaraldehyde vapour at 30°C for 30 min. After the immobilization, the enzyme electrode and the reference electrode were immersed in 10 mM HEPES buffer (pH 7.4) containing ferrocene monocarboxylic acid (FCA), with stirring, at 30°C. The potential was set at 350 mV for the mediator oxidation and at −300 mV for the oxygen electrode. After the current output had reached a steady state, glucose or galactose solution was injected into the buffer solution and the current changes were measured. The measurement was carried out under air

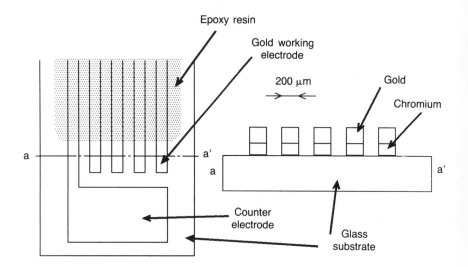

Figure 40.11   Structure of the integrated micro-electrodes

bubbling conditions. The glucose concentration was determined at various galactose concentrations. The response toward glucose decreased with the increase in the galactose concentration because GAO reacts not only with oxygen but also with the mediator. The response of the glucose electrode, therefore, depended on the galactose concentration. By this scheme simultaneous determination of both glucose and galactose was possible. First, galactose was measured using the oxygen electrode and then the glucose concentration was determined by choosing an appropriate calibration curve for glucose in the presence of galactose.

40.7.3 Glucose Determination under Unknown Oxygen Tension

Accurate monitoring of glucose is very important in such fields as bedside monitoring and direct implantation of a micro-biosensor. In the sensors described in the previous section, a different potential was applied to the each working electrode and different modes of enzymatic reaction employed for the simultaneous determination of two substrates. The same technology was applied for the precise determination of glucose (Yokoyama *et al.*, 1989b).

GOD was immobilized on the polypyrrole film on the working electrode. DMFe was incorporated by dipping the glass substrate into DMFe dissolved in acetone solution. One of the working electrodes was used for the mediated glucose sensor, the other being used as an oxygen-based glucose sensor. An Ag/AgCl electrode was used as the reference electrode. The same counter and reference electrode were employed for the glucose sensors. The working potentials of the mediated and the oxygen-based glucose sensors were 0.1 V and that of the oxygen-based sensor was −0.3 V versus the Ag/AgCl electrode.

The characteristics of the glucose sensors were evaluated simultaneously in 0.1M phosphate buffer (pH 7.0, 30°C). The response time was approximately 1 min when 50 µl of 10 gdl^{-1} glucose sample solution was injected into 30 ml of phosphate buffer. Under several dissolved oxygen concentrations, glucose concentration was determined from the response of the mediated glucose sensor, and corrected on the basis of the oxygen-based glucose sensor.

40.7.4 Direct Photo-patterning Method

The direct photo-patterning method (Hanazato *et al.*, 1986) uses direct application of lithography for enzyme immobilization. The key point is use of a water-soluble photo-cross-linkable polymer which does not cause deactivation of enzymes. The fabrication of this glucose sensor was as follows. The surface of the ISFET was treated by 3-APTES to improve the adhesion strength between the surface and the polymer membrane. Then 0.1 ml of polymer solution, consisting of polyvinylpyrrolidone (PVP) and 2, 5-bis (4'-azido-2'-sulphobenzal) cyclopentanone (BASC) (PVP:BASC:WATER = 100:3:1000) containing 10 mg

of GOD and 10 mg of BSA, was spin-coated on the surface of the integrated
ISFET, on which three ISFET elements were formed, at 2000 rpm for 2 min, to
make the thin film. A limited area of the membrane around the gate surface of
one of the ISFETs was exposed to ultraviolet irradiation to polymerize and entrap
the enzyme. The ISFET was immersed in water to develop the GOD membrane.
Thus, the GOD membrane was formed on a limited area of the gate surface. By
repeating this procedure, a lipase membrane and BSA membrane were also
formed on the other ISFET elements.

The ISFETs were immersed in 3% glutaraldehyde solution and then in 0.1 M
glycine solution. The glucose sensor was found to be useful for the determination
of glucose concentration in the range 0–5 mM. When the lipid sensor was
employed to measure triolein concentration, triolein could be detected up to 3
mM. Through thorough verification of the interference between the glucose and
triolein responses of the integrated ISFET biosensor, this multi-biosensor can be
shown to measure both glucose and triolein concentration simultaneously.

40.7.5   Micro-pool Method

In the micro-pool method (Kimura *et al.*, 1986) a small amount of pre-gel enzyme
solution is poured into a small well fabricated on the gate of ISFET. The fabrication process of this biosensor is shown in Figure 40.12 and consists of the
following steps.

(a)   The silicon layer of the SOS wafer was anisotropically etched to form
      'islands' of silicon.

(b)   The source, drain and gate region were doped. The SiO_2 layer was thermally
      oxidized. The polysilicon layer for the MOSFET gates was deposited by
      chemical vapour deposition (CVD). Finally, to control the threshold voltage,
      phosphorous ions were implanted into the gate surface.

(c)   The Si_3N_4 layer was deposited by CVD as a protecting membrane against
      saline water. It also works as a pH-sensitive membrane. Contact pads were
      created, followed by aluminium deposition. A gold layer was sputtered on
      the whole surface.

(d)   A film-resist photopolymer layer was adhered to the wafer surface and
      patterned as shown in Figure 40.12(d). This photopolymer was used to make
      micro-pools for four types of the membranes. It also acts as a physical
      protection membrane for the ISFET surface.

(e)   The wafer was cut into individual sensor chips, which were placed on a
      flexible printed circuit board. Electric connections were achieved by alumi-

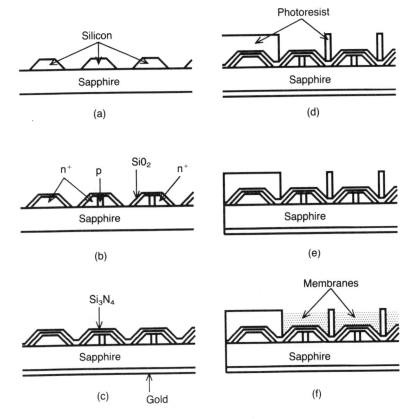

Figure 40.12 Micro pool method fabrication process (from Kimura *et al.*, 1986).

nium bonding between the ISFET and the flexible printed circuit board. The bonding area was moulded by heat-hardening adhesion.

(f) Membranes which are sensitive to potassium, glucose and urea were formed, together with a reference membrane.

(1) *Potassium-sensitive membrane.* A potassium-sensitive membrane, which needs the high-temperature process, was formed. First 10 mg of valinomycin was dissolved in 120 µl of dioctyl adipate with 800 mg of negative photoresist polymer (OMR 83) added. After stirring well, a small amount of the mixture was dropped into one pool of the sensing area using a microsyringe. The volume of one drop is approximately 0.03 µl. Then, after prebaking at 80°C for 30 min, another drop was applied to the membrane surface. After a second pre-baking and ultraviolet irradiation, the membrane was hardened by baking.

(2) *Glucose-sensitive and urea-sensitive membranes.* Here 5 mg of GOD or 1 mg urease was dissolved in 15% BSA solution. A small drop (0.05 µl) of this

mixture was poured into one of the ISFET sensing areas with a microsyringe. After drying at room temperature for 10 min, 0.05 µl of 25% glutaraldehyde solution was poured onto the dried solution to immobilize the enzyme. After 10 min the device was immersed in water.

(3) *Reference membrane.* The role of this membrane is to protect the ISFET surface against physical destruction and to prevent bubble formation, while maintaining the pH sensitivity. The process of fabricating this membrane is the same as that for the enzyme membranes. However, this membrane excludes any sort of enzyme.

A gold layer works as a pseudo-reference electrode and a potential of +1.6 V was applied to the gold electrode. The urea and potassium-sensitive FETs showed rapid responses to urea and potassium, respectively. The urea-sensitive FET showed no response to glucose. The glucose-sensitive FET showed slowly increased response to glucose and a small response (less than 1 mV) to urea. The

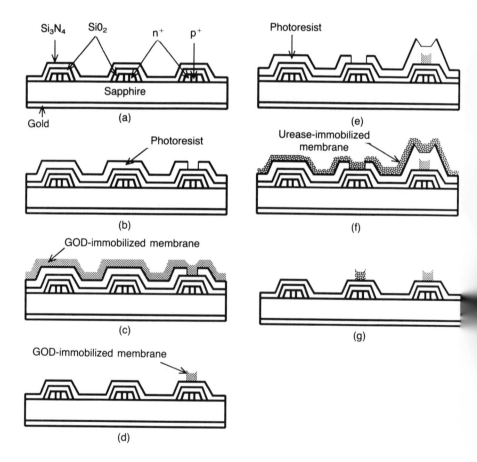

Figure 40.13   Lift-off method fabrication process (from Kimura *et al.*, 1988a).

influence of diffusion of chemical species to the neighbour electrode surface was smaller than predicted. Glucose concentration could be determined between 1 and 50 mg/dl. Urea concentration could be determined in the range 1–100 mg/dl. The potassium differential output shows a linear relationship between 1×10^{-3} and 2×10^{-2} m with a sensitivity of 60 mV/pK. This method makes it quite easy to create the membrane and to control the amount of the immobilized enzyme. Further, it requires only a small amount of enzyme for immobilization and, hence, is an economical method. However, the major portion of the membrane has the tendency to be located towards the film-resist side wall. As a result, the membrane centre is rather thin. The thickness uniformity needs to be improved to get a large and rapid response. By controlling the micro-syringe automatically a more precise and uniform enzyme membrane will be formed.

40.7.6   Lift-off Method

A lift-off method (Nakamoto et al., 1988; Kimura et al., 1988a) is one of the IC fabrication processes for precisely preparing thin films on a small area. This method is suitable for preparing an enzyme membrane in mass production of integrated multi-biosensors.

Fabrication steps of the lift-off method for making biosensors are shown in Figure 40.13. In this case, GOD and urease membranes were formed on different ISFETs.

The steps are as follows:

(a)   1% 3-APTES aqueous solution is spin-coated on the wafer and is heated at 110°C for 5 min to modify the wafer surface, with amino groups undergoing the reaction with glutaraldehyde.

(b)   A positive photoresist is spin-coated onto the wafer and then pre-baked, followed by patterning using ultraviolet light exposure through the photomask, and development to expose the gate region on which the GOD immobilized membrane is deposited (Figure 40.13(b)). The photoresist layer is 1.8 μm thick.

(c)   Next 0.25 ml of 28% BSA solution in 50 mM PIPES-NaOH (pH6.8), 0.15 ml of 23% GOD aqueous solution and 0.1 ml of 5% glutaraldehyde aqueous solution are mixed and spin-coated onto the wafer at 1500 rpm. The wafer is held at room temperature for 2 hours to complete the cross-linking reaction. As a result, the wafer is covered with the GOD immobilized membrane (Figure 40.13(c)).

(d)   Ultrasonic vibration is applied to the wafer placed in acetone. The photoresist between the GOD immobilized membrane and the wafer dissolve in the acetone. The GOD immobilized membrane on the photo-resist is then lifted off. A precisely patterned GOD immobilized membrane for placement on the gate region is obtained. The resulting wafer is washed in water and dried with nitrogen gas.

(e) The positive photo-resist is spin-coated onto the wafer again (Figure 40.13(e)). To prevent the enzyme inactivation, due to heat from baking, the photo-resist is dried in an evacuated desiccator for 30 min. The wafer is patterned and developed by the same process as that in (b) to expose the gate region on which the urease-immobilized membrane is deposited.

(f) The urease immobilized membrane is deposited onto the gate surface area in the same way as that in step (c). The enzyme solution containing 0.25 ml of 28% BSA, 0.15 ml of 5% urease and 0.1 ml of 5% glutaraldehyde is spin-coated on the wafer (Figure 40.13(f)).

(g) The urease immobilized membrane is lifted off by the same method as that in (d).

By repeating the fabrication processes described above, the GOD and urease immobilized membranes were precisely deposited on each gate surface of the wafer. The membranes are 100 μm wide and 400 μm long. The thickness is approximately 1 μm.

After the enzyme solution, containing urease, was spin-coated, the GOD-immobilized membrane was covered with photo-resist. Thus, no chance exists to contaminate urease with the GOD-immobilized membrane. It was considered possible that presence of the photo-resist would cause reduction of the GOD activity. However, the sensor showed the same response to glucose as without the photo-resist treatment.

In the case of the glucose sensor, a linear relationship between the differential output (under steady state) and final glucose concentration up to 90 mg ml^{-1} was observed. The slope of the output began to decline at 90 mg ml^{-1} and a limit was reached at 200 mg ml^{-1}. In the case of the urea sensor, the linear relationship between the logarithm of urea final concentration and differential output was in the range 1–100 mg ml^{-1}.

Among the other multi-biosensors, a glucose sensor based on amperometry and a urea sensor based on potentiometry are integrated. This was realized by the complementary use of ISFETs and micro-patterned planar gold electrodes. A GOD-immobilized membrane, urease immobilized membrane and a BSA cross-linking membrane were deposited in appropriate positions by the lift-off method described above. The output of this sensor, to glucose, increased linearly with the logarithm of glucose concentration in the range 1–100 mg dl^{-1}. The output begins to decline at 100 mg dl^{-1}. The response to urea showed an S-shape curve against the logarithm of urea final concentration. Glucose concentrations in the range

5–100 mg dl^{-1} were able to be measured. This multi-biosensor employed the complementary use of amperometric and potentiometric principles and could effectively measure glucose and urea concentration in solution with high selectivity.

40.7.7   Ink-jet Nozzle Method

The ink-jet nozzle was originally developed as a printing device where it is used to deposit drops of printing ink onto the paper. By controlling the movement of the ink-jet nozzle against the paper and the timing of the ink jet, any kind of character can be formed on the paper.

The method explained here (Kimura *et al.* 1988b), employs this device as a precise deposition unit to place enzyme onto an ISFET.

The ink-jet nozzle construction is shown in Figure 40.14. Two kinds of immobilization methods were carried out using this equipment. One of them is gas phase immobilization. This fabrication process is as follows.

(a)   10 drops of the enzyme solution are emitted by the ink-jet nozzle onto different positions of the ISFET gate area.

(b)   To carry out preliminary immobilization, the sensor device is placed in a vapour chamber containing 25% glutaraldehyde solution for 5 min.

(c)   By repeating processes step (a) and (b) five times, 50 drops of the enzyme solution are immobilized on the ISFET gate area.

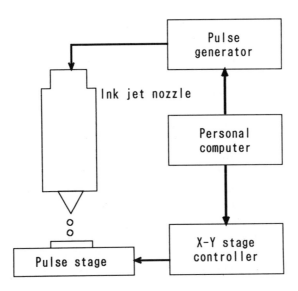

Figure 40.14   Ink-jet nozzle system (from Kimura *et al.* 1988b. Reproduced by permission of Elsevier Science Publishers Ltd.)

(d) To complete the immobilization, this sensor device is immersed in 1% glutaraldehyde solution for 5 min.

Using sensors, glucose was determined in the range 2–100 mg dl^{-1} and urea was determined in the range 1–1000 mg dl^{-1}.

In this method, only ten drops of enzyme solution covered a large part of the gate region. The direction of the enzyme emission was precisely determined by employing an X–Y stage controlled by a personal computer. However, in the gas phase immobilization method it takes one hour to emit and immobilize enzyme solution in one cycle, and the method is thus unsuited for mass production.

An alternative method is liquid phase immobilization. This fabrication process is almost the same as the micro-pool method as described in section 40.7.5. The enzyme solution is emitted by the ink-jet nozzle into micro-pools made by the photo-resist. However, in this method, special skill is necessary to deposit an immobilized membrane on a small sensing area.

With the ink-jet nozzle method loss of enzyme is minimal and an expensive enzyme can be used efficiently for fabricating biosensors. This method can easily distinguish an enzyme among several enzymes and rapidly and reliably determine the enzyme emission position. Thus many kinds of multibiosensors, having different enzyme combinations, can be realized on one wafer.

40.8 IMAGING SENSOR

40.8.1 Introduction

Many clinical analyses are based on the determination of soluble marker substances added to body fluids such as blood and urine. Direct analysis in cells or tissue is of great important in clinical diagnosis. In the case of cancer detection, highly sensitive, rapid, detection methods for abnormal cells are required. Cell diagnosis is carried out mainly through visual inspection by trained experts or with the use of a flow cytometer.

Recently, much attention has been focused on image analysing systems composed of an imaging sensor followed by a digital image processing system. In this section, application of image analysis systems, composed of a CCD (charge-coupled device) imaging sensor and digital image processing system, to automated cell and tissue diagnosis is described.

The CCD is an integrated semiconductor chip composed of photodiode arrays and charge transfer circuits. There are many advantages of a solid-state CCD imaging sensor as compared with conventional vidicon. They are compactness in size, high sensitivity, little distortion, no after-image, low power consumption and a long operational life.

A phase-contrast, transmitted light, microscope was used to observe the cyto-
lysis of tumor cells. A CCD video camera was mounted vertically onto the
microscope. Images of cells were displayed on a video monitor. The video
display was fed into an image memory board connected to a 16-bit personal
computer. Some examples of its use are now described.

40.8.2 Detection of Tumour Cells

The first example is detection of tumour cells (Suzuki *et al.*, 1987 a, b). The
specific detection of Line-10 hepatocarcinoma (L-10) cells is based on the com-
plement-mediated cytotoxic reaction with monoclonal antibody 3C4. When cells
were observed through a phase-contrast microscope, the normal cell looked
brighter than the damaged cell. This is because the normal cell retains its cell
membrane structure, which produces a phase-lag of transmitted light due to the
difference of optical refractive index between intracellular and extracellular
space, while the damaged cell loses its membrane structure. Calculating the
decrease in brightness of lysed cells by a two-dimensional imaging sensor and
image processor, specific and non-specific cells are distinguished. The experimental procedure is as follows. 100 µl of cell suspension containing Line-10
cells and non-specific L2C cells, 100 µl of 3C4 antibody solution and 50 µl of
rabbit serum (complement source) were mixed. After 30 min incubation at 37°C,
the sample was transferred into a hemacytometer and analysed using the imaging
sensor system.

L-10 cells can be detected quantitatively in the range 10–100%. A linear
relationship between the L-10 content and image area are obtained in the same
range. This process requires no cell-washing step, therefore making it possible
to automate the whole system during the reaction stage. However, the detection
limit of this system is insufficient for cancer diagnosis. To overcome this problem, target cells were stained directly. By using a DNA (or RNA) specific
fluorescent dye, propidium iodide and this detection system, L-10 cells can be
detected quantitatively when present at above 2%.

40.8.3 Latex Agglutination Test

A turbidimetry-based latex agglutination test in immunoassay has been applied
in practice, although the sensitivity is still insufficient (Matsuoka *et al.*, 1987;
Tamiya *et al.*, 1988c). It was formerly known that specific and non-specific
reactions differed in their initial reaction rate. Therefore, precise measurement
of the turbidity change at the early stage of the reaction would improve the
sensitivity of turbidimetry. Use of an imaging sensor is a suitable technique for
such purposes. Application of linear (one-dimensional) imaging sensors to immunoassay based on latex agglutination is now described.

The experimental procedure is as follows. Human IgG solution and an Ab-L (antibody bound latex beads) suspension were mixed in a test tube and immediately transferred to a glass capillary (0.8mm internal diameter by 70mm length). The glass capillary was then set on the sample holder of the imaging sensor, placed in the dark box and illuminated with a fluorescent lamp through the slit. Light scattered by the capillary at a scattering angle of 120° was passed through the lens and focused exactly on the sensing area of the imaging sensor. Optical intensity was adjusted with an iris. The progress of agglutination was tracked using the imaging sensor. As Ab-L suspended in a glass capillary gradually precipitated, the linear image pattern changed. This linear imaging sensor system enabled immunoassay for human IgG in the range $5 \times 10^{-8} - 5 \times 10^{-5}$ g ml^{-1}.

In latex agglutination immunoassay, an electric pulse accelerates the specific immunoreaction between Ab-L and Ag by increasing the contact frequency of the small particles. Such acceleration can improve the sensitivity of this assay method. By using the pulse immunoassay method, human IgG was determined in the range 6.7×10^{-8} to 6.7×10^{-6} g ml^{-1}. Combination of the imaging sensor system and pulse immunoassay method has the potential to yield rapid and precise immunoassays.

40.9 PIEZOELECTRIC CRYSTAL BIOSENSORS

40.9.1 Introduction

It is known that the resonant frequency of a piezoelectric resonator changes according to the surface mass change. The degree of change Δf of frequency f_0 is written as

$$\Delta f = \Delta m / m$$

where m is the mass of the resonator, Δm is surface mass change and f_0 is the basic resonant frequency. Because it is easy to measure resonant frequency to a precision of 9 digits, it is possible to detect very small mass changes at the surface of a piezocrystal resonator. Employing this principle, the piezoelectric resonator has been applied for detection of gases, ions, microbes, biopolymers and electrochemical mass changes. It has also been reported that modification of the crystal surface with organic compounds or enzymes that bind a particular gaseous substrate can provide enhanced detection specificity. The properties of piezoelectric devices immersed in water or organic solvents have recently been studied by various researchers and several theoretical equations governing their behaviour in liquids have been derived. Methods employing a coupled gravimetric and electrochemical assay have also been used for the determination of electrodeposited metal ions.

Application of piezoelectric crystals as biosensors is now described.

40.9.2   Detection of Pathogenic Micro-organisms

The first application of a quartz crystal to an immunosensor (Shons et al, 1972) involved measurement of anti-BSA antibody using a BSA-immobilized quartz crystal. Leucocytes, immunogloblin and other substances have also been detected using a quartz crystal. Muramatsu *et al.* (1986) report detection of Candida albicans, determination of which has gained importance in clinical analysis. *C. albicans* is found in the human body even under normal conditions, but an ncrease in cell population can induce infection and disease. It is conventionally assayed by visual inspection of antibody-antigen (*Candida* species) aggregation. This method, however, requires technical skill and cannot be automated. Furthermore, it gives only a semi-quantitative assessment of concentration. Matsuoka *et al.* (1987) showed that potentiometry can be applied to its immunoassay. In this method, negatively charged *C. albicans* is adsorbed onto the membrane-bound antibody, inducing a considerable change in the membrane potential (linear in the range 10^4 to 5×10^5 cells cm^{-3}). Another immunoassay, based on an electric pulse technique, was also proposed and applied to *C. albicans* immunoassay (linear in the range 10^7 to 6×10^7 cells cm^{-3}).

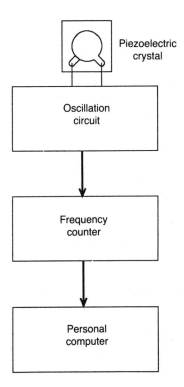

Figure 40.15   Micro-organism detection using a piezoelectric crystal

Piezoelectric crystals coated with immobilized anti-Candida antibody have been applied to immunoassay of *C. albicans*. This appears to be the first application of piezoelectric crystals to the detection of micro-organisms. The piezoelectric crystal used was AT-cut quartz, 8 mm × 8 mm × 0.18 mm, with a basic resonant frequency of 9 MHz. Silver electrodes were formed on the crystal by vacuum deposition and were plated with palladium. The palladium electrodes were treated by anodic oxidation. The electrodes were then treated by 3-APTES and glutaraldehyde, after which anti-Candida antibody was immobilized on the electrodes. The treated piezoelectric crystal was dipped in the microbial suspension for 30 min to allow the immunoreaction to occur between the immobilized antibody and the microbe. The crystals were then rinsed with 0.5 M NaCl and water and dried in air.

The schematic diagram of this sensor system is shown in Figure 40.15. The resonant frequency was measured and the frequency shift was calculated. The correlation between the cell concentration and the resonant frequency shift resulting from the binding of *C. albicans* was obtained in the range 10^6 to 5×10^8 cells cm^{-3}. The number of cells adsorbed on the electrode was also measured by microscopy. This confirmed the magnitude of Δf to be related to the actual number of adsorbed microbes.

40.9.3   Determination of IgG Subclass

Recently much attention has focused on the application of quartz crystals in liquids. Muramatsu et al. (1987) reported the application of AT-cut piezoelectric crystals to the determination of human IgG concentration and IgG subclasses in solution. They immobilized protein A on the surface of a piezoelectric crystal as a recognition element for IgG because of its high binding specificity toward IgGs.

The experimental procedure was as follows. The crystal electrodes were modified with 3-APTES and glutaraldehyde and then protein A was immobilized on the electrodes. After immobilization of Protein A, the remaining aldehyde groups were blocked with glycine. The determination procedure of IgG concentration was as follows. The Protein A-modified piezoelectric crystal was placed inside the flow cell. First, the resonant frequency was monitored at a constant flow rate. The steady resonant frequency (f_2) was obtained. The water was drained off and a solution of human IgG was then injected into the cell. After incubation, the crystal was rinsed with 0.5 M NaCl solution to remove any nonspecifically adsorbed IgG, followed by remeasurement of the steady resonant frequency f_2 in constantly flowing water. IgG bound to Protein A was removed with glycine-HCl buffer, allowing the successive measurement of different human IgG concentrations.

Determination of the IgG subclass was as follows. Mouse γ-globulin or human γ-globulin solution was injected into the cell and the steady resonant frequencies, f_1 and f_2, were measured as described above. The crystal was then rinsed with a stepped gradient of phosphate–citric acid buffer (from pH 7 to pH 2.5) each pH step being 0.5 pH units. After the cell was rinsed at each step, the steady resonant frequency f_{2pH} was remeasured in flowing water.

The steady resonant frequency decreased with injection of γ-globulin solution. By rinsing with a stepped gradient of phosphate–citric acid buffer from pH 7 to pH 3, the steady resonant frequency increased. The pattern of the overall resonant frequency shift, during each of the separate rinsing steps, corresponded to the result that was obtained by affinity chromatography. The peaks at pH 6.5, pH 5.5–4.5, and pH 3.5–3.0 indicated the presence of IgG1, IgG2a and IgG2b, respectively. The pattern of the overall resonant frequency shift during each of the separated steps for the case of bound human γ-globulin clearly showed difference from that of mouse γ-globulin, and corresponded to the results previously reported in which IgG2 and IgG4 were eluted at pH 4.7 and IgG1 and IgG4 were eluted at pH 4.3. This system can be applied to various analyses through the application of protein–protein affinity reactions.

40.9.4   Determination of Pyrogen

Many applications of piezoelectric crystals are based on its mass detection. Likewise, in the studies of quartz crystals in contact with liquids, the resonant frequency shift and the electrical resistance (included in the electric equivalent circuit) of the quartz crystal were shown to be a function of viscosity and density of the contacting liquid. For this the change Δf in resonant frequency of the crystal is given by

$$\Delta f = -f_0^{3/2} (\eta_L \rho_L / \pi \mu_Q \rho_Q)^{1/2}$$

where η_L and ρ_L are the absolute viscosity and density of the liquid, respectively, and μ_Q and ρ_Q are the elastic modulus and density of the quartz, respectively.

Muramatsu et al. (1988) reported that the electrical resistance of the quartz crystal shows an excellent linear relationship to the viscosity of the liquid and has applied this fact to endotoxin measurement. Endotoxin causes gelation of Limulus amebocyte lysate, in accordance with viscosity change. Endotoxin is a kind of fever-inducing pyrogen. It is produced by gram-negative microbes and widely found in water samples. Endotoxin determination is of great importance in medical products. Blood from the horseshoe crab coagulated by the addition of endotoxin has enabled the gelation of Limulus amebocyte lysate (LAL) to be used for determination of endotoxin; this is called the Limulus test. The conventional method is a manual inversion method, in which a mixture of LAL and a sample solution is incubated in a test tube for 1 hour at 37°C after mixing. Determination of endotoxin is based on visual observation of the gel settling. Therefore, the conventional method is semi-quantitative. An experimental procedure for the piezoelectric sensor was as follows. The sample solution (0.2 ml) was added to solute lyophilized LAL. The mixture was poured into the well-type cell attached to one side of the quartz crystal and the resistance and resonant frequency were measured at 37°C. By monitoring the viscosity change, endotoxin was determined in the range 1 to 10^5 pg ml^{-1}.

40.9.5   Odorant Detection and Recognition

The piezocrystal resonator has also been applied for odorant recognition and the measurement of odorant concentration, (Muramatsu *et al.*, 1990). When a crystal is covered with lipid, the resonant frequency change is sensitive to concentrations of odorants such as n-amyl-acetate citral, β-ionone and menthone, because these compounds are absorbed into the lipid layer on the piezocrystal causing mass increase on the surface. The procedure was as follows.

A 9 MHz AT-cut quartz resonator was used and an impedance analyser employed to estimate the resonant frequency of the device. The lipid film was formed over the resonator using a chloroform solution (5 mg ml^{-1}). After film formation, nitrogen gas was allowed to flow onto the resonator until the resonant frequency stopped drifting. The vessel containing the resonator was closed and odorant concentration was calculated from the volume ratio of vaporized odorant to the volume of the vessel. The concentration of odorant was controlled by diluting it in diethyl ether. When the resonant frequency stopped drifting the frequency was recorded. Before the next injection, the vessel was purged with nitrogen gas. An asolectin-coated quartz crystal showed a detection limit of ca. 1 ppm, for β-ionone, 1–10 ppm for citral, 10 ppm for menthone and 100 ppm for n-amyl acetate.

In the case of odorant recognition, because the resonant frequency was a function of both the amount of the coating film and the injected gas concentration (odorant), it was necessary to normalize the response by two steps: first to decide the film amount and second the gas concentration. The first normalization was performed by dividing the frequency shift caused by ethanol gas by the frequency shift caused by film coating. The second normalization was performed by dividing each of the first normalized values by the sum of that value for all odorants. The resultant values showed a pattern which corresponded to a specific odorant.

40.10   SOLID ION ELECTRODE

40.10.1   Introduction

The measurement of potassium and sodium ions in body fluids is routine in the clinical situation. There is high demand for rapid, easy and simultaneous measurement of these analytes. The conventional flame photometric method provides superior reliability. However, is not easy to handle and is not suitable for rapid measurement. Glass membrane ion selective electrodes suffer from limitation of miniaturization because of their inner electrolytic solution. In this section, new sodium and potassium ion concentration determination methods using a solid ion conductor are described.

40.10.2 Sodium and Potassium Sensor

NASICON (natrium super ion conductor) is a solid-state ion conductor which has selectivity for the sodium ion. If one side of a NASICON membrane is immersed in a solution containing sodium ions, electrochemical equilibrium is attained when a proper sodium ion concentration profile is formed inside of the membrane. NASICON exclusively conducts sodium ions. In this condition excess sodium ions cause a potential difference between the bulk solution and the NASICON membrane. This potential is proportional to the logarithm of bulk sodium ion concentration with Nernstian slope RT/F.

The experimental procedure of Tokumoto *et al.* (1990) is as follows. A NASICON membrane was attached to a metal electrode and fixed with an O-ring inside an electrode housing (Figure 40.16). The constructed electrode was immersed in buffer solution and the electrode potential measured (by a high input impedance voltmeter/electrometer versus a reference electrode, such as Ag/AgCl.

The electrode system shown in Figure 40.16 was able to measure sodium ion concentration in the range of 1×10^{-3} to 1 M. The sensitivity was almost the same as predicted by the Nernstian relationship. Interfering ions such as potassium, lithium, ammonium and calcium had less effect on sensor response.

For the potassium ion, the NASICON membrane was coated with poly (vinyl chloride) (PVC) including valinomycin. The response to potassium ion was as quick and stable as when bare NASICON was used for sodium ion detection. Measurement using this potassium sensor could be performed in the linear range

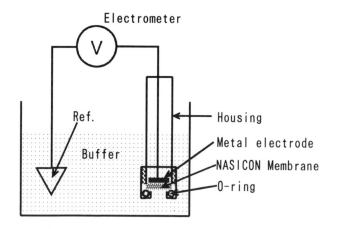

Figure 40.16   Sodium sensor using a NASICON membrane

of 5×10^{-4} to 1 M, the slope being a little less than the Nernstian prediction. The selectivity for potassium was also high against interfering ions (except for the ammonium ion).

These NASICON membranes were arranged on one plate for simultaneous measurement of sodium and potassium ion concentration in a control serum. The performance of both sodium and potassium ion detection with these electrodes was in good agreement with that of the standard reference values for the controlled serum.

40.11   SOURCES OF FURTHER ADVICE

The reader is referred to the following textbooks and journals for further information on biosensors.

Books

Buck, R. P., Hatfield, W. E., Umana, M. and Bowden, E. F. (1990) *Biosensor Technology* Marcel Dekker, New York.

Cass, A. E. G. (ed.) (1991) *Biosensors: A Practical Approach* Oxford University Press, Oxford.

Hall, E. A. H. (1990) *Biosensors* Open University Press, Milton Keynes.

Turner, A. P. F., Karube, I. and Wilson, G. S. (eds) (1987) *Biosensors: Fundamentals and Applications* Oxford University Press, Oxford.

Wise, D. L. (ed.) (1989) *Applied Biosensors* Butterworths, Stoneham, MA.

Journals

Biosensors Elsevier, UK.

IEEE Transactions on Biomedical Engineering IEEE, USA.

Journal of Biomedical Engineering Butterworth Scientific, UK.

Sensors and Actuators B—Chemical Elsevier Sequoia SA, Lausanne.

REFERENCES

Bergveld, P. (1970). *IEEE Trans. Biomed. Eng.*, **BEM-17**, 70–1.

Caras, S. and Janata, J. (1980). *Anal. Chem.*, **52**, 1935.

Cass, A.E., Davis, G., Francis, G.D., Hill, H.A.O., Aston, W.J, Higgins, I.J, Plitkin, E.V., Scott, L.D.L. and Turner, A.P F. (1987). *Anal. Chem.*, **56**, 667.

D'Costa, E.J.D., Higgins, I.J. and Turner, A.P.F. (1986). *Biosensors*, **2**, 71.

Foulds, N.C. and Lowe, C.R. (1988). *Anal. Chem.*, **60**, 2473.

Gotoh, M., Tamiya, E., Seki, A., Shimizu, I. and Karube, I. (1988). *Anal. Lett.*, **21**, 1785.

Gotoh, M., Oda, S., Shimizu, I., Seki, A., Tamiya, E. and Karube, I. (1989), *Sensors and Actuators*, **16**, 55–65.

Hanazato, Y., Nakako, M., Maeda, M. and Shiono, S. (1986). *Proc. 2nd Int. Meeting Chem. Sensors*, 576.

Ikeda, T., Hamada, H. and Senda, M. (1986). *Agric. Biol. Chem.*, **50**, 883.

Karube, I., Tamiya, E., Dicks, J.M. and Gotoh, M. (1986). *Anal. Chim. Acta*, **185**, 195.

Karube, I., Wang, Y., Tamiya, E. and Kawarai, M. (1987). *Anal. Chim. Acta.*, **199**, 93.

Karube, I., Yokoyama, K., Sode, K. and Tamiya, E. (1989). *Anal. Lett.*, **22**, 791.

Kimura, J. Kuriyama, T. and Kawana, Y. (1986). *Sensors and Actuators*, **9**, 373–87.

Kimura, J., Murakami, T., Kuriyama, T. and Karube, I. (1988a). *Sensors and Actuators*, **15**, 435–43.

Kimura, J., Kawana, Y. and Kuriyama, T. (1988b). *Biosensors*, **4**, 41.

Matsuo, T. and Wise, K.D. (1974). *IEEE Trans. Biomed. Eng.*, **BEM-21**, 485.

Matsuoka, H., Tanioka, S. and Karube, I. (1987). *Anal. Lett.*, **20**, 63.

Muramatsu, H., Kajiwara, K., Tamiya, E. and Karube, I. (1986). *Anal. Chim. Acta.*, **188**, 257.

Muramatsu, H., Dicks, J.M., Tamiya, E. and Karube, I. (1987). *Anal. Chem.*, **59**, 2760.

Muramatsu, H., Tamiya, E. and Karube, I. (1988). *Anal. Chim. Acta.*, **215**, 91.

Muramatsu, H., Suda, M., Ataka, T., Seki, A., Tamiya, E. and Karube, I. (1990). *Sensors and Actuators*, **A21-3**, 362–8.

Nakamoto, S., Ito, N., Kuriyama, T., and Kimura, J. (1988). *Sensors and Actuators*, **13**, 165–72.

Shons, A., Dorman, F., Najarian, J. and Biomed, J. (1972), *Mater. Res.*, **6**, 565.

Sibbald, A., Covington, A.K. and Cater, R.F. (1984). *Cli. Chem.*, **30**, 135.

Suzuki, M., Tamiya, E., Kataoka, T., Tokunaga, T. and Karube, I. (1987a). *Clin. Chem.*, **33**, 558.

Suzuki, M., Tamiya, E. and Karube, I. (1987b). *Anal. Lett.*, **20**, 337.

Suzuki, H., Tamiya, E., Karube, I. and Oshima, T. (1988a). *Anal. Lett.*, **21**, 1323.

Suzuki, H., Tamiya, E. and Karube, I. (1988b). *Anal. Chem.*, **60**, 1078.

Suzuki, H., Kojima, N., Sugama, A., Takei, F., Ikegami, K., Tamiya, E. and Karube, I. (1989). *Electroanalysis*, **1**, 305.

Tamiya, E., Karube, I., Kitagawa, Y., Ameyama, M. and Nakashima, K. (1988a). *Anal. Chim. Acta*, **207**, 77.

Tamiya, E., Seki, A., Karube, I., Gotoh, M. and Shimizu, I. (1988b). *Anal. Chim. Acta.*, **215**, 301.

Tamiya, E., Watanabe, N., Matsuoka, H. and Karube, I. (1988c). *Biosensors*, **3**, 139.

Tamiya, E., Karube, I., Hattori, S., Suzuki, M. and Yokoyama, K. (1989). *Sensors and Actuators*, **18**, 297–307.

Tokumoto, J., Kato, T., Tamiya, E. and Karube, I. (1990). Extended abstract of the *First World Congress on Biosensors '90*, 274.

Turner, A.P.F., Karube, I. and Wilson, G. S. (1987). *Biosensors—Fundamentals and Applications*, Oxford University Press.

Umana, M. and Waller, J. (1986). *Anal. Chem.*, **58**, 2979.
Updike, S.J. and Hicks, G.P. (1967). 'The enzyme electrode'. *Nature*, **214**, 986–8.
Yokoyama, K., Sode, K., Tamiya, E. and Karube, I. (1989a). *Anal. Chim. Acta.*, **218**, 137.
Yokoyama, K., Tamiya, E. and Karube, I. (1989b). *Anal. Lett.*, **22**, 2949.

PART **3**

Advances in Design and Manufacturing Techniques

Chapter

41 N. H. HANCOCK

Computer Aided Engineering of Instrumentation

Editorial introduction

The number of measurement systems being set up each day grows at a gigantic rate. Each new application is a custom task because the sensing interface must be created to harmonize with the knowledge gathering need of the application. Users expect their use to be simpler, their performance to be better and their cost to be less. As with all other fields of design of engineering artefacts computer aided design is highly relevant. This chapter shows just how much design support exists for the measurement systems designer and suggests where these tools are heading.

41.1 INTRODUCTION

Use of computers to assist in all manner of tasks across the breadth of engineering has become commonplace during the last decade. This radical change has been fuelled by the increased ease of access to computers by engineers, in particular via the low-cost 'personal computer' and, more recently, the much more powerful 'engineering workstation'. The potential advantages including speed, accuracy, record keeping and the possible reduction in human errors are well appreciated: in addition, engineers simply enjoy using computers!

Three main component parts of the engineering process are design, analysis and manufacture and hence there exists *computer aided design (CAD)*, *computer aided analysis (CAA)* and *computer aided manufacture (CAM)*. Other essential dimensions of engineering including specification, planning, financial analysis and control,

Handbook of Measurement Science, Volume 3
Edited by P. H. Sydenham and R. Thorn
© 1992 John Wiley & Sons Ltd

quality assurance and documentation may be similarly computer aided, to a greater or lesser extent. The term *computer aided engineering (CAE)* encompasses all of these facilities but is now generally understood to imply also integration of and interaction between them. (In production engineering the term computer integrated manufacture (CIM) is commonly used to emphasize this aspect.)

Unfortunately, and perhaps unavoidably, the application of computers to engineering tasks has been patchy and largely unstructured. Programs, and suites of programs, (software 'packages' or 'tools') have been developed on an *ad hoc* basis, according to need and according to the capabilities of the computers (the hardware) and the languages etc. (the software) available at that moment. As a result integration of these various tools into coherent and convenient ('user-friendly') packages remains a problem. Hence, while some CAD tools are well developed (e.g. mechanical design, printed circuit board layout) and also some general-purpose analytical tools (e.g. finite element analysis), it is clear that CAE in general is still at an early stage of development.

CAE applied to instruments

In principle at least the CAE of an individual instrument, or of a complete instrumentation system, should be no different from the CAE of any other device or product: a specification would be drawn up, an appropriate technology chosen, a design produced and manufacture undertaken. Hence a large proportion of available CAE tools (CAD, CAA and CAM) are applicable to and very useful across the breadth of instrument and instrumentation system design and manufacture—these are reviewed in Section 41.4.

Some CAE tools have been produced specifically to aid the creation of measurement systems. The need for these tools has arisen for two main reasons. The first is the cross-disciplinary nature of the design skills and techniques needed to produce almost any modern measuring instrument (for example, see Figure 41.18). This requires the use of a range of different and essentially non-compatible CAE tools for each of the different technologies involved. Whilst this is workable for the professional instrument designer it is unsatisfactory for the occasional or non-specialist user. Hence some integrated packages have been produced with features tailored to the design of particular classes of instrument.

The second reason arises from the need to adequately define the user's measurement requirement and instrument specification. This task is often far from straightforward. CAE tools have been written to assist the thinking and communication processes involved in conceiving and then specifying the instrument that will properly meet the user's measurement requirement.

Both of these types of package employ not only the well-developed calculation and modelling techniques traditionally associated with engineering design but also the newly emerging computing techniques of knowledge engineering. In consequence most of these tools are still in developmental form. They are examined in Section 41.5.

A computer provides capabilities for organizing, manipulating and storing information. Combined with the practice of engineering in the form of software 'tools' a range of advantages may be realized. There are also, of course, potential disadvantages. The possible requirements of a particular CAE system are next considered in the light of the service the user might require from the particular system.

41.2 NATURE AND SCOPE OF CAE

41.2.1 User Considerations

There are essentially two types of user who can derive benefit from the CAE of instruments. These are first the professional instrument designer and manufacturer, whose task it is to create new, non-standard and improved instruments; and second the instrument user who wishes to accomplish a particular measurement task but is not professionally skilled in instrumentation. From the point of view of the product (the desired measurement system) these may be labelled the supplier's domain and the user's domain respectively as set out in Peuscher (1983).

The CAE system appropriate to service user needs in each of these domains is likely to be very different, even in the same narrow area of application.

CAE for instrument suppliers

Faced with a request to supply or build a new instrument the instrumentation professional would undertake the task as a number of interacting stages which would commonly proceed as follows:

- confirmation of the actual measurement required and agreement on a firm specification;

- design or selection of appropriate sensing principle(s) and/or actual sensors and transducers;

- design of appropriate signal processing and its implementation, normally by means of electronic circuitry with embedded software;

- simulation of the system performance, in parts or as a whole;

- final design and manufacture of each part of the measurement system;

- assembly, test and calibration; and

- documentation, including calibration and maintenance procedures, etc.

CAE software is available to assist in many of these separate tasks and these tools are briefly surveyed in Section 41.4. Clearly these tools span a range of technologies and techniques and hence they are usually independent. In all cases the instrumentation expertise must be supplied by the user and the success of the design depends critically on this expertise.

CAE for instrument users

In contrast to the above the instrument user often does not possess the instrumentation expertise required and hence any CAE system must provide this skill if it is to be of use. Nor is it usually convenient to use the range of separate software tools necessary for the single task of obtaining one measurement system.

Clearly it will be very difficult, if not impossible, to incorporate the range of knowledge, experience and skill of the measurement professional into a CAE system.

'Customized' instrumentation

It may be observed that much of the demand for measurement system design arises because off-the-shelf units are not quite suitable. Hence the measurement requirement may be seen as non-standard rather than original and may often be achieved by 'customizing' an existing design.

Again professional measurement knowledge, experience and skill is required. But for a given (strictly limited) class of measurements the task of adequately encapsulating the required part of this knowledge, experience and skill becomes more feasible. (Such software packages have been produced and these are reviewed in Section 41.4.4.)

41.2.2   Requirements for a CAE System

According to the user requirements a CAE system must have a range of features and capabilities which will include most or all of the following.

- *Information storage.* This may be in a number of forms including:
  — numeric (sizes, part numbers, etc.),
  — symbolic (labels, etc.);
  — graphical (schematics and drawings, etc.);
  — heuristic (practical limitations, 'rules of thumb', etc.); or
  — conceptual (which may be in the form of natural language text).

- *Calculation* according to set procedures (algorithms).

- *Decision making and logical processing* based not only on numerical results but also on heuristic information.

- *Modelling*, which is essentially a structured combination of the above features. It may also be required to formulate and construct models as well as make use of existing models.

● *An appropriate human interface.* This is often referred to as I/O (input/output) interface and the hardware required may include keyboard, mouse, digitizing tablet, with a screen for text, a separate, high-resolution screen for graphics, plotters and printers. Adequate software to support these facilities is obviously essential.

● *External interfaces.* The CAE system will often be required to communicate with other computer-based systems, for example those involved with management, financial and project control and also those controlling production. A comprehensive CAE system will usually have these facilities integrated within it.

● *Expertise.* To some extent the CAE system (or at least the CAD part of the system) may be required to display direct design capability. Quasi-human capability is therefore expected (without, of course, the human propensity for error). Whilst this is a particularly demanding requirement such systems are now becoming available (see Section 41.5).

41.2.3 Some Advantages and Benefits

These may be summarized as follows.

● *Engineering time, and hence direct costs are reduced.* This is perhaps the most significant advantage of and justification for CAE. These saving may be quite considerable: Fox (1986), for example, reports a saving of two thirds of overall engineering time for the instrumentation of a large chemical plant.

● *Structure and discipline is imposed on the engineering process.* Whilst this can be helpful to the individual it is a particular advantage where large teams are involved.

● *Reporting of the status of the work is rapid, accurate and convenient.* Clearly this will aid management of the overall project and may be made virtually automatic.

● *Documentation of the design procedure is fully available.* The job of producing complete and error-free final documentation is tedious and often much resented by engineering staff. This burden is greatly eased, particularly when convenient word-processing with fully annotated diagrams and graphs ('report-ready' output) is available directly from the computer system.

● *Comprehensive filing and archival systems may be conveniently established.*

- *The physical space required for information storage is dramatically reduced.* This also represents real cost savings and encourages the retention of design notes and alternatives which might have to be scrapped for lack of space.

- *Previous designs are always available for easy replication, modification and re-use.* Repetition of design work can thus be avoided and this, of course, includes previously abandoned or inappropriate designs. Major cost saving can result. In consumer type electronic products some 60% of the electronic files are reused in a subsequent product.

- *Standardization of designs and procedures is encouraged.* The foregoing re-use of designs and designs shared between team members on the same CAE system often results in *de facto* standards emerging.

- *Engineering experience accumulates within the CAE system.* Hence this valuable resource is not lost when staff leave the project.

- *Project familiarization may be accelerated.* New staff joining the project will often find it easier to appreciate the current status of the project via the CAE system than via reports and drawings.

41.2.4   Some Disadvantages and Difficulties

The advantages and benefits outlined above do not come without cost and like any other machine the computer and its software have deficiencies and limitations. CAE systems are similarly restricted in capability. The possible disadvantages, difficulties and costs associated with the implementation and use of CAE may be summarized as follows.

- *Installation and maintenance costs may be substantial.* The purchase of hardware and software, often with maintenance contracts to ensure serviceability, may represent a substantial overhead.

  With respect to CAD the major cost items are the high-resolution screens and plotters required for full mechanical design, analysis and draughting (Section 41.4.2). However, the mechanical design required in instruments can often be adequately serviced by more modest facilities based on the larger personal computer which can represent a major cost saving.

  Similarly the expense of fully computer-controlled production facilities to implement CAM may be difficult to justify.

- *The hardware components and software packages required may be difficult or impossible to integrate.* This problem arises particularly in the CAE of instrumentation due to the range of different technologies involved in the design and manufacture of even quite modest measuring systems.

● *Design and production may be halted by the breakdown of the CAE system.* It must be recognized that CAE based on a single, central computer system is particularly vulnerable. Although this architecture is becoming less common in favour of a number of networked computers ('workstations') it is still common to use a central ('fileserver') machine to control the bulk storage of files, without which individual machines may not be able to perform much useful work.

● *Introduction of CAE may involve considerable staff retraining.* Although this represents an additional installation cost it may also be seen by staff as a vote of confidence in the future.

● *Staff may perceive their long-practised manual skills to be downgraded or valueless.* This is one of a range of morale problems that may occur, particularly with older, highly skilled staff (see, for example, Francis, 1983). This is a potentially serious problem as the instrumentation industry has traditionally relied on the skill of experienced staff to maintain quality in both design and manufacture (Goulding, 1988) and hostility toward CAE can negate many of the possible benefits.

This potential problem is part of a wider personnel issue caused by the major changes to engineering design practice as a result of computer-based tools (discussed in Section 41.4.2). These changes impose a fundamental revision of the roles of the engineer and the technician, particularly in the practice of electronic engineering (Goulding, 1988).

In addition to these general considerations the nature of the instrumentation industry appears to militate against the rapid development of a specific CAE of instruments. Both the production volume, the rate of sale and, with only a few exceptions, the unit value of a particular instrument model is likely to be small in comparison with many other industrial or consumer products. Hence instrument designers and manufacturers do not appear to represent a large potential market to fund the cost of development of comprehensive CAE installations. As a result it is not surprising to find that specific CAE tools to support the design of instruments is in its infancy.

41.2.5 Discussion

The perceived advantages of CAE will be to a large extent permanent and enhanced by cheaper (and improved) computer hardware and improved software tools.

In contrast it appears that the disadvantages are largely due to the present costs, technical limitations (hardware and particularly software) and personnel implications. It appears reasonable to expect these to diminish, perhaps quite rapidly, in future years.

41.3 COMPUTER AIDED DESIGN

41.3.1 The Design of Instruments

Computer aided design (CAD) is central to the concept of CAE and to a large extent dominates today's CAE installations. The determination and design of useful CAD tools, appropriate to the CAE of measurement systems, requires firstly a review of the theory, methodology and practice of design and secondly an appreciation of the capabilities (particularly with regard to software) of computer based systems.

The anatomy of a simple, modern instrument may be set out as illustrated in Figure 41.1. A number of observations may be made about the design requirements for such a device.

● It is likely that a range of technologies will need to be employed to realize the instrument. These will depend primarily, of course, on the part of the physical world to which it is to be connected, the particular phenomenon being sensed and the transducing techniques deemed appropriate to yield the measurement information required. In addition electronic technology will (almost invariably) be required to process the information.

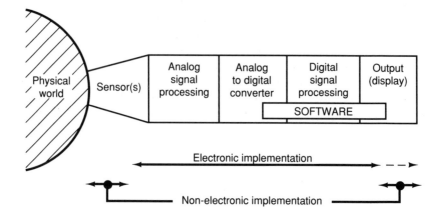

Figure 41.1   The typical anatomy of a modern measuring instrument

- In common with the great majority of other areas of engineering design the processes employed and functions implemented within the instrument are essentially algorithmic, that is based on formulae resulting from mathematical models of the component parts of the instrument. The design of these parts is therefore likely to be amenable to computer aid.

- Both analog (continuous) and digital (discrete, quantized and sampled) representations of the measurement information will be employed in different parts of the instrument.

- To a large extent the functionality and performance of the device may be determined in software. The design of software therefore assumes corresponding importance.

- In contrast, functionality at the sensing interface cannot be realized in software—a physical sensor must be produced and possess appropriate properties to convey the measurement information required.

This combination of electrical and non-electrical technologies, analog and digital signals and embedded software usually required even in a simple, basic instrument can be seen to distinguish instrumentation from the broad range of other engineering products. This has major implications for the establishment of appropriate CAD and CAE systems (Barwicz and Morawski, 1989).

41.3.2 The Place of the Computer in Design

The potential advantages to the user and the present limitations of computer application have been set out in Sections 41.2.3 and 41.2.4 respectively. However, realization of these advantages (and acceptance of the limitations) can profoundly affect both the process and the practice of design.

Figure 41.2 sets out the traditional design process followed in the creation of an instrument (This is, of course, essentially the same procedure as that used for the creation of a great majority of electronic, mechanical and other products.)

Computer-based *analysis tools* can greatly aid the evaluation process within the design loop shown. However, where computer-based *simulation tools* are available the need for prototype construction and physical testing may be greatly diminished. This change can yield major savings in convenience, design time and cost. However, as has been discussed in Section 41.2.4, it may also have a profound effect on the requirements for, and activities of, personnel involved in the design process.

In cases where prototyping is essentially impossible (e.g. integrated circuit design) simulation techniques have to be relied upon entirely until the final product is produced. Similarly in many other areas simulation tools are becoming so well proven that prototyping, whether inconvenient or not, is being dispensed with entirely (Goulding, 1988).

However, design is still generally regarded as an essentially human activity. Hence the appropriate use of the computer is strongly influenced by its abilities (and lack of abilities) relative to the human designer. Table 41.1 sets out the traditional comparison. It is often used to determine which aspects of the design

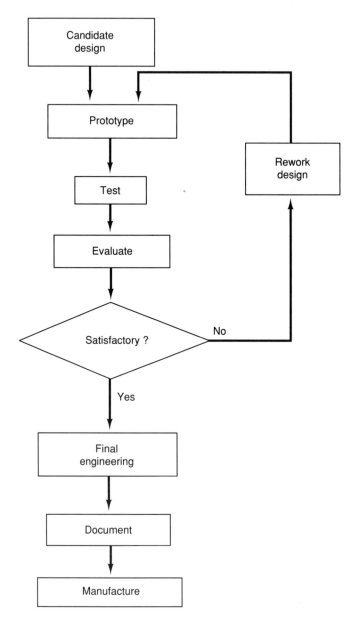

Figure 41.2 Traditional design methodology

process are best performed by the computer and which by the design engineer. In the major areas of CAD application such as mechanical design (see, for example, Besant and Lui, 1986) this view has dominated the methods of the application of the computer.

Table 41.1   A traditional comparison of the characteristics of man and computer (reproduced from Besant and Lui, 1986, Chapter 1). Area where the capabilities of computer are commonly perceived to be limited or deficient are shown boxed.

	Man	Computer
1. Method of logic and reasoning	Intuitive by experience, imagination, and judgement	Systematic and stylized
2. Level of intelligence	Learns rapidly but sequential. Unreliable intelligence	Little learning capability but reliable level of intelligence
3. Method of information input	Large amounts of input at one time by sight or hearing	Sequential stylized input
4. Method of information output	Slow sequential output by speach or manual actions	Rapid stylized sequential output by the equivalent of manual actions
5. Organization of information	Informal and intuitive	Formal and detailed
6. Effort involved in organizing information	Small	Large
7. Storage of detailed information	Small capacity, highly time dependent	Large capacity, time independent
8. Tolerance for repetitious and mundane work	Poor	Excellent
9. Ability to extract significant information	Good	Poor
10. Production of errors	Frequent	Rare
11. Tolerance for erroneous information	Good intuitive correction of errors	Highly intolerant
12. Method of error detection	Intuitive	Systematic
13. Method of editing information	Easy and instantaneous	Difficult and involved
14. Analysis capabilities	Good intuitive analysis, poor numerical analysis ability	No intuitive analysis, good numerical analysis ability

However, recent research in a range of fields has demonstrated that the perceived deficiencies of the computer (boxed in Table 41.1) are very much open to challenge.

The techniques associated with artificial intelligence (AI) in particular have yielded packages which can provide a computer with (following the order of Table 41.1):

- self-learning capability;

- assistance in the organizing of information;

- ability to analyse natural language input and extract information;

- some tolerance of erroneous and inconsistent information; and

- simple and convenient editing.

Only with regard to the final deficiency, that of intuitive analysis, does it not appear likely that computers can provide significant assistance.

41.3.3   Theory and Methodology of Design

It has been shown that the process of instrument design may be usefully and formally structured (Bosman, 1978; Sydenham, 1985a,b) and therefore might be to some extent automated. Hence a *theory of design* is required. Work in this area has a substantial literature (see, for example, Warman and Yoshikawa, 1980; Finkelstein and Finkelstein, 1983) which assumes increasing relevance with the expanding role for the computer in the design process. Warman (1990), for example, has recently set out the objectives of the theory of design as:

(a)   to clarify the human ability to design in a scientific way;
(b)   to produce practical, useful knowledge about design methodology; and
(c)   to frame design knowledge in a certain way suitable for its implementation on a computer.

Clearly this work has real practical importance for the user of CAD systems: for the designer of CAD systems these objectives will be key guidelines (Warman, 1990).

Closely related to the theory of design is the science of methods of design or *methodology of design*. This field has been reviewed by Finkelstein and Finkelstein (1983) and Barwicz and Morawski (1989) with particular reference to the design of measuring systems.

As an example a typical design methodology incorporated within a CAD package is set out in Figure 41.3.

In recent years several design methodologies have been proposed and (to some extent) implemented. Van Biesen *et al.* (1989) and Rudolph *et al.* (1990) are significant. At a greater level of generality Barwicz and Morawski (1989) present the design methodology illustrated in Figure 41.4. Here the design task is conceived as being decomposed into smaller and smaller entities until each entity

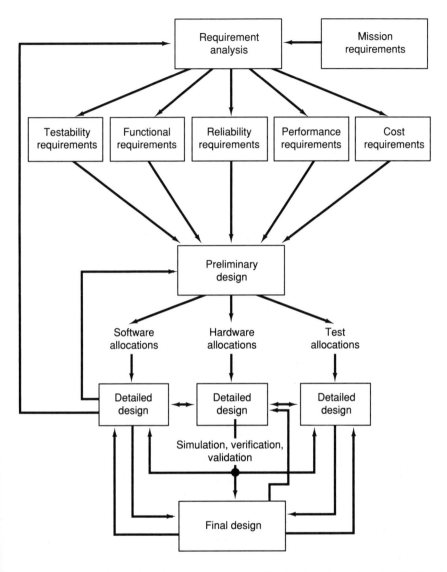

Figure 41.3   The design methodology followed by the 'Test Engineer's Assistant', a CAD tool to facilitate the design of adequate testability within digital electronic systems (Hallenbeck *et al., 1989)*

can be synthesized directly. The concept of successive decomposition dates back to philosophical writings on the structure of knowledge published in the 16th century. Clearly such an approach can form the basis of computer aided and automated design procedures if adequate (software) libraries can be provided for each of the four thesauri specified in the figure.

41.3.4   Knowledge Engineering

Although CAD is now well established in many fields of engineering Warman (1990) states:

Much of commercially available CAD is still at the drafting end of design and many of today's systems have, at their core, code and ideas of some 20 years ago.

And despite the major investment (by both vendors and users) in existing packages there is a growing recognition of their deficiencies, shortcomings which are implicit in the view of computer capabilities set out in Table 41.1 given earlier.

Knowledge engineering is a discipline concerned fundamentally with the capture, representation and manipulation of knowledge within a computer. This may be seen in contrast to the view which regards computer capability as that of acquiring and processing data, which is normally numeric (but may also be symbolic).

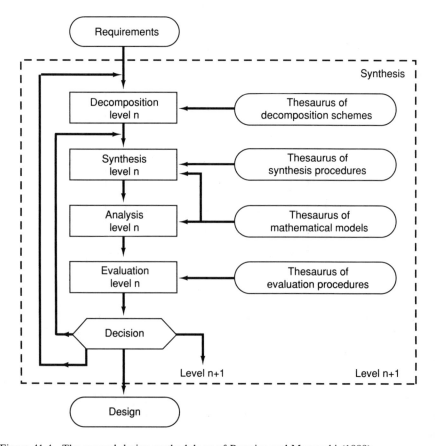

Figure 41.4   The general design methodology of Barwicz and Morawski (1989)

Knowledge engineering uses the same techniques and ideas as those used in the pursuit of *artificial intelligence* (AI) but the former is often preferred because it avoids the philosophical dabate about the nature of intelligence. Hence 'AI techniques' are not distinguished from 'knowledge engineering techniques'.

Devices and systems constructed from a computer and using knowledge engineering techniques are commonly labelled *knowledge-based systems* (KBS). Early versions were often labelled intelligent knowledge-based systems (IKBS).

The representation of knowledge in a KBS is normally by means of *frames*. These are data structures capable of representing not only the information itself but also the relationships between it and other items of knowledge. A frame will also contain *attributes* associated with the item of knowledge and frames will be organized in a *hierarchy* with more general knowledge (e.g. concerning the attributes of classes of objects) set higher up than more specific, detailed knowledge. Bench-Capon (1990), for example, provides a recent introduction to the theory and programming of KBS.

Mirza *et al.* (1990) have applied this structure to the classification of instrument subsystems and a grouping according to the common functional characteristics of each is presented in Figure 41.5. This provides the knowledge base for a KBS currently under development (by Mirza *et al.*, 1990) for the computer-aided generation of design concepts for instrument systems.

41.3.5 Expert Systems

One major area of application of KBS (or AI techniques) is in the construction of expert systems, so-called because they seek to mimic the thought processes, and therefore (hopefully) the performance, of the human expert. Hence each will have a particular and usually limited domain of expertise.

The potential advantage of the expert system lies in areas where knowledge, procedures and techniques cannot be readily expressed in the form of data and algorithms. Many aspects of design fall into this category including much associated with the practical design of instruments.

The success (or otherwise) of an expert system depends critically on two things: firstly the adequacy and completeness of the knowledge obtained from the (human) expert, and secondly an appropriate encapsulation of this knowledge within the computer. The first of these remains a key problem area requiring (at present) innovative solutions on a case-by-case basis (see, for example, Edmonds, 1989). The second has benefited greatly from the development of knowledge engineering.

The core of an expert system is its *knowledge base*. This will contain the knowledge acquired from the appropriate human expert in the form of related facts and rules and usually stored in a frame structure. Hence expert systems are also known as *rule-based frame systems (RBFS)* (see, for example, Barber *et al.*,

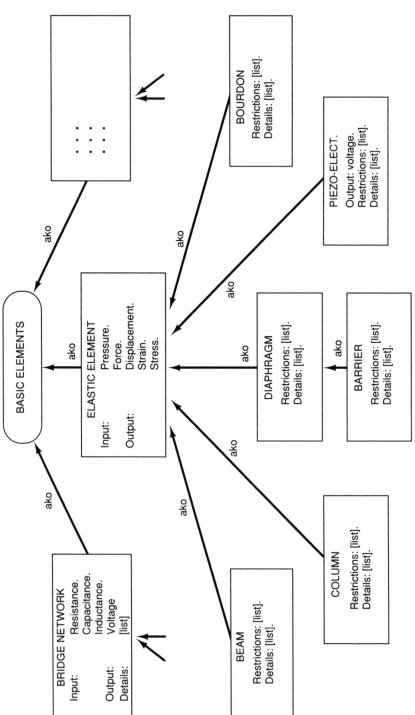

Figure 41.5   The organization of a knowledge base by means of grouping together instrument subsystems according to their common functional characteristics (ako = a kind of) (Mirza *et al.*, 1990)

1988). These rules will often be *heuristic* rather than determinate in nature, that is practical 'rules of thumb' appropriate to the domain of expertise. Bench-Capon (1990), Chapter 9, for example, provides an introduction.

The expert system 'solves' a particular problem by means of an *inference engine* which draws upon and manipulates the knowledge base. Rules contained in the knowledge base are used (or 'fired') in sequence using forward- or backward-chaining as appropriate to reach a logical conclusion.

Examples of expert system implementations are numerous. In the area of instrumentation Barwicz and Morawski (1989) have formulated requirements for a system to provide the CAD of general measurement systems, and its structure is illustrated in Figure 41.6.

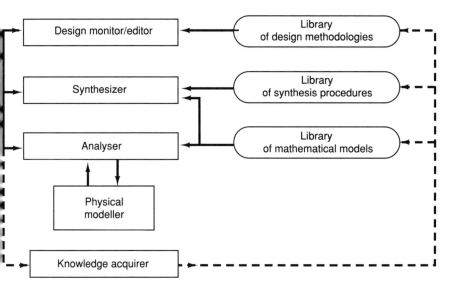

Figure 41.6 The general structure of an expert system for CAD of measuring systems proposed by Barwicz and Morawski (1989)

Freiling *et al.* (1986) have provided a theoretical analysis of an inference engine used for electronic instrument troubleshooting: other expert systems are considered as specific CAE tools in Section 41.5.

41.3.6 Object-oriented Programming

A recent theoretical concept which offers major potential advantages for CAD in general is that of object-oriented programming (OOP).

The essential feature of *object-orientated systems* (usually abbreviated as *OOPS*) is the representation of every item of information and knowledge as an 'object', that is an entity which (as in the normal meaning of the word) has properties of 'being' or 'doing'. Each object is a block of (program) code which may contain simply data or may include a set of variables and procedures. Objects may then be classified (within a given OOPS) according to the properties they have in common.

Objects communicate and relate to one another, and to the external environment, by passing messages. These messages specify the type of interaction required and may be commands or queries: the response to a message is itself an object. Hence an OOPS has its functionality (message structure) totally separated from its implementation (objects). This results in several major advantages particularly when implemented in large software structures (see, for example, Warman, 1990).

Perhaps the most significant advantage stems from the inherent modularity of object-oriented structures. Information is represented as a set of related but independant items (objects) which may be labelled an 'extensional' representation. In contrast non-OOPS use 'intentional' representation, wherein strongly related items are represented together as a single entity (e.g. a CAD drawing of an assembly of mechanical parts). In consequence modification of an OOPS simply involves the addition (or deletion) of further objects rather than the fundamental revision of the entire structure. Clearly this is a particular advantage for large software structures.

The field of OOPS is rapidly developing an extensive literature. Rudolph *et al.* (1990) have applied OOPS techniques to the automation of electronic measurement system design. Other contributions significant to the CAD of instruments (and elsewhere) include a discussion of the major advantages for the design of (mechanical) CAD systems (Warman, 1990), overcoming knowledge representational problems in expert system shells (Leung and Wong, 1990) and advantages in the analysis and specification of software systems (Kurtz *et al.*, 1989).

41.4   CURRENT GENERAL-PURPOSE CAE TOOLS

This section provides a brief overview of some of the software tools or classes of tools currently available to the instrument designer and manufacturer. They are general purpose in the sense that none were created with the specific purpose of assisting instrumentation design—those tools which have been so created are considered in more depth in Section 41.5. Also covered are some software packages that might be useful to the instrument owner or user in such areas as testing, training and maintenance.

Instrumentation design and operation spans the whole range of engineering practice and so it is likely that a large proportion of the software packages for engineering design presently available will have relevance to the design and manufacture of measuring instruments. As these tools are of such general application throughout commerce and engineering only a cursory mention need be made here.

Although computer aided engineering software has been readily available for less than two decades the range of packages applicable to the CAE of instruments is already very large and continues to increase. In this section the overview of software cannot attempt to be exhaustive nor remain up to date. Mention of a particular commercial product should not be taken as an endorsement nor any omission as a criticism.

41.4.1 Mathematics Software

Mathematics packages are available for formulating and solving the equations which arise during any stage of the design process. These tools are conceived as alternatives to the use of a general-purpose programming language such as FOR-TRAN and seek to provide a much more convenient and rapid completion of the task. MathCAD and Matlab are examples of numerical equation solvers which will solve families of linear and quadratic equations including derivative and integral terms and provide two-variable or three-variable graphical output. In contrast Chico Solver and Derive are examples of mathematics packages which solve equations by symbolic manipulation and to this extent may said to be more general purpose. However, at present, neither approach is free of limitations nor the software free of potential difficulties for the occasional user (Hines, 1990).

Much of engineering design results in families of differential equations constrained by a set of boundary conditions. Examples are the stress/deflection relationship of a mechanical transducer, the heat flow in the sheath (the thermowell) of a temperature sensor and the electromagnetic field distribution within a capacitive transducer. From the mathematical point of view these may be considered together as *boundary value problems*. There are three numerical techniques commonly available within software packages which can produce accurate and reliable solutions to boundary value problems and to initial value problems which may be regarded as having time-varying boundary values in time.

The *finite difference method (FDM)* involves the choice of an appropriate *grid* of points within the space of the object or region. Simple algebraic equations derived from the differential or integral equations can then describe the difference between values at each point on the grid and may be solved directly or iteratively.

The *finite element method (FEM)* involves the choice of a *mesh* which divides the object or region into an assembly of simpler, interacting objects or subregions known as *elements*. Approximate but adequate numerical calculation can then be performed on each of these individual elements to produce the overall solution. This method has found major application in mechanical engineering stress analysis despite the limitation that complex objects or regions will usually require a three-dimensional mesh with a very large number of elements and hence a powerful computer.

The *boundary element method (BEM)* utilizes the fact that integral equations relating to the space within the object or region may be transformed into equations existing solely at the boundary of the object or region. This has the immediate advantage of transforming, for example, a three-dimensional problem into a two-dimensional problem and may greatly reduce the amount of computation required to yield an adequate solution.

A useful comparison between the methods is provided by Cookson and El-Zafrany (1987). Application of these methods to the modelling of electromagnetic fields involves the solution of Maxwell's equations, usually in three dimensions, and in a recent review Cendes (1990) reports that the necessary algorithms and automatic grid and mesh generation routines have been refined to the extent that useful simulations can now be run on a modest personal computer. In contrast the application of FEM in mechanical stress analysis usually requires more substantial computing resources due to the complexity of the object. Mechanical design is now considered in more detail.

41.4.2 Mechanical Design Software

Across the breadth of engineering, perhaps the area of greatest impact of CAE techniques is in the CAD (computer aided design) and increasingly the CAM (computer aided manufacture) of mechanical engineering parts and assemblies.

Mechanical design still plays a major role in the creation of instruments and instrumentation systems. The techniques of fine mechanics involving gears and springs, levers and linkages are now only seldom used for signal processing and have diminishing application in displays (pointers, meter movements, etc.) and in data logging (chart recorders, etc.). However, all instruments still have a case and/or a chassis which must meet shape and weight restrictions, installation and environmental requirements and mechanical strength specifications (shock, vibration, etc.).

Likewise all transducers have mechanical installation requirements and there are areas in which the mechanical design of the transducer is critical. Obvious examples are the transducers of mechanical quantities (mass, force and pressure, strain and displacement, acceleration and vibration, flow, etc.). Mechanical CAD techniques are, of course, applicable independently of scale and are also being applied to the micromechanics of quartz and silicon transducers at the nanometer scale—see, for example, Grattan (1990). Electrical measurements or signal processing involving radio and microwave frequencies also have specific mechanical design requirements.

Graphical capabilities

Central to the practice of mechanical design is the drawing board. Use of a computer to aid design in any role beyond that of a calculator depends principally on the extent to which the computer display may be used as a drawing board. Graphics capabilities of computer systems, large and small, have developed greatly in recent years and now provide a comprehensive data entry, manipulation, storage and display facilities based on industry-wide standards such as the Graphics Kernel System (GKS)—see for example, Besant and Lui (1986).

The display device is still, almost universally, the cathode ray tube (CRT), and a high spatial resolution (typically 1024 by 1024 pixels) is required if it is to provide the detail to approach that of the traditional pen-and-paper drawing. A 'window' or 'zoom' is provided whereby only a small portion of the complete drawing is displayed over the whole screen and to a large extent this compensates for the poorer spatial resolution. Substantial memory is required to store each drawing, typically one megabyte or more, and hence a reasonably powerful computer is needed to manipulate the drawings. A high-resolution plotter is then required to produce the final drawing on paper and the result is often better, both in accuracy and presentation, than the equivalent manual drawing. In consequence full professional computer hardware for mechanical CAD is expensive and may not, at present, be cost-effective for the casual or small scale user.

Modestly priced CAD packages (for example, AUTOCAD) are available for the larger personal computers with relatively high-resolution screens. These make heavy use of the 'window' facility and are, at present, generally restricted to computer aided drafting applications. The 'engineering workstation' size of computer networked to a bulk storage facility ('fileserver') is a common configuration for full mechanical CAD.

Computer-aided drafting

The lowest level of computer aid to mechanical design is two-dimensional computer aided draughting (also, unhelpfully perhaps, referred to as CAD). Software for this purpose serves to assist and automate the production and the maintenance of standard engineering drawings. Computer-aided draughting has several major advantages with respect to conventional pen-and-paper draughting. The ability to replicate or relocate parts of a drawing can greatly speed up the production of both new and modified drawings. Automatic dimensioning can help avoid transcription errors and inconsistencies between drawings. Storage on disk rather than in a drawing cabinet can save valuable space. The requirement for handskills in drawing are also eliminated.

An enhancement of two-dimensional draughting is $2\frac{1}{2}$-D draughting, in which the extra 'half dimension' refers to height or thickness information stored with each drawing. This enables a limited three-dimensional view of the object to be presented, usually as an orthogonal view—see, for example, Haigh (1985).

Computer aided design

To use the full power of computer aided design requires software to form a three-dimensional geometrical model of each mechanical part or assembly within the computer. There are three main types of model (see, for example, Besant and Lui, 1986). The simplest is the wireframe model, which provides a transparent outline of the part as if it were made from a frame of thin sticks or wires (see Figure 41.7(a)). Whilst this is helpful to visualize the part, the view may be

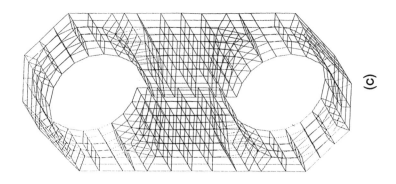

(c)

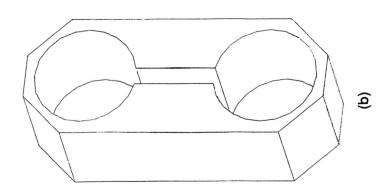

(b)

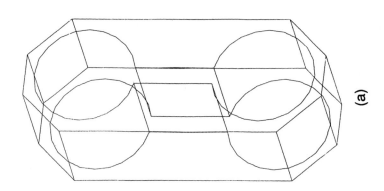

(a)

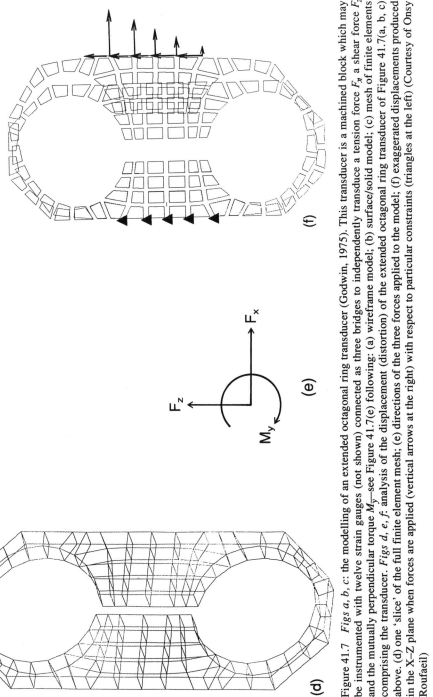

Figure 41.7 *Figs a, b, c*: the modelling of an extended octagonal ring transducer (Godwin, 1975). This transducer is a machined block which may be instrumented with twelve strain gauges (not shown) connected as three bridges to independently transduce a tension force F_z, a shear force F_x, and the mutually perpendicular torque M_y—see Figure 41.7(e) following: (a) wireframe model; (b) surface/solid model; (c) mesh of finite elements comprising the transducer. *Figs d, e, f*: analysis of the displacement (distortion) of the extended octagonal ring transducer of Figure 41.7(a, b, c) above. (d) one 'slice' of the full finite element mesh; (e) directions of the three forces applied to the model; (f) exaggerated displacements produced in the X–Z plane when forces are applied (vertical arrows at the right) with respect to particular constraints (triangles at the left) (Courtesy of Onsy Roufaeil)

ambiguous (for example, front and back are not distinguished) and no geometrical properties can be automatically deduced.

Much more useful is the *surface model* in which each edge, face and surface of the object is stored as a mesh of coordinate points. Hence surface intersections and areas may be automatically calculated. More sophisticated versions will remove hidden lines and edges and can shade the visible surfaces according to angle so that the object appears as the observer would actually see it under illumination. The view provided may be simply orthogonal (with parallel edges shown parallel) or may be corrected for perspective. The object can be rotated such that the viewpoint may be changed at will.

Whilst the surface model is adequate to take sections, produce standard two dimensional drawings and derive instructions for machining and manufacture analysis of the mechanical properties of the object is limited. A full *solid model* comprises a complete description of the object in terms of the coordinate space which it occupies. Hence in addition to the surface properties, the *mass properties* of the object may be automatically calculated, namely mass, centre of gravity moments and products of inertia. These quantities are essential for vibration and other dynamic analyses.

Figure 41.7(a), (b) and (c) illustrates a wireframe, surface/solid model and the preparation for analysis (respectively) for a simple instrumentation transducer.

Computer aided mechanical analysis

The value of computer aided mechanical design is greatly enhanced if the performance of the object may be determined without the need to manufacture and test each design. Whilst the static properties (mass, volume, centre of gravity, etc.) may be derived from the solid model the calculation of mechanical strength, deflections and distortion under the action of external forces, modes of vibration and natural frequencies requires further modelling. Direct calculation of these properties is possible for only the simplest geometrical shapes, and of the various techniques available for more complex geometries the general purpose *finite element method (FEM)* is now widely available as part of, or as a companion program to, most CAD packages. (FEM is considered in Section 41.4.1: see also Haigh, 1985.)

Successful application of finite element (FE) analysis depends critically on an appropriate shape, size and arrangement of the interacting elements into which the volume of the object is subdivided. The surfaces separating the elements constitute a *mesh* and the choice of this mesh is very much one of engineering judgement. Hence as part of the interface to the FE package the more useful CAD systems provide an *automatic mesh generation* facility which has embedded rules of good mesh geometries for particular mechanical configurations and may vary according to the directions of the stresses to be applied to the object. The mesh is then displayed in elemental form and the designer can add, delete or modify elements as desired. The facility also exists to undertake a partial or complete manual layout of the mesh for very unusual geometries.

Figure 41.7(c) illustrates the mesh required for the analysis of a relatively simple object. The fairly coarse mesh shown (which would normally be inadequate for instrumentation purposes because it will yield only low precision results) contains 304 elements. Figure 41.7(d) shows one 'slice' of this mesh containing 76 elements.

The outputs of a finite element analysis are also best displayed graphically. For mechanical analysis, forces and constraints are applied and the distortion of the shape may be displayed directly, for example the 'exploded view' of Figure 41.7(f). Similarly, if required, the magnitudes and concentrations of stress in the model may be shown by contouring, differential shading or colouring of sections through the model.

41.4.3 Electrical and Electronic Design Software

Within the great majority of today's measurement systems signals are represented electrically and processed electronically. It follows that electronic design, and increasingly digital electronic design, comprises a very large proportion of instrumentation design effort.

A very wide range of software is available to aid the electronic design engineer. Again these tools and packages provide not only aids to the calculation and analysis involved in design but also simulation of the performance of the design under chosen circumstances and aids for the management of the overall design process. Such tools seek to entirely replace the pencil, paper, hand-held calculator and even the data book on the designer's desk, and various stages of 'breadboard' and prototype on the designer's bench. It is no exaggeration that the complete computer based design of many electronic systems from conception to production is now both possible and affordable: Kaplan (1990) provides a recent, practical overview. Jain and Harpas (1988) covers PC forms. All of these tools and packages are of value in the design of instruments according to circumstance.

A wide range of *analog circuit simulation packages* is available and many require only a modest personal computer. One of the most commonly used is SPICE (simulation program with integrated circuit emphasis) which is available in various forms (PSpice, HSpice, InSpice, SpicePlus and others)—see, for example, Tuinenga (1988). The facilities provided typically include d.c. properties, frequency response and transfer function, transient response, noise analysis, distortion analysis and large signal effects. Some of these packages have a 'probe' or 'instrumentation interface' facility which can provide simulated oscilloscope and spectrum analyser displays of the simulated performance of the circuit.

Electromagnetic interactions causing interference (EMI, RFI) are common in instrumentation systems due to the very low signal levels produced by most sensors and the physical separation, often many metres, between sensor and amplifier. Associated design problems in low-level analogue circuits include grounding, shielding and guarding, cabling and p.c. layout, interference from adjacent digital circuits and power supply distribution. Software is becoming available to analyse these and other potential electromagnetic compatibility/control (EMC) problems and to evaluate possible solution schemes (for example, Program 7700 of Interference Control Technologies Inc., USA).

Components chosen for instrumentation, particularly for high-performance systems, often include custom circuits manufactured in thick-film or thin-film technology, or custom integrated circuits or *application specific integrated circuits (ASICs)*. The design of these components relies very heavily on specially developed CAD tools plus data about the manufacturing process (Haskard, 1990).

Closely related to circuit design and analysis packages are *component layout and printed circuit board (p.c.b.) design packages*. Again the choice is considerable and many may be run on modest personal computers. Whilst early versions were little more than specialized two-dimensional draughting tools, current packages often include microstrip and stripline models to cope with transmission line phenomena occurring in high-speed circuits. The use of such tools for schematic drawing and p.c.b. layout is now almost universal in electronic manufacture and with the increasing use of surface mounting rather than leaded, through-hole component types the computer-generated layout can interface directly with the 'pick and place' robots used for board assembly. The current trend is towards increased provision of autoplacement of components and autorouting of tracks (Kaplan, 1990) although skilled technicians still tend to intervene during autorouting to produce a more efficient use of board space.

Larger, more comprehensive systems, provide an *integrated design environment* with more comprehensive libraries of components, more detailed descriptions of their properties and hence require larger computing facilities. Hines (1990) provides a useful survey.

With this proliferation of independently developed software the communication of data between different stages of design can be a major problem. To facilitate general and non-standard data transfer a translation system *EDIF (electronic design interchange format)* has been written by a consortium of electronic design automation and semiconductor vendors. Although still developmental, the current version (EDIF 2.0.0) has proved successful and been incorporated in many of the larger design packages (Eurich and Roth, 1990).

'Building block' approaches

All of the above design aids, and most readily available CAE tools, are general purpose in the sense that they commence with 'a clean sheet of paper'. The designer is required to know not only what is required to be done but also, in most cases, how to do it, and must be ready to enter into the system details of the block diagram and schematic circuitry.

An alternative approach to the computer-aided design of electronic circuits is advocated by Sydenham and Jain (1988). This is set out in a developmental package of software labelled CAENIC (CAE of electroNICs) in which the user has a range of pre-designed 'building blocks' from which to construct the required electronic system. For example, upon requesting a second-order, low-pass filter with a particular cut-off frequency—the user is presumed to know what a filter is and what it does, but not how to design one in electronics—the user is presented with a series of screens such as that of Figure 41.8. This example is taken from the original AJITA library of building blocks (Jain and Harpas, 1988) implemented on a modest personal computer. Subsequent screens display the characteristics of this circuit, list the components required for construction and an appropriate printed circuit layout for these components.

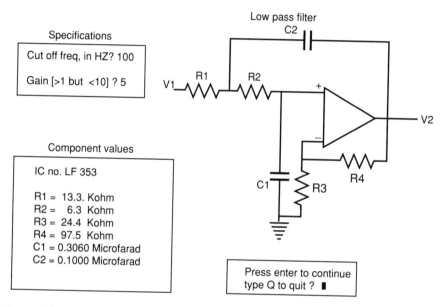

Figure 41.8 A second-order, low-pass filter building block presented to the user as part of CAENIC software

Further building blocks available for analog electronics include amplifiers and d.c. power supplies (to energize transducers and circuits). Hence the engineer or scientist, untrained in electronic design, can assemble a complete, albeit modest, electronic system as long as the design requirements (gains, frequencies and voltages, etc.) are known. In general the design produced will be neither elegant nor optimal, but it will work to specification and may be immediately manufactured. It is produced very rapidly (around one minute) and without the need to consult other persons or sources of information.

Clearly the design produced is strictly limited by the range of building blocks available and to be generally useful across the spectrum of instrumentation system design a wide variety of blocks are required. In this context CAENIC may be seen as simply one part of a wider building-block concept to construct complete instruments (CAEINST, which is considered further in Section 41.5 below). At present CAENIC software has not been released commercially and there is an obvious requirement to provide the user with some assistance in selecting and arranging appropriate blocks for all but the simplest systems. Harris *et al.* (1991) describe recent developments including progress with the incorporation of artificial intelligence techniques.

41.4.4   Software to Aid the Writing of Software

The use of software for operating the internal behaviour within instruments and instrumentation systems has increased greatly in recent years: a bench test instrument designed in 1990 will contain typically three microprocessors and a million lines of code (Terry, 1990). The design effort required in writing and verifying software within most electronic products, including instruments, has increased accordingly and is now recognized to be typically twice that required for the design of the electronic hardware. A range of computer-based packages have emerged to assist in these tasks and they are generally referred to as *CASE (computer aided software engineering)* tools. A useful review of current CASE packages is provided in Kaplan (1990).

Central to CASE software are the traditional tools and techniques of computer engineering. 'High-level' languages and associated compilers, interpreters and assemblers are available such that the programmer does not need to work directly at the binary machine-code level. Cross-compilers provide translation between languages and linkers and loaders are used to create the final software, then referred to as 'executable code' or 'object code', ready for implementation in the hardware.

Tools for the error detection and correction of the object code ('debugging') are also provided. The user may set (and remove) breakpoints within the code to halt operation at particular points, or may 'single-step' through portions of the code to observe its performance. Various levels of trace may be set to record details of the operation of the code. Dis-assemblers can re-translate machine code back to a readable source-code form.

History has provided some painful and expensive lessons with regard to the unmanaged evolution of software. It is now universally recognized that the creation of successful software is much more than the writing of correct code: an unmanaged, unstructured approach will result in the generation of software which is difficult to follow, dangerous to modify and very difficult to maintain or repair when faults occur. Hence a rigorous 'top-down' approach is adopted commencing from a detailed *software specification*. A modular construction of the required code is promoted by the use of the more recent 'structured' languages such as Pascal, C and Ada. A major advantage of this approach is the ability to re-use modules of code for different purposes in different products. CASE packages provide library facilities for the storage and cataloguing of modules which can greatly increase the speed and efficiency of software production. Major cost savings can result from this approach.

CASE packages also provide support for the rigorous documentation of software, documentation which is essential if the software is to be maintained. Maintenance is required as well-hidden errors and deficiencies are eventually discovered and when upgrades or enhancements are required. It is now recognized that the use of the occasional 'comment' statement in an otherwise unbroken block of source code is grossly inadequate and that writing complete, unambiguous documentation is a major task for the programmer. Software engineering is now a recognized professional area.

Digital signal processing (DSP) software

One major reason for the greatly increased volume of software in modern instruments is the use of digital signal processing (DSP) techniques. DSP may be implemented within an instrument on a general-purpose computing subsystem, such as a microprocessor, or by means of specialized, programmable DSP integrated circuits when high speed is required. In both cases specialist software development tools have recently become available (Mather, 1990).

General DSP support is provided by packages which provide signal acquisition and manipulation plus general mathematics capabilities. Support of DSP integrated circuits requires specialized packages which provide the appropriate programming environment including a high-level language user interface and error detection and correction ('debugging') facilities. In both cases an essential feature is the library of standard algorithms and functions such as fast Fourier transform (FFT), correlation and convolution. Filter design is a very common DSP application and standard routines are provided to design a range of finite impulse response (FIR) and infinite impulse response (IIR) filters. Any chosen design is analysed and its characteristics and performance displayed to assist in selecting the optimum design. Where high-speed DSP integrated circuits are to be programmed a realistic 'real-time' simulation can be achieved using a specialized DSP co-processor card fitted to the personal computer on which the design is being undertaken (Mather, 1990).

41.4.5 Software to Aid Manufacturing

Automation of manufacturing methods and processes has advanced greatly in the last decade, driven by the need for lower production costs, greater speed of production and better quality control.

These considerations apply to instrument manufacture, particularly with volume production, just as they do to any other area of engineering. In contrast, perhaps, to other areas of engineering the manufacture of an instrument often involves a range of engineering techniques and technologies from sensors through to the display. Whilst mechanical and electronic technologies may dominate the construction of most instruments there are also optical, chemical, nuclear and other technologies which may need to be incorporated.

Computer aided manufacture (CAM) is perhaps most well developed in the area of mechanical manufacturing. Numerically controlled (NC) machine tools and computer numerically controlled (CNC) machine tools follow a program of instructions to create a mechanical part by turning, milling, grinding, etc. as appropriate without operator intervention. Industrial robots may be similarly programmed to undertake assembly of parts, welding, painting or other tasks which require general movement in three dimensions.

With suitable software these manufacturing facilities may be directly interfaced with a computer aided design system and major benefits accrue when the two are conceived and implemented together. Hence the abbreviation CAD/CAM is often used. The solid model of the part, or assembly of parts, may then be used to determine the path of the cutting tool or the movement of robot arm. Although comprehensive systems can determine a path entirely automatically there often remain matters of engineering judgement and experience which can produce a better product or a more efficient use of material. Hence the automatic facilities provided for tool path generation or robot movement are normally subject to intervention by the design or production engineer. Comprehensive software usually provides libraries about available materials and cutting tools for particular machines.

Clearly major advantages are to be gained by the smooth combination of design, analysis and production. The overall process is referred to as *computer integrated manufacturing (CIM)* in which, ideally, the computer will also control material requirements, job scheduling, quality assurance and inventory. Experience has shown that the design and factory-floor layout of manufacturing facilities is important to realize the benefits of CIM.

A further consideration is the length of the production run of the product and the lifetime of the product in the marketplace. With 'high technology' products, which will include the great majority of modern instruments, production runs will often be measured in weeks and months rather than years. The ease and speed with which the manufacturing facility or production line can be reconfigured for another product is then of paramount importance and the adoption of a *flexible manufacturing system (FMS)* layout will be required. In contrast to a production line an FMS seeks to provide an appropriate general arrangement of production facilities and machines linked by automatic transport of materials and partially completed product.

Introductions to CAD/CAM, CIM and FMS are provided by Haigh (1985) and Besant and Lui (1986).

While CIM remains the goal of manufacturing companies there may be significant problems of compatibility between the software controlling different parts of the process. For this reason multi-industry cooperation has led to the definition of a common format known as *manufacturing automation protocol (MAP)*. Park and Talley (1990) outline one current implementation of MAP. MAP is based on the Reference Model for Open Systems Interconnection (OSI) of the International

Standards Organisation (ISO). See Chapter 42 for further details on communications protocols.

41.4.6   Equipment and Software for Automatic Testing

Testing, and quality control in general, represent a significant proportion of production costs across the breadth of engineering. It has become clear that customer expectations with respect to quality and reliability are continually rising and that quality is a major and sometimes dominant factor in purchase decisions. This is particularly so in the instrument and test equipment manufacturing industries (Terry, 1990).

The activities of production testing and inspection are traditionally labour-intensive, requiring skilled staff and a heavy investment in training. The work may also be very repetitive and hence tedious and prone to human error. Sample or batch testing is often adopted in order to contain the high costs involved. Unfortunately sample and batch testing imply that some substandard product will reach the customer and the manufacturer will have to rely on the warranty period to detect and replace the defective goods. In general this approach is not acceptable to customers in the instrumentation industry (Terry, 1990).

Automated testing and quality control offers an alternative to the tedium of repetitive testing and the risk of human error. It may also be much more rapid, such that 100% testing of all product becomes feasible with its obvious advantages in the marketplace. However, the establishment costs may be very high, both as regards capital equipment and the time and effort required to implement the required test sequence. In addition, and particularly in the manufacture of instruments, production runs may be small, the tests complex and the available time-to-market for a new product may be only a few months.

In consequence most automated test facilities are computer-controlled and will be to some extent general-purpose. The facility may be rapidly reconfigured to suit changing requirements and this will, as far as possible, be under the control of software. Ease of use of this software and its integration as part of the overall CAE of the product are obviously critical.

As implied from the foregoing general remarks the great majority of automatic, computer-controlled testing is *production testing* and it is in this area that major benefits may be immediately realized. Here the tasks of testing and conformance to specification may be set out in a fixed sequence. The result of each test is either 'pass/fail' or the instruction (to the operator) to make a specific adjustment.

In contrast the task of *diagnostic testing* occurs only when a unit malfunctions and the scheme of appropriate measurements to make and interpretation of the results varies according to the nature of the fault. This area is traditionally the area of expertise of the service technician and the human qualities of intuition and experience can make a major contribution to the rapid repair of complex systems. In consequence automatic diagnostic testing is far less well developed and less frequently employed. At present techniques commonly employ 'brute force' methods such as the complete sequential retesting of a subsystem or of the complete unit. However with the rapid growth of rule-based, self-learning computer systems it is likely that automated diagnostic testing will find greater application in future years.

Automated mechanical testing

Automated mechanical testing involves the inspection and measurement of manufactured parts. Positions and dimensions may be sensed by instrumented mechanical feelers or (increasingly) by non-contact techniques using ultrasonic or laser probes. These sensors may be installed as stand-alone instruments, or as part of an assembly line, and be programmed with the expected measurement and tolerance to provide 'pass/fail' information to the computer control of the product flow. Alternatively the sensors may be incorporated as one or more of the tools within a computer-numerically controlled (CNC) machine and are usually referred to as *gauging probes*. Use of these tools during the manufacturing process provides feedback to the cutting tools to ensure that the part is manufactured within the specified tolerance.

In both cases it is highly desirable to incorporate these testing facilities within the overall computer-integrated manufacturing (CIM) process. The controlling software for these measurement tools and the specifications against which the measurements are being compared will be part of the CIM software.

Automated electronic testing

Electronic testing is required for individual components, for assembled circuit board and for complete systems. *Automatic test equipment (ATE)* may be custom designed, but is often assembled from units of general-purpose bench test equipment under the control of a small computer. A typical ATE system comprises a *controller* (or an instrument with control capability), one or more *stimulus instruments* configured to provide the required input signals and one or more *response instruments* to make the performance measurements. Programmable power supplies, signal switching facilities and a *bed of nails* board to probe the tracks of a printed circuit card may also be incorporated. Ibrahim (1988) provides an introduction to application of ATE. Figure 41.9 shows a modest, general-purpose ATE facility.

Testability in electronic circuits and systems

Modern electronic instruments, like other electronic systems, are of increasing complexity, and even small subsystems such as microprocessor boards possess a very wide range of functional capability. Exhaustive testing of even one unit is therefore a major undertaking, both to determine the test sequence and to actually perform the tests. The situation is potentially even worse with the increasing use of integrated components such as gate arrays and custom or application-specific integrated circuits (ASICs). The result is that to a greater or lesser extent the resulting system may be untestable.

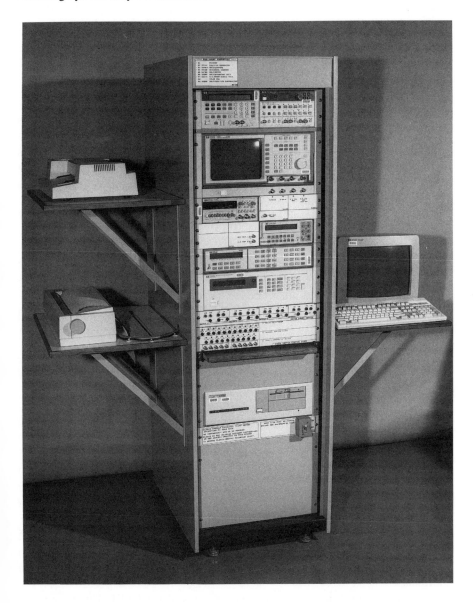

Figure 41.9 A mobile, general-purpose, rack-mounted ATE system. The facility comprises (from the top) two signal generators, a digital storage oscilloscope, a counter/timer, a digital voltmeter, a signal switching unit and a programmable power supply. These instruments, plus a printer and plotter (left-hand side), are interconnected via the GPIB such that each may be entirely controlled by a personal computer (the lowest unit, with the monitor on the right-hand side)

Fischer (1984), for example, stresses the need to specifically design for testability. At the physical level it is necessary to provide adequate and appropriate access to particular signals within the system by means of probe points, buffered as necessary. At the conceptual level it is necessary to design with regard to the ATE system and facilities available and hence avoid incomplete or ambiguous test results. In short the testability of circuits should be integrated into the overall CAE design and production process.

To assist the designer meet test and diagnosis requirements Hallenbeck *et al.* (1989) report the development of a set of CAD software tools called Test Engineer's Assistant (TEA). By interfacing with other general-purpose CAD systems for electronic circuits and systems TEA seeks to address testability at all stages of the design process and at all levels of the eventual system hierarchy—factors inhibiting unambiguous test are identified, recommendations for eliminating hard-to-test circuit features are provided, the locations of appropriate test points are determined and the hardware (and hence cost) overhead is assessed.

The General-purpose interface bus (GPIB)

Clearly a communications system is required between the component parts of the ATE system. (Another overview is given in Chapter 46, where the emphasis is somewhat different.) Whilst binary-coded decimal (BCD) format or the commonplace RS-232 serial communications standard may be employed to transfer data in very simple two-unit (instrument plus controller) systems this is normally inadequate. In virtually all cases the application of ATE requires several instruments plus a controller (usually a computer) and the capability not only to transfer measurement data from the response instruments but also to set up and modify the control settings of each instrument.

To meet this requirement a communications system generally referred to as the *General-purpose Interface Bus (GPIB)* has been almost universally adopted. Developed by the Hewlett Packard Company it is also known as the *Hewlett Packard Interface Bus (HPIB)* and has been endorsed by the Institution of Electrical and Electronic Engineers (IEEE) and adopted by the American National Standards Institute (ANSI) as *Standard ANSI/IEEE 488(1978)*. It has been further adopted by the International Electrotechnical Commission (IEC) as *Standard IEC 625-1*. These four names are synonomous and define a 16-signal, 24-wire hardware connection for up to fifteen instruments in one system. The standard fully defines the electrical and mechanical aspects of the connection such that equipment from different manufacturers may be interconnected with confidence.

Figure 41.10 illustrates the sixteen signal lines of the GPIB. These comprise an eight-line data bus used for the passage of data and commands in a byte-parallel form at speeds up to one megabyte per second; a three-line transfer bus which provides a 'handshake' between instruments to synchronize the transfer of data; and a five-line management bus to effect management and control.

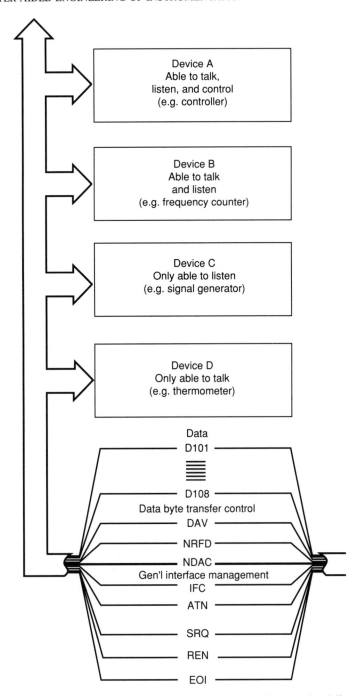

Figure 41.10 The General Purpose Interface Bus (GPIB) showing the signal lines and bus structure

Figure 41.10 also illustrates four classes of instrument which may be connected to the GPIB distinguished according to whether each can act as a 'controller' (of which there can only be one), as a 'talker' which is placing data on the bus (and hence there can only be one enabled at any instant) or as a 'listener', receiving data (as many as required). The status and action of each instrument is managed by the controller using commands uniquely addressed to that instrument and communicated along the data bus. A number of texts, for example Ibrahim (1988), present details of the GPIB and Loughry (1983) introduces the formal ANSI/IEEE Std.488-(1978) and recommended software practice.

The implementation of GPIB capabilities within a test instrument may vary and provision of a GPIB interface does not guarantee that all the instrument functions are controllable via the bus. Older instruments which were not specifically designed for GPIB control or have had their interfaces retro-fitted will often have certain key controls which can only be operated from the front panel. Common examples are frequency-band switches and range selection. In certain cases this may be a significant disadvantage as an operator must then be prompted (often from the screen of the controller) to make the adjustment.

While IEEE 488-(1978) and IEC 625-1 specify the overall mechanical and electrical properties and the low-level protocol of the transfer bus and the management bus, the command structure and software is largely undefined. Hence the programming conventions adopted by different manufacturers vary significantly. To assist the test engineer assemble the strings of commands necessary to set up each instrument, to execute the test routine and to acquire and analyse the results a range of software tools have proliferated. Hewlett Packard Interactive Test Generator (ITG) and Tektronix Test Management System (TekTMS), for example, provide a full pictoral emulation of each instrument front panel. The code required to set up the instrument is generated automatically as the on-screen 'controls' of the emulated instrument are manipulated. At a more modest level various data acquisition, analysis and processing packages such as ASYST and LabWindows provide libraries to assist programming of GPIB operations. Nevertheless the incorporation of instruments from different manufacturers in a single ATE system can still cause substantial programming inconvenience.

To overcome this problem recent development of the GPIB has resulted in the additional *standard IEEE 488.2 (1987)* which defines data structures, a syntax of message exchange and also some common commands and queries between the controller and the instruments. However specific messages between particular devices connected to the GPIB are still left to be defined by the device designer. One general-purpose language based on IEEE 488.2 is Test and Measurement System Language (TMSL) developed by the Hewlett Packard Company (Nemeth-Johannes, 1990).

Recent developments use expert systems to select and run the appropriate system. The IMMI concept, Van Biesen *et al.* (1990), switches the signal to the appropriate software analysis package. An alternative (Rudolph *et al.*, 1990) switches the signal to the appropriate instrument unit.

'Instruments on a card'—the VXIbus

A recent development in ATE architecture is the recognition that most of the front panel facilities—controls and displays—are not required in an ATE application and that the measuring circuits of each instrument may often be constructed on a single printed circuit card. Hence a complete ATE system comprising controller, stimulus instruments and response instruments may be arranged as modules within a single chassis and communicate via a common backplane. The result is a more compact system offering the possibility of a more comprehensive set of interconnections and the potential for much faster communications between instruments.

A consortium of instrument manufacturers have adapted the well-established international 'Eurocard' format and VMEbus backplane standard to define the *VME Extensions for Instrumentation (VXIbus)*. This standard permits different instruments designed by different manufacturers to be interconnected and to communicate in a manner similar to the GPIB. To date it has not been endorsed by any national or international standards authorities.

Each VXIbus instrument meets Eurocard mechanical standards (extended to include larger C and D card sizes) and plugs into the VXI backplane which provides various buses for high-speed data transfer, clock, trigger, interrupt and power supply buses via three multi-pin sockets. Mechanical aspects of the VXIbus standard also provide for improved cooling capacity and electrical aspects provide for the control of possible electromagnetic interference between adjacent modules (instruments). The standard defines a limited system protocol which includes a command hierarchy and a 'handshake' between communicating devices.

Every chassis requires a module to occupy 'Slot 0' and provide controller functions, 10 MHz and 100 MHz clock and other system signals, and usually an interface to the GPIB. From the perspective of the user the system and the software required are very similar to those of a GPIB-interconnected ATE system. Software support tools such as the previously mentioned TMSL are applicable to both systems. Jessen (1989) provides an overview of the VXIbus and Narciso (1990) considers the implications for the instrument designer.

41.4.7 General CAE Support Software

There exists a wide range of engineering planning and project control packages appropriate and applicable to instrumentation design and installation ranging from elementary guides (for example Bacon, 1989), simple applications of databases (for example Bownds and Gillett, 1984) to those which can incorporate heuristic information and employ knowledge engineering techniques. Examples of these advanced packages are PIPPA (Professional Intelligent Project Planning Assistant) and XPERT (eXpert Project Expedition Reasoning Toolkit) reviewed by Barber *et al.* (1988).

Other general tools include word processors, graphics and 'desktop' publishing packages which are indispensable for producing the high standard of documentation that is expected to accompany instrumentation.

At the large-scale level the integration of these project control and support packages can become a major problem unless they are all purchased at the same time and from the same manufacturer. Hence a common data interchange and communication format labelled *Technical and Office Protocol (TOP)* is currently under development and like the Manufacturing Automation Protocol (MAP), Section 41.4.5, is based on the Reference Model for Open Systems Interconnection (OSI) of the International Standards Organisation (ISO).

41.4.8   Operator Training Software

In parallel to the increased use of computers with instrumentation and even within stand-alone instruments in recent years there has been a corresponding increase in complexity of the equipment. Many additional facilities are provided (for example even a simple voltmeter may now provide maximum, minimum, averaging and other statistical functions) although the basic measurement capability of the instrument is unchanged. Accordingly ease-of-use, front panel layout and the anticipated operator 'learning curve' have become increasingly important factors in any purchase decision. Complex instruments are actively marketed as 'user-friendly' and manufacturers often provide assistance under the headings of 'Getting Started' or 'Tutorial'. Experienced users are then provided with an alternative 'Fast Track' familiarization.

It is important to note that the type of training required for modern computer-based instruments—even simple voltmeters—is therefore different from that required for 'traditional' types of instrument. The manual skills required to make good measurements with modern instruments have diminished, whereas the knowledge required of the facilities and controls has increased considerably. Training requirements which are *knowledge based*, rather than skills based, lend themselves particularly to automated, computer based tutorial techniques.

Access to a personal computer or a computer terminal is so commonplace in 1991 that the instrument designer or vendor may confidently presume its availability. Hence the documentation that accompanies any instrument may be usefully supplied in software as well as written form. It is particularly convenient to have the Operator's Manual 'online' so that it may be consulted by the various users at any time. Upgrades and amendments can be easily added and the risk of a critical section of the document disappearing from the company library is avoided.

For simple instruments the documentation will be supplied as disks and then installed on personal computers or the company central computer facility: for computer-based instrumentation systems, or instruments with substantial internal computers, the documentation will be built-in. Furthermore the common techniques from software engineering practice such as selection by menu and the provision of a 'help' key will usually be provided, and if the documentation is internal to the instrument the 'help' facilities may be interlocked with the instrument functions such that only that information from the manual relevant to the immediate problem is presented to the operator.

For complex or extensive systems simple, unmonitored tutorial facilities may not suffice and a more formal training programme may be required with some evidence of operator proficiency. The traditional approach to this requirement is to provide an intensive short course using professional lecturers, course notes and written assessments. Such courses are expensive to run, disruptive to the other duties of staff and are usually available only on a "one-off" basis.

Computer Aided Instruction

There exist alternative computer-based approaches to supplement or replace the traditional instrument training course. The simplest is *computer aided instruction (CAI)*, which is an extension of the on-line tutorial. In addition to the factual material presented in the tutorial, tests are provided requiring the user to respond via a keyboard or perhaps via the instrument front panel. The answer provided determines whether the user progresses to the next stage of the tutorial or is required to repeat the previous section until a satisfactory response is obtained. Hence CAI provides useful feedback to the trainee user of the instrument.

Computer Managed Learning

A more complete control over the training of operators may be achieved by the techniques known collectively as *computer managed learning (CML)*. These techniques are currently being developed to be applicable in essentially any area of education and training and may encompass a range of teaching media as appropriate to the training task (Barker, 1988a). They may include on-screen CAI, the viewing of a videotape, the use of an interactive video disk, particular exercises on the instrument itself (or on a software simulation of the instrument) or simply direction to study certain parts of the published manual. The key word in CML is 'managed': the task of the computer is not primarily to provide the instruction or training but rather to supervise, monitor and control the training process.

The potential benefits of CML to the owner of the instrumentation are considerable. The manager responsible for personnel can monitor the progress of particular operators (identified as each person logs-on for training sessions) whilst the engineer responsible for the equipment can identify common areas of difficulty which may indicate a deficiency in the training program or in the equipment itself. Furthermore the training software may also be interlocked with the instrument to deny access to particular facilities or capabilities for safety or security reasons.

As with CAI the benefits to the trainee include the opportunity to receive direct training feedback, to repeat instruction in difficult or seldom used areas of operation and to use the facility whenever there is free time available with equipment.

Commercial packages for the implementation of CAI and certainly CML are few. Hence most are written by the particular instrument manufacturer or developed 'in-house' by a large-scale user. McSherry (1987) reports that the teaching design and animation package AUTHOR has been used to develop effective CAI in a range of commercial as well as educational environments, although not specifically in the area of instrumentation training. General-purpose intelligent tutoring systems are assessed as being in their infancy (Lesgold, 1987) and similarly CML appears to require substantial knowledge engineering to be of general applicability (Barker, 1988b).

Simulators

Large-scale, comprehensive instrumentation training facilities are often labelled simulators. Examples include those specifically developed for instrumentation within aircraft and other vehicles, within power stations, chemical process plant and military systems. They often employ the full range of computer aided and computer managed techniques—see, for example, Short (1988). Simulators are of particular value where a high degree of training is essential for safety or security reasons. Realism of the simulation is important; hence they are usually specific to one type or class of instrumentation system. Design of the simulator hardware and the appropriate training programs have developed from operational experience and typically involve considerable expense.

41.5 SPECIFIC TOOLS AND TECHNIQUES FOR THE CAE OF INSTRUMENTATION

Some CAE packages have been specifically developed to assist in the instrument design process. These deal with matters either unique to, or of particular importance to, the creation and operation of measurement systems. Some of these packages are now seen to have much wider application (for example, the Specriter package, Section 41.5.2) than solely the design of instrumentation.

41.5.1 Establishing the Measurement Requirement

The small number of such dedicated software tools is in part a recognition that instruments are engineering products just like any other and hence general-purpose CAE tools are applicable; and in part that quite advanced computing techniques are required, notably those associated with knowledge engineering and artificial intelligence (as introduced in Section 41.3). In addition there exist software packages developed in-house by instrument manufacturers for private use only; these are not discussed here.

A software package called MINDS (Measurement INterface Design System) has recently been produced to assist the user to establish precisely and unambiguously what needs to be measured for a stated knowledge gathering need (Sydenham *et al.*, 1990; Terwisscha van Scheltinga, 1991).

Establishing the measurement requirement is not the same as drawing up a specification for the required instrument and will normally precede the formal specification stage (Section 41.5.2). The need for such a package arises from the experiences and frustrations of instrument designers who can easily fail to achieve customer satisfaction (at least in initial designs) because the actual measurement requirement turns out to be rather different from that initially perceived by the customer or instrument designer. To this extent the package acts as a clarification and communications interface tool between customer and designer prior to the drawing up of a formal instrument specification.

In some cases the measurement requirement may be obvious—for example, the user requires to measure a temperature at a particular location with a particular range, discrimination, output type etc.—and the appropriate method (or choice of methods) may be obvious. If so the user can write a full specification directly and the instrument required can be designed in detail and manufactured.

However, there is a wide range of circumstances for which details of the measurement requirement cannot be immediately determined by the intending user. Hence full specification, detailed design and manufacture of the appropriate instrument cannot occur with any confidence until the measurement requirement is further refined. Examples range from what may be termed 'slightly non-standard' (such as the requirement to measure temperature in a particular inaccessible place which may require unusual sensing and transducing methods) to the obviously complex (such as the measurement of efficiency of a machine or quality of a product).

For these circumstances the usual next step is to 'consult an expert'. However, seldom will the expert be able to produce an instant answer; rather the procedure followed will be one of exhaustive examination of the system of interest, its properties and environment. The MINDS software package seeks to mimic this procedure and may be labelled an 'expert system'. The present version contains no measurement expertise allowing it to recognize when a particular aspect of the measurement requirement is a potential measurand, that is when it may be directly measurable. However, the package does require the user to provide basic metrological information and then seeks to organize that information in an appropriate manner in a procedure similar to that which might be followed by the expert.

Although the approach of the MINDS software appears to be unique with regard to measurement science and instrumentation design, this approach and similar methodologies are applied in other fields to determine user and customer requirements. Berlin (1989), for example, sets out a methodology to determine the user's real requirements in the area of applications software and human–computer interfacing.

Operation of the MINDS package

The MINDS package is an implementation of the first part of the *measurement process algorithm (MPA)* as set out by Sydenham (1985a) and illustrated in Figure 41.11. The MPA is the formal presentation of a systematic and structured approach to instrument design (Sydenham, 1985b) and is applicable to the specification of any measurement system.

The package uses an interactive question-and-answer format which leads the user through the procedure illustrated in Figure 41.12.

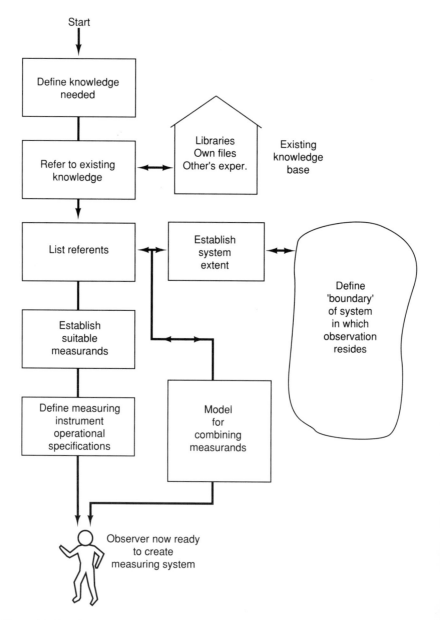

Figure 41.11   The measurement process algorithm. (Reproduced from Sydenham, 1985a)

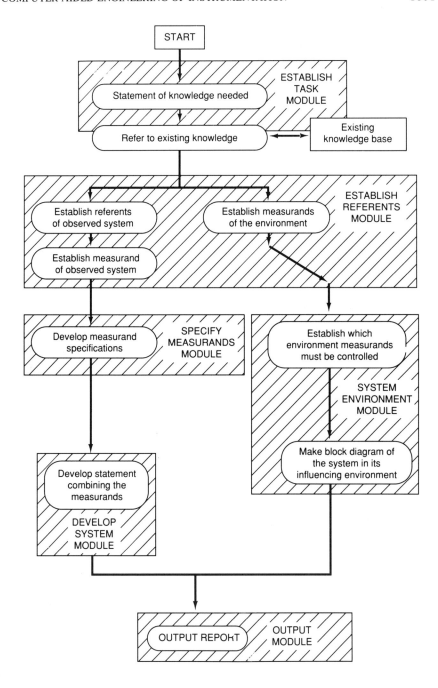

Figure 41.12 The structure and operation of the MINDS software package. (Reproduced from Sydenham *et al.*, 1990)

Of the six software modules set out in Figure 41.12 the first, the establish task module, is perhaps the most critical. This requires the user to state as simple English-language sentences:

- the particular system/phenomenon/object to be measured;

- the information related to the knowledge desired about the particular system/phenomenon/object; and

- the physical environment in which the particular system/phenomenon/object exists and in which the measurement system will have to operate.

At this stage it is vital that the user does not attempt to guess the measurands required or presume a particular measurement technique—it is the task of MINDS to ensure that the user systematically deduces the measurands required as the analysis proceeds. Prejudging these matters (perhaps subconsciously) by failure to keep an open mind seriously handicaps the effectiveness of the software.

Having determined a satisfactory statement of the task (usually after some iteration and revision) the Establish Referents module then requires the user to identify *referents*, that is those attributes of the system/phenomenon/object which are of relevance to the information required. MINDS then assists the user to transform all referents into measurands. Some referents are recognized immediately by MINDS as measurands (e.g. mass, temperature) whereas others (e.g. size, heat) are ambiguous or not directly measurable and are referred back to the user for further development. Hence each referent may be refined or 'unravelled' (size, for example, as certain characteristic dimension(s) or perhaps as a volume) and the progress is logged as a 'tree diagram' where each branch or sub-branch eventually must end as a measurable quantity. Figure 41.13 is this part of a final report from the PROMINDS version. The shading in a box indicates the type of environment of that parameter.

Subsequent modules of MINDS prompt the user to determine the metrological parameters for each separate measurand, to consider the measurement environment and the effect of influence variables and to combine the measurands into an overall measurement system. A report, in the form of a log of the progress to date, may be produced at any stage during the use of the package. The original version was written in the procedural PASCAL language. This was replaced with PROMINDS which is written in Quintus-PROLOG to give it the advantages of frame-based knowledge representation and the declarative programming feature of user entry at any point.

User experience

MINDS has been subjected to trials for a range of 'difficult to measure' applications where a fundamentally new instrumentation system appears to be required. (Two examples are reported and analysed in detail in Sydenham *et al.*, 1990.)

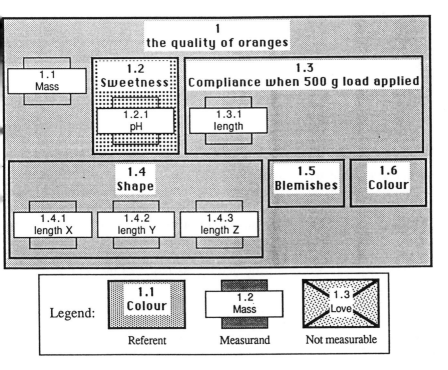

Figure 41.13   System architecture part of a PROMINDS report on a system for measuring the quality of oranges. (From Terwisscha van Sheltinga, 1991)

Initial reaction to the output produced has sometimes been that it was 'obvious', but in many cases the requirement was far from obvious (at least to the instrument designer) at the outset and the actual measurement requirement turned out to be significantly different from that which was envisaged at the start.

In some cases the results obtained may be considered totally successful in that a clear, unambiguous statement of the measurement requirement has been produced in terms of fully specified measurands: others have been less successful as one or more essential referents could not be adequately transformed into measurands. However in these cases the exercise was still considered valuable because critical areas of inadequate understanding or incomplete knowledge were clearly identified. If the system of measurable parameters cannot be indentified then a real sensing system cannot be realized.

It is recognized by Sydenham et al., (1990) that deficiencies remain. The user interface in particular was judged as 'unfriendly' by those unskilled in instrumentation. A more serious problem is that of encouraging the user to set aside preconceived solutions and think laterally with respect to potential referents. These issues were better overcome using the knowledge based methods, those leading to PROMINDS (Terwisscha van Scheltinga, 1991).

The value of the MINDS package as a communications tool has been proven. Whether the MINDS concept will lead to the design of an instrument which would not otherwise have been designed remains an open question—the package was conceived as an aid to, and not a substitute for, human ingenuity. Its purpose is to provide a sound foundation for further automation of instrument design via CAE by capturing the systemic parts of the initial steps of the design process.

41.5.2 Specifying the Instrument

Having determined the measurement requirement the next task is usually to produce a formal, detailed specification. This is required for a number of reasons which may include management evaluation, cost analysis and the issue of formal invitations to tender for the design, manufacture or direct supply of the required system. Production of this document in adequate detail and free of omissions, ambiguities and inconsistencies can be a lengthy and difficult task—unfortunately it is also a tedious and usually unwelcome task, and therefore particularly prone to human error.

The program Specriter 1 has been developed (Cook, 1988) to assist users produce a standard specification document for any single measuring sensor. The output is in a report format which meets the internationally accepted US military specification MIL-STD-490A (USDD, 1985). Major section and subsection headings are set out in Figure 41.14.

Specriter uses a question-and-answer format to obtain information from the user, information which may be provided in one of three forms as appropriate. These are:

● as a number, for example the value or limit of a particular parameter;

● as a selection from a number of alternatives presented by the program; or

● as text, for example a name, title or descriptive paragraph.

In addition an entry of the universally used 'TBD' (to be determined) may be made or the default entry of 'TBS' (to be specified) accepted such that the user is not stalled by incomplete information.

The information supplied is held in an *attribute file* and may be edited as required at any time. The output specification document is then produced by a merging of direct user input text and standard paragraphs customized according to the contents of the attribute file. The structure of the program is illustrated in Figure 41.15.

Whilst Specriter 1 is an effective communications tool and will always produce a complete and correctly formatted specification the resulting document will obviously only be as good as the information supplied by the user. Hence it comes as no surprise that this software usually only produces a good specification when

```
1  SCOPE
2  APPLICABLE DOCUMENTS
3  REQUIREMENTS
   3.1 Instrument Definition
   3.1.1 General Description
   3.1.2 Interface Definition
   3.1.2.1 Electrical Interface
   3.1.2.1.1 Power Interface
   3.1.2.1.2 Communications Interface
   3.1.2.1.3 Electromagnetic Compatibility
   3.1.2.2 Mechanical Interface
   3.1.2.2.1 Sensor
   3.1.2.2.2 Data Processor
   3.1.2.2.3 Display
   3.1.2.3 Thermal Interface
   3.1.2.3.1 Sensor
   3.1.2.3.2 Data Processor
   3.1.2.3.3 Display
   3.2 Characteristics
   3.2.1 Performance
   3.2.1.1 Range
   3.2.1.2 Discrimination
   3.2.1.3 Repeatability
   3.2.1.4 Hysteresis
   3.2.1.5 Drift
   3.2.1.6 Dynamic Response
   3.2.1.7 Measuring Error
   3.2.1.8 Power Consumption
   3.2.2 Physical Characteristics
   3.2.2.1 Sensor Physical Characteristics
   3.2.2.2 Data Processor Physical Characteristics
   3.2.2.2.1 Data Processor Housing
   3.2.2.2.2 Data Processor Mass
   3.2.2.3 Display Physical Characteristics
   3.2.2.3.1 Display Housing
   3.2.2.3.2 Display Mass
   3.2.3 Reliability
   3.2.4 Maintainability
   3.2.5 Environmental Conditions
   3.3 Design and Construction
   3.4 Documentation
   3.5 Logistics
   3.6 Personnel and Training
4  QUALITY ASSURANCE
5  PREPARATION FOR DELIVERY
6  NOTES
```

Figure 41.14 The paragraph heading of the instrument specification document produced by the program Specriter 1. (From Cook, 1988)

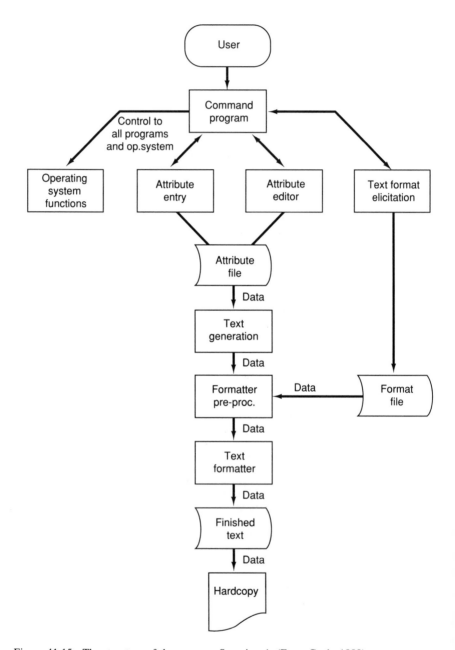

Figure 41.15   The structure of the program Specriter 1. (From Cook, 1988)

the user fully appreciates the application and has a degree of instrument engineering experience (Cook, 1988).

To improve the usefulness of the tool for non-specialists in instrumentation Cook (1990) reports Specriter 2 and Specriter 3. These include instrumentation expertise within the software package and the latter, Specriter 3, employs artificial intelligence computing techniques. The package contains a *knowledge base* comprising rules and heuristic information about the meaning of and relationships between the various items of the specification (i.e. those of Figure 41.14) and *reasoning mechanisms* to apply this knowledge to the input provided by the user. The software can then:

- check consistency and reasonableness of input information;

- generate appropriate default values and limits for parameters of the specification; and

- provide selective, 'context-sensitive' help information to the user.

41.5.3 Designing and Modelling the Sensor

At present the design of the sensor is not well supported by readily available CAE tools. Where the sensor or transducer required is essentially standard, or at least employs standard and well-understood sensing techniques, some support is available but this is normally incorporated in the wider CAE of a complete instrument type. These packages are considered separately in Section 41.5.4.

Where the sensor required is novel instrument designers, like inventors, are essentially 'on their own'. However, having determined a physical effect and the possibility of a sensing technique, the design of an optimal sensor or transducer depends on an adequate understanding and description of the physics of the effect. Whether this understanding is empirical, theoretical or a combination of both, the necessary analysis will require a formulation as a mathematical model. Only then can a design be optimized with respect to the energy exchange at the sensing interface to determine sensor performance such as loading effects, input and output impedances, bandwidth and dynamic characteristics. At present software support for sensor and transducer modelling is almost entirely by means of general- purpose *mathematics software* (as described in Section 41.4.1.) Finite element methods have been applied in this way, examples being Grattan (1990) and Tran Tein (1990).

The MEDIEM package

MEDIEM (Multi-Energy Domain Interactive Element Modelling) is a simulation and modelling tool specifically designed to describe and analyse energy flow at the sensing interface and through a chain of transducers. The program implements the systematic scheme of instrument description presented by Finkelstein and Watts (1983) in which basic transducer types are classified and treated as functional blocks according to their energy-processing role, namely sourcing, conversion, interconnection, storage and dissipation.

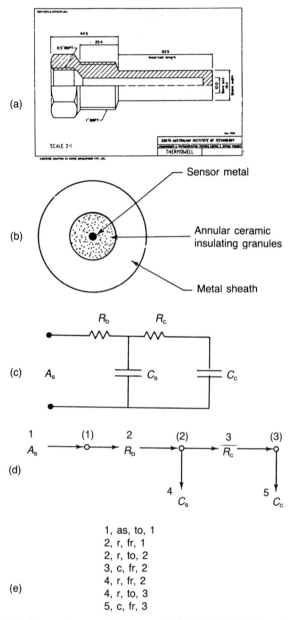

Figure 41.16 Modelling a thermowell system with MEDIEM: (a) given physical shape and size; (b) cross-sectional model according to heat-flow considerations; (c) electrical equivalent circuit; (d) structure graph; (e) MEDIEM instruction list. (From Sydenham *et al., 1989*)

A proposed sensing system to be simulated is presented to the package as a 'structure graph', a graphical representation of the energy flow which may be readily derived from a block diagram or schematic of the system (Liebner and Abdullah, 1981). Appropriate state equations are then automatically set up according to the choice of blocks, input parameters and structure graph and then solved to yield the system performance. Figure 41.16 shows the steps needed to set up a MEDIEM exercise for modelling a temperature sensor placed in a thermowell.

MEDIEM has been used to model various instrument sensing systems for research and industrial purposes, for example electromechanical transducers (Abdullah *et al.*, 1977).

41.5.4 Design Shells according to Measurand

As indicated in Section 41.3 computer aided design of an instrument will commonly involve the sequential use of a range of general-purpose CAD packages. At each stage the user—the designer—is required to conceive and create the item (transducer, signal processing circuit, etc.) and possess the knowledge and experience to produce a sensible and satisfactory final design.

An alternative approach is the use of a single *design shell*. The shell is an integrated suite of programs for a specific purpose—in this case the design of an instrument—containing the total knowledge and procedures required for the purpose and in a form which is independent of the application. For an instrument design shell this should (ideally) cover all phases of the design process between specification and final manufacture.

Such a tool is intended for the non-expert user, that is, for the engineer or scientist who is not an instrumentation professional but understands the particular measurement application. It is also useful for the routine designer who wishes to keep updating a design. The starting point is a clear definition of the measurement requirement and the result is normally in the form of a comprehensive report containing:

● a full explanation and justification of the design;

● complete mechanical drawings for the manufacture and installation of the sensor assembly;

● electronic schematic and p.c.b layout for the energizing, signal processing and output (interfacing or display) circuitry;

● a complete bill of materials and parts list; and

● a statement of calibration procedures, maintenance requirements and any operational restrictions.

Software that might fulfil this requirement will almost certainly be both large and complex, as it seeks to encapsulate the expertise of the instrumentation professional. Hence for any given package the scope of the professed design capability will have to be severely restricted. An obvious first restriction is normally that of limiting a package to one particular measurand and then to a limited range of design archetypes: further restrictions include sensing technique, area of application (for example, environment) and output interface format.

ThermoShell

ThermoShell is one example of a measurand-specific design shell which has been written not only to meet an industry need but also to explore the practicalities of implementing the ideas set out above.

The package provides the complete custom design of thermocouple thermometers installed in pipelines as used in the process industries (Harris and Hancock, 1991). Such thermometers are commonplace and many instruments are available off-the-shelf. However, in practice there is great variation in individual measurement and installation requirements such that non-standard, custom designs are often required. It is therefore likely that the instant design capability provided by a package like ThermoShell would be of considerable value to process control engineers.

Figure 41.17 shows the limited design scope of the package and in addition there are limitations on the form of the electronic output available. Figure 41.18 illustrates the output documents provided.

ThermoShell is not a CAD system (in the normally understood sense of the term) as it does not require the user to create the design. Rather the package commences by presenting a standard default design and invites the user to *cus-*

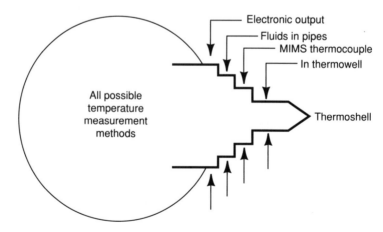

Figure 41.17   Temperature-measuring instruments that may be designed using ThermoShell. (MIMS—Metal Insulated Metal Sheath.) (From Harris and Hancock, 1991)

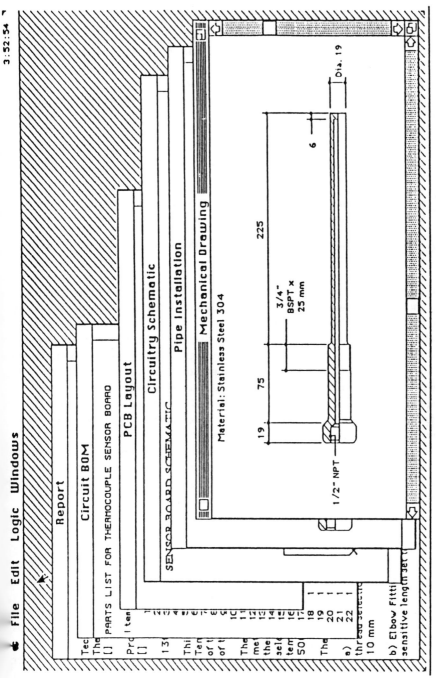

Figure 41.18   A composite computer screen illustrating the six output documents produced by ThermoShell. (From Harris and Hancock, 1991)

tomize this design by modifying the design constraints, including installation geometry, pipe size, fluid type, flow rate and temperature range. At each change the entire design is reworked by the software and where conflicting requirements have been specified (for example, the optimum thermowell size cannot be accommodated in the required pipe diameter) the user is informed and, where possible, alternative configurations are suggested.

Within the software the design process involves a combination of traditional computing techniques and knowledge-based methods more commonly associated with artificial intelligence. These include:

- calculations using standard design algorithms (for example to determine the resonant frequency of the thermowell);

- practical design experience and other heuristic information in the form of a hierarchy of interacting rules; and

- a database of standard materials and their properties, standard parts and sizes (for example screw threads).

All of this information and processing is invisible to the user but is reported at the end of the design process. (Harris and Hancock, 1991, set out an example of the interaction of these techniques).

Harris and Hancock (1991) also report that the limited user experience and reaction to date has been positive and encouraging. Criticisms have been limited to the design restrictions which were imposed at the outset, particularly with regard to the electronic design and output interfacing, and indicate a need for wider options (for example, use of alternative electronic parts and circuits). Being a knowledge-based system written in the Macintosh environment, addition of knowledge, in many forms, is straight forward.

One unexpected reaction has been the considerable interest of the professional instrumentation designers. This group, of course, do not require the embedded expertise of the system but found the speed of design, quality of reporting and potential standardization might offer major cost savings.

The 'reverse engineering' approach

The use of knowledge- and rule-based computing techniques permits an alternative to the direct design approach illustrated by ThermoShell. If there exists a range of working (and hopefully optimal) designs for a particular type or class of sensing system it may be possible to compare and analyse these designs and derive simple numerical relationships between parameters. These relationships and other design rules may then be used predictively to design similar systems.

This approach is known as 'reverse engineering', and is inferior to the proper mathematical modelling of the class of sensing systems because any design produced cannot be independently assessed nor its performance optimized. However, the approach is useful where sensor modelling is incomplete or inadequate. Clearly the design rules and parameter relationships may be incorporated in software and customized designs produced as required. It is essential, however, that the user not be permitted to extrapolate the parameters of the design beyond the range from which the design rules were derived.

Sydenham (1987a) reports the application of this approach to the computer based design of LVDT (linear variable differential transformer) displacement transducers for which direct design involves complex and lengthy mathematical modelling. The numerical relationships and design rules required were derived from manufacturer's catalogue data and yielded acceptable results for non-standard sizes and configurations.

Design shells under development for other measurands including pressure (by diaphragm displacement), electrical conductivity (of liquids) and dissolved oxygen (in natural and waste waters) have been reported (Sydenham, 1987b). These employ a combination of direct design (via algorithms), heuristic and reverse engineering approaches as appropriate and as available. The ThermoShell tool, mentioned above, was constructed making use of an earlier system (Sydenham et al., 1988b).

41.5.5 Calibrating the Instrument

Specialist software has been developed to extend the advantages of CAE to the calibration and certification of instruments. Poulter (1985) reports the advantages realized and problems experienced from the use of computers in calibration work associated with engineering metrology. Three levels of computer application may be distinguished, as follows:

(1) Direct entry of manually acquired measurement data. This permits immediate (on-line) data analysis, provides feedback to the calibration engineer and hence (manual) quality control of the calibration procedure can also be used to impose a strict measurement sequence. Problems can arise, however, when the computer is given decision-making powers, for example the power to reject out-of-bounds or (apparently) erroneous measurements.

(2) Direct computer acquisition of the calibration measurements. The transfer from manual to computer data acquisition is a major step as it usually involves a substantial redesign of the calibration equipment. This will involve considerable expense in both hardware and software but can yield major improvements in the speed of calibration and the virtual elimination of human errors. A less fortunate result is that the operator has been to a considerable extent removed from the calibration process and hence when anomalous results or faults occur they can be more difficult to rectify.

(3) Computer control of the complete calibration procedure. This is the final stage in automation and fully computer-controlled metrology has become a necessary part of computer integrated manufacturing (CIM). The quality and reliability of software required here is very high and hence the development cost will be substantial. A research example, Dianos (1988), demonstrates how complete set-up, initial calibration and life-long calibration can be achieved for an example sensor—a thermistor thermometer—using a personal computer.

Software for engineering metrology

A major requirement of engineering metrology is geometrical measurement, and software for automated dimensional metrology will contain at least the following five major features:

- definition of geometric form;

- specification of the measurement procedure;

- a mathematical model of the measurement(s);

- mathematical algorithms appropriate to the geometric form; and

- a validation procedure,

as set out by Cox and Jackson (1983). Production of adequate geometrical definition and associated mathematical description is critical to provide the necessary level of software reliability and hence user confidence (Anthony and Cox, 1984). Nonetheless problems of measurement interpretation can occur (Poulter, 1985) and Figure 41.19 illustrates how the error in straightness might be quoted differently according to the definitions of tolerance adopted and the statistical procedures employed. A useful account of such a system is given in Seiffert (1990).

41.5.6   Maintaining and Troubleshooting the Instrument

With the increased volume and complexity of electronic test equipment in use there exists a need for the automation of diagnostic testing ('troubleshooting'). Faults occur less frequently due to improved quality control and inherent reliability and hence it is difficult for technical staff to acquire adequate real (as opposed to simulated) fault-finding experience on a particular item of instrumentation.

Diagnostic testing clearly involves a procedure dominated by chains of IF. . . THEN. . . steps following the logical flow of signals through the instrument. However, the procedure required would normally be very different for every instrument type.

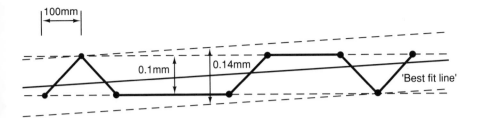

Figure 41.19   An example of how differing measures of straightness error (0.10 mm or 0.14 mm) are obtained for different choices of datum. Both pairs of parallel lines shown contain all the measurement points but only the wider pair (yielding 0.14 mm) is referred to the 'best-fit line'. (Reproduced from Poulter, 1985)

To avoid the need to write a separate routine for each instrument a knowledge-based approach may be adopted. Here the unique properties of the device are encapsulated in a database and used with a conventional expert system shell. Prevost and Laffey (1985) report the design of such a system for the diagnostic testing of a commercial digital voltage source (DVS), an item of bench test equipment judged to be of typical size (9 p.c. cards, 39 circuits and 300 components), moderate complexity and for which significant fault-finding experience was available. The system uses the general-purpose Lockheed Expert System (LES) (Perkins and Laffey, 1984) with knowledge derived directly from human experts (experienced serviced technicians). The knowledge is represented in three forms, namely:

● a frame structure of factual information which effectively models the instrument;

● a hierarchical set of IF...THEN... rules which describe how to identify faults; and

● a set of WHEN...THEN... rules which control the reasoning process.

The WHEN...THEN... rules are used to implement the judgement of the human expert and operate by modifying the priorities of various fault-finding (IF...THEN...) paths possible. The operation of the system is thus 'expert-like' in that it appears to mimic the thought processes of an experienced service technician.

Prevost and Laffey (1985) report greatly improved acceptance of this system by technical staff. These staff are required to conduct tests on the faulty instrument as directed by the system and they expect to be required to conduct only those tests necessary to find and rectify the fault. Previous computer- controlled diagnostic systems have often met with a hostile reaction as they required (apparently) unnecessary tests to be conducted of no obvious relevance to the problem at hand.

41.5.7   CAE for Modular Instrument Systems

A number of software tools have been developed to assemble and configure modular instrumentation systems, that is those created largely from standard instruments or modules. The software available to support GPIB-interconnected test equipment is an example of this and is discussed in Section 41.4.3. However, more general tools are also available with application wider than automatic testing (ATE systems).

Modular instrumentation are almost invariably computer controlled and the ease and speed with which a system may be created is of obvious importance. The direct approach of traditional programming (for example in BASIC, the most commonly employed language for systems controlled by a personal computer) requires that a working program be created before any measurement can be taken: the user must possess programming skills.

The 'virtual instrument'

An alternative approach is that of the 'virtual instrument' in which the user develops the required system block diagrams, instrument front panel controls and output displays by means of specialized graphics on the computer screen. Thus a virtual instrument may be defined as an instrument whose entire functions and capabilities are determined in software.

The concept of providing a human interface appropriate to measurement engineers and instrument users was first implemented by National Instruments Corp. as the LabVIEW (Laboratory Virtual Instruments Workstation) package (Santori, 1990). The software is designed to enable users to create their own virtual instruments by developing a set of (specialized) block/flow diagrams which describe the measurement functions and procedure. An appropriate 'front panel' for the instrument using a library of controls, indicators and displays is indicated in Figure 41.20.

The virtual instrument may then, of course, be connected to the physical world and used to make measurements, for example via a signal acquisition card in the computer itself or via modules in a VXI chassis. It is then very much a 'real' instrument. This style of instrument is under continuous development, a term being used is the *surrogate instrument.*

Database systems

For large and extensive instrumentation systems management of the design and documentation of the system becomes a major task and is obviously amenable to computer assistance.

A modest system running on a personal computer is reported by Bownds and Gillett (1984) as an aid to, and resource for, process instrumentation designers. Koskinen (1985) reports the development of a larger package labelled as a

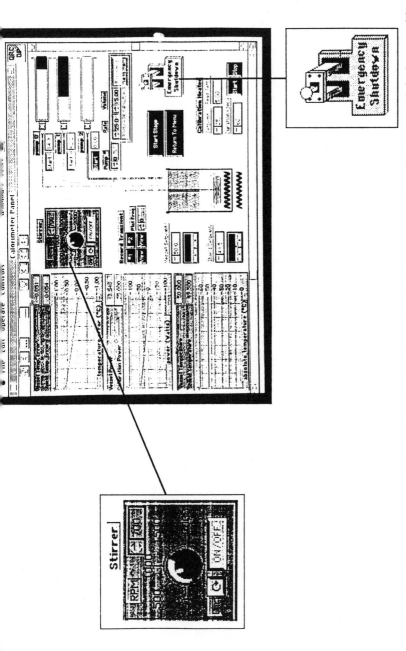

Figure 41.20 A virtual instrument screen for a chemical reaction thermal analyser with detail of two of the 'controls', both operated by the computer's mouse, are shown enlarged. The virtual instrument was developed by Columbia Scientific Industries, Austin, Texas and is based on LabVIEW 2 software by National Instruments Corp. to run on a Macintosh II computer. (From Santori, 1990)

computer aided instrumentation design system which handles also cabling, alarms and interlocks. Major process control system vendors now provide software to assist setting up the plant process and in doing so automatically produce the ordering and installation specifications.

In both examples major advantages are realized by the maintenance of a comprehensive, universal data base as part of the package. Although set up primarily to service information requirements for design, the data base is also directly applicable to the production of standardized instrument specifications, to procurement and to the management of maintenance programmes.

Both report very positive user experience and major cost savings due in large measure to the availability of system information which is reliable, single-valued and error-free.

Fox (1986) describes a very large installation (a chemical manufacturing plant) involving over 3000 instruments for which six interrelated data bases were established, namely:

- the process and instrument drawing (P&ID) data base (containing the overall instrumentation design derived from a CAD system);

- the instrumentation inventory data base;

- the instrumentation specification data base;

- the purchasing data base;

- the I/O configuration and construction data base; and

- the process control computer configuration data base.

Overall Fox (1986) reports a saving in total system engineering time of two thirds compared with management using manual procedures.

Expert systems control of data acquisition systems

Setting up the interface parameters of a data acquisition plug-in board for personal computers requires programming expertise not always available to the user. To overcome this limitation products and published reports show how an expert system can be used to make the task feasible by non-programmers. ANNIE is such a system based on the CRYSTAL expert system shell. The same shell has been used to construct a small data collection and controller system (Harris et al., 1990).

Computer aided selection of instruments

Several projects have sought to streamline the selection of proprietary sensors. The THESAC project (Warren Spring Laboratory, UK) attempted to cover just

73 variables found in the process industry. This led to the identification of over 100 000 products to be included. Such services are yet to find acceptance, major barriers being that users are not willing enough to pay for the development of such systems and the proliferation of sensor products and their currency as a saleable product provides an element of change that militates against the service ever catching up. More progress in the formalized classification of sensors may eventually lead to an efficient methodology but that is still to be researched at a useful level.

41.6 CONCLUDING REMARKS

Clearly a considerable range of useful tools exists for the CAE of instrumentation. However, at present most are 'general purpose' and drawn from disciplines other than measurement science.

The need for and potential advantages of specialized tools for the CAE of instruments is also clear and recent work has shown it is possible to integrate the tools available into useful CAE packages. The measurands derived from the use of MINDS (Section 41.5.1), for example, could lead directly into Specriter (Section 41.5.2) which would invoke the appropriate design shells (Section 41.5.3) to complete the design of measurement system—as illustrated in Figure 41.21. This possibility has led to the concept of CAEINST (Sydenham, 1987b).

Furthermore, if the system can be made complete and sufficiently robust the non-specialist engineer or scientist might reasonably expect to be able to specify, design and manufacture their own measurement system—a 'do-it-yourself' approach as set out by Sydenham et al. (1988a).

Whilst this might appear fanciful the success and positive reactions to the demonstrator package ThermoShell (Section 41.5.3), which integrates sensor knowledge and existing tools demonstrates that it is both possible and of value to the user. Considerable further development is obviously still required and will be heavily dependent on the techniques of knowledge engineering. It is likely also that applications will be restricted to well-trodden paths of instrument design technique.

It is certain that the range of general, computer-based tools applicable to the CAE of instruments will continue to increase: these tools will be of immediate benefit to the professional instrument designer in the rapid creation of new measurement systems which will continue to increase in complexity at a rapid rate.

41.7 SOURCES OF FURTHER ADVICE

The reader is referred to the following textbooks and journals for further information on the CAE of instrumentation.

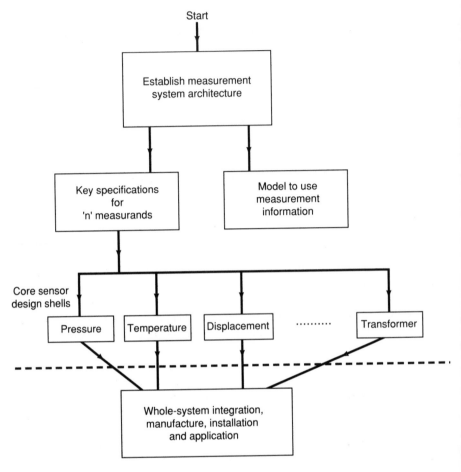

Figure 41.21　An integrated approach to the design of a complete measurement system—the concept of CAEINST. (From Sydenham, 1987b)

Books

O'Reilly, W. P. (1986) *Computer Aided Electronic Engineering* Van Nostrand Reinhold, New York.

Gero, J. S. (ed.) (1988) *Artificial Intelligence in Engineering Design* Elsevier, Amsterdam.

Dimarogonas, A. D. (1989) *Computer Aided Machine Design* Prentice-Hall, New York.

Vlach, J. and Singhal, K. (eds) (1987) *Selected Papers on Computer-Aided Design of Analog Networks* IEEE Press, New York.

Greenbaum, J. *et al.* (1988) *Analysis and Design of Electronic Circuits Using PC's* Van Nostrand Reinhold, New York.

Journals

Computer Aided Design Butterworths Scientific, UK.

Computer-Aided Engineering Journal IEE, UK.

Measurement IMEKO, UK.

REFERENCES

Abdullah, F., Finkelstein, L. and Rahman, M.M. (1977). 'The application of mathematical models in the evaluation and design of electromechanical instrument transducers', *J. Appl. Sci. Engng*, **A2**, 3–26.

Anthony, G.T. and Cox, M.G. (1984). *The Design and Validation of Software for Dimensional Metrology*, Report DITC 50/84 (October, 1984), National Physical Laboratory, Teddington, UK.

Bacon, J.M. (1989). *Instrumentation Installation Project Management System*, Instrument Society of America.

Barber, T.J., Marshall, G., and Boardman, J.T. (1988). 'Tutorial—a philosophy and architecture for a rule-based frame system: RBFS', *J. Eng. Applic. of AI*, **1**, 67–86.

Barker, P. (1988a). *Multi-Media Computer Assisted Learning*, Kogan Page, London.

Barker, P. (1988b). 'Knowledge engineering for CAL', in *Multi-Media Computer Assisted Learning*, Barker, P. (ed.), Kogan Page, London.

Barwicz, A. and Morawski, R.Z. (1989). 'Requirements for tools of computer-aided design of measuring systems', *IEEE Instrument and Measurement Technology Conf.*, Washington DC, April 1989, 150–2.

Bench-Capon, T.J. M. (1990). *Knowledge Representation*, Academic Press, London.

Berlin, L.M. (1989). 'User-centered application definition: a methodology and case study', *Hewlett Packard J.*, **40** (5), 90–7.

Besant, C.B. and Lui, C.W.K. (1986). *Computer-Aided Design and Manufacture* (3rd edn), Wiley, Chichester.

Bosman, D. (1978). 'Systematic design of instrumentation systems', *J. Phys.E: Sci. Instrum.*, **11**, 97–105.

Bownds, A.F. and Gillett, J.S., (1984). 'Computer applications in instrument design', *ISA Conf. No. 39: Advances in instrumentation*, Instrument Society of America, Houston, Texas, USA.

Cendes, Z.J. (1990). 'Electromagnetic simulators', in 'Special guide to software—systems, packages and applications', Kaplan, G., *IEEE Spectrum*, **27** (11), 47–102.

Cook, S.C. (1988). 'Automatic generation of measuring instrument specifications', *Measurement*, **6** (4), 155–60.

Cook, S.C., (1990). 'Knowledge-based generation of measuring instrument specifications', *IMEKO Symposium on Knowledge-Based Measurements, Karlsruhe*, October, VDI Verlag, Book number 856, 145–52.

Cookson, R.A. and El-Zafrany, A. (1987). 'State of the art review of the Boundary Element Method', in *Advances in the Use of the Boundary Element Method for Stress Analysis* I. Mech. E., Mechanical Engineering Publications, London.

Cox, M.G. and Jackson, K. (1983). *Algorithms and Software for Engineering Metrology: a Statement of Need*, Report MOM65 (June 1983), National Physical Laboratory, Teddington, UK.

Dianos, A. (1988), *Sensor Electronic Testing Using a PC*, Internal Report, Measurement and Instrumentation Systems Centre, University of South Australia.

Edmonds, E.A. (1989). 'Intelligent measurement and sensing systems: speech knowledge externalisation', *Australasian Instrumentation and Measurement Conference*, Adelaide, Australia, **AIM-89**, 22–5.

Eurich, J.P. and Roth, G. (1990) 'EDIF grows up', in 'Special guide to software—systems, packages and applications', Kaplan, G., *IEEE Spectrum*, **27** (11), 47–102.

Finkelstein, L. and Finkelstein, A.C.W. (1983). 'Review of design methodology', *Proc. IEE*, **130** (4), 213–22.

Finkelstein, L. and Watts, R.D., (1983). 'Fundamentals of transducers: description by mathematical models', in *Handbook of Measurement Science*, Vol.2, Sydenham, P.H. (ed.) John Wiley and Sons, Chichester, p.p.747–95.

Fischer, J. (1984). 'PC-based instrumentation merges with computer-aided engineering' *IEEE Westcon 84 Conference Record*, Anaheim, California, 6/5/1–3.

Fox, W.A. (1986). 'Instrumentation systems using computer databases', *Proc. 5th Annual Conf. on Control Eng., Rosemount, IL, USA*, 165–70.

Francis, A. (1983). 'The social effect of CAE in Britain', *Electronics and Power*, **29** (1), 72–4.

Freiling, M.J., Rehfuss, S., Alexander, J.H., Messick, S.L. and Shulman, J.S. (1986). 'The ontological structure of a troubleshooting system for electronic instruments', *Conf. on Applications of AI in Engineering*, April, Southampton, UK 609–20.

Godwin, R.J. (1975). 'An extended octagonal ring transducer for use in tillage studies', *J. Agric. Eng. Research*, **20** (3), 347–52.

Grattan, K.T.V. (1990). 'Photothermal excitation of resonant structures in fibre optic sensors: application of computer modelling techniques', *Sensors and Actuators*, **A21–3**, 1146–9.

Goulding, F.S. (1988). 'The professional realities of modern electronic instrument design', *IEEE Trans. on Nuclear Science*, **35** (3), 1017.

Haigh, M.J. (1985). *An Introduction to Computer-Aided Design and Manufacture*, Blackwell, Oxford.

Hallenbeck, J.J., Cybrynski, J.R., Kanopoulos, N., Markas, T and Vasanthavada, N. (1989). 'The Test Engineer's Assistant', *IEEE Computer*, **22** (4), 59–68.

Harris, D.D. and Hancock, N.H. (1991). 'ThermoShell: an instrument design system for the non-expert user' *Computer-Aided Eng. J.*, **8** (5), 195–9.

Harris, D., Zhang, F. and Sydenham, P.H. (1990) 'Low cost, real time, expert system based, monitoring and control system', *Proc. Conf Control 90*, Inst. Meas. Control, New Zealan, paper 22.

Harris, D.D., McNiell, T. and Sydenham, P.H. (1991). 'Pre-schematic electronic designer', *Proc. Conference on Artificial Intelligence in Engineering*, May, Oxford, UK.

Haskard, M.R. (1990). *Application Specific Integrated Circuits*, Prentice Hall, Sydney.

Hines, J.R. (1990). 'Analog design', in 'Special guide to software—systems, packages and applications', Kaplan, G., *IEEE Spectrum*, **27** (11), 47–102.

Ibrahim, K.F. (1988). *Instruments and Automatic Test Equipment*, Longman, UK.

Jain, L.C. and Harpas, P. (1988). 'PC software for electronic circuit design in printed circuit form', *CAE J.* **5** (4), 148–52.

Jessen, K. (1989). 'VXIbus: a new interconnection standard for modular instruments', *Hewlett Packard J.* **40** (2), 91–5.

Kaplan, G. (1990). 'Special guide to software—systems, packages and applications', *IEEE Spectrum*, **27** (11), 47–102.

Koskinen, K. (1985). 'Computer aided design system for instrumentation engineering', *Proc. 3rd IFAC/IFIP Symp. on Computer Aided Design in Control and Engineering Systems*, Lyngby, Denmark, 311–14.

Kurtz, B.D., Ho, D. and Wall, T.A. (1989). 'An object-oriented methodology for systems analysis and specification', *Hewlett Packard J.* **40** (2), 86–90.

Lesgold, A., (1987). 'Intelligent tutoring systems; practice opportunities and explanatory models', *Conference of the Australian Society for Computers in Learning, 1987*, Sydney, Australia.

Leung, K.S. and Wong, M.H. (1990). 'An expert-system shell using structured knowledge', *IEEE Computer*, **23** (2), 38–47.

Liebner, R.D. and Abdullah, F. (1981). *User's Guide to MEDIEM*, Doc. DSS/RDL/FA/226, The City University, London.

Loughry, D.C. (1983). *IEEE Standard Digital Interface for Programmable Instrumentation and IEEE Recommended Practice for Code and Format Conventions for use with ANSI/IEEE Std.488–1978*, The Institution of Electrical and Electronic Engineers, New York.

McSherry, A. (1987). 'Development of computer-based teaching using AUTHOR', *Conference of the Australian Society for Computers in Learning, 1987*, Sydney, Australia.

Mather, B.C., (1990). 'Needed DSP software emerges', in 'Special guide to software—systems, packages and applications', *IEEE Spectrum*, **27** (11), 47–102.

Mirza, M.K., Neves, F.J.R. and Finkelstein, L. (1990). 'A knowledge-based system for design-concept generation of instruments', *Measurement* **8** (1), 7–11.

Narciso, S. (1990). 'The VXIbus from the instrument designer's perspective', *IEEE Instrumentation and Measurement Technology Conference, IMTC/90*, San Jose, California, February, 280–5.

Nemeth-Johannes, J. (1990). 'A standardised instrument programming language based on IEEE Std.488.2', *IEEE Instrumentation and Measurement Technology Conference, IMTC/90*, San Jose, California, February, 306–10.

Park, C.Y.W. and Talley, B.J. (1990). 'HP Manufacturing Automation Protocol 3.0', *Hewlett Packard J.*, **41** (4), 6–14.

Perkins, W.A. and Laffey, T.J. (1984). 'LES: a general expert system and its applications', *Proc. SPIE Technical Symposium East—Applications of Artificial Intelligence Applications*, May, Arlington VA, USA, 46–57.

Peuscher, F.G. (1983). 'Design and Manufacture of Measurement Systems', in *Handbook of Measurement Science*, Vol.2, Sydenham, P.H. (ed.), Wiley Chichester, pp.1209–49.

Poulter, K.F. (1985). 'Computer-aided dimensional metrology, advantages and disadvantages', *Proc. 7th Intl. Conference on Automated Inspection and Dimensional Metrology*, Birmingham, UK, 37–44.

Prevost, M.P. and Laffey, T.J. (1985). 'Knowledge-base diagnosis of electronic instrumentation', *IEEE 2nd Conference on Artificial Intelligence Applications*, Miami Beach, USA, December, 42–8.

Rudolph, C., Schwetlick, H. and Filbert, D. (1990). 'Knowledge-based automation of measurement system design', *IMEKO Symposium on Knowledge-Based Measurements*, Karlsruhe, October, VDI Verlag, Book number 856, 233–40.

Santori, M. (1990). 'An instrument that isn't really', *IEEE Spectrum*, **27** (8), 36–9.

Seiffert J.P. (1990). 'Statistical process control—TESA's concept' *Proc. Conference 'Control 90'*, Inst. Meas., Control, New Zealand, Paper 23.

Short, R.F. (1988). 'Training of deck officers—electronic navigation simulator', in *Multi-Media Computer Assisted Learning*, Barker, P. (ed.) Kogan Page, London.

Sydenham, P.H. (1985a). 'Structured understanding of the measurement process—Part 2: Development and implementation of a measurement process algorithm', *Measurement*, **3** (4), 161–8.

Sydenham, P.H. (1985b). 'Structured understanding of the measurement process—Part 1: Holistic view of the measurement system', *Measurement*, **3** (3), 115–20.

Sydenham, P.H. (1987a). 'Design shells for common measuring sensors', *International Conference on Modelling and Simulation*, Melbourne, Australia, October.

Sydenham, P.H. (1987b). 'Computer-aided engineering of measuring instrument systems' *Computer-Aided Engineering J.* **4**(3), 117–23.

Sydenham, P.H. and Jain, L.C. (1988). 'CAENIC—user-characterised, electronic systems development software', *CAE J.* **5** (5), 200–5.

Sydenham, P.H., Skinner A. and Beijer, R. W. (1988a). ' "Do-it-yourself" measurement and control package', *J. Measurement and Control*, **21**, 69–75.

Sydenham, P.H., Morgan, N. and Anfiteatro, M. (1988b). 'Thermocouple temperature sensor CAD tool', *CAE J.*, **5** (5), 206–10.

Sydenham, P.H., Hancock, N.H. and Thorn, R. (1989). *Introduction to Measurement Science and Engineering*, John Wiley, Chichester, UK.

Sydenham, P.H., Harris, D.D. and Hancock, N.H. (1990). 'MINDS—A software tool to establish a measuring system requirement', *Measurement*, **8** (3), 109–7.

Terry, W.E. (1990). Executive Vice-President, Hewlett Packard Company, 'Emerging measurement technologies', the Algie Lance Keynote Address to the *IEEE Instrumentation and Measurement Technology Conference, IMTC/90*, San Jose, California, February 1990. Published in abridged form in *IEEE Instrumentation and Measurement Society Newsletter*, Summer 1990, 10–11.

Terwisscha van Scheltinga, J.A.S. (1991). *PROMINDS in Apple Macintosh*, Internal Report, Measurement and Instrumentation Systems Centre, University of South Australia.

Tran Tein, L.F. (1990). 'Toward a rational conception of transducers', *Abs. Conference Eurosensors*, Karlsruhe, October.

Tuinenga, P.W. (1988). *SPICE—A Guide to Circuit Simulation and Analysis using PSpice*, Prentice Hall, Englewood Cliffs, N. J.

USDD (1985). *MIL-STD-490A, Military Standard—Specification Practices*, United States of America Department of Defense, June.

Van Biesen, L. and Schoukens, J. (1990). 'IMMI: A concept and model for the design of knowledge based instrumentation', *IMEKO Symposium on Knowledge-Based Measurements*, Karlsruhe, October, VDI Verlag, Book number 856, 161–71.

Van Biesen, L.P., Schoukens, J., Barel, A.R.F., Renneboog, J., Pintelon, R.M. and van den Bossche, M. (1989). 'The IMMI concept of the VUB, Department of Fundamental Electricity and Instrumentation, as a CAD tool for intelligent instrumentation', *Australasian Instrumentation and Measurement Conference, AIM-89*, Adelaide, Australia, November, 32–8.

Warman, E.A. (1990). 'Object-oriented programming and CAD', *J. Eng. Design*, **1**(1), 37–46.

Warman, E.A. and Yoshikawa, H. (1980). *Design Theory for CAD*, North Holland, Amsterdam.

Chapter

42 J. R. JORDAN

Communication Standards for Measurement and Control

Editorial introduction

There is a strong tendency to ignore, or be unaware, of the importance of communication in the modern multi-sensing system. As can be seen by the content of other chapters in this Handbook—especially Chapters 35, 36 and 44 where multi-sensor systems are essential in implementing machine monitoring across a large plant, controlling pollution of large areas of land and in setting up quality control systems on machine tool centres,—communication technology needed to become specialized for these inanimate sensor-based systems. Chapter 13, in Volume 1, provides an account of the theory and practice of data transmission. This chapter is concerned with the increasingly important topic of communication standards in measurement and control.

42.1 INTRODUCTION

A very large international activity has driven standards development to the point where a detailed knowledge of standards documentation is seen to be an essential part of product development. In fact product definition and standards development are now seen as parallel activities and standards can appear before products reach the market place. Users will benefit from a standards driven approach to products because a detailed study of published standards will provide information that will reliably define the functionality of future systems. In this chapter the communication standards applicable to measurement and control will be described and particular consideration will be given to the use of serial data high-

Handbook of Measurement Science, Volume 3
Edited by P. H. Sydenham and R. Thorn
© 1992 John Wiley & Sons Ltd

ways to interconnect field devices. This contrasts with information given in Chapter 13, which related to telecommunications applications.

It is important to recognize that standards bodies form committees whose members are usually volunteered by industrial organizations at their own expense. The international standards groups hold meetings all over the world so travel expenses are large, and therefore small, innovative companies may find it difficult to participate. Benefits that can be expected to arise from an involvement with standards development include increased sales of standard based products, an increased understanding of user needs and an indication of industrial competitive trends. It is clearly important for equipment users to participate at an early stage but it is likely that this will be difficult to economically justify. The combined expert knowledge of each committee is distilled into a standards document that defines a concept which industry is expected to interpret to enable products to be manufactured in conformity to the standard. Hardware and implementation details are not specified by a standard. Hence testing for conformance of a product to a particular standard is important. An objective of communication standards is the facilitation of the construction of multi-vendor systems and therefore testing for interoperability of equipment (which is not guaranteed by conformance) will become increasingly important.

The importance of standards based products has been emphasized by a 1985 European Community Ministers agreement on a 'New approach to technical harmonization and standards' to facilitate the free movement of goods in the single European market. Essential requirements which must be met before products may be sold in the community are defined by European standards and products meeting the requirements will carry the 'CE mark' (DTI, 1989). Standards will be prepared by the European Standards bodies, CEN (European Committee for Standardization) and CENELEC (European Committee for Electrotechnical Standardisation) which bring together the national standards bodies of the Community and EFTA. Appendix 42.1 lists the names and position of standards bodies related to communication systems used in measurement and control.

A large number of groups contribute to the communication standards work of the USA. The American National Standards Institute (ANSI) coordinates the writing and approval of domestic standards. The Electronic Industries Association (EIA) is responsible for the RS series of standards. The National Bureau of Standards (NBS), now the National Institute of Science and Technology (NIST), is a US government organization which carries out research and development of scientific measuring methods for standards purposes. The Instrument Society of America (ISA) represents the US process control Industry—it originated the 4–20 mA current loop standard and it is heavily involved in the development of the field bus standard—see Section 42.4.1. The Institute of Electrical and Electronic Engineers (IEEE) is involved with standards in many areas of electrical technology and acts through the US national committee of the IEC and through ANSI for the ISO.

Two key international standards bodies are involved with communication standards of relevance to measurement and control. The International Electrotechnical

Commission (IEC) is involved with standards covering the whole range of electrotechnology and it is currently coordinating the field bus activity. The International Organisation for Standardisation (ISO) is particularly involved with the Open System Interconnect (OSI) for information and data processing. The term 'open' was chosen to signify that OSI standards would allow systems to be constructed that would enable communication with other systems conforming to the same standard.

The process control industry currently uses the twisted pair with 4–20 mA signalling range as a standard method for coupling remote instrumentation in a star (point-to-point) configuration to a central control area. A standard for this analog connection method (Mathews, 1986) was established in 1970 (BS 3586) and is now supported by a wide range of international equipment suppliers. Attempts to use a digital communication link were initiated soon after the first digital computers were in practical use (see, for example, Keefe et al., 1967 and Collins, 1968). However, developments in silicon circuit technology and communication protocols were required before significant progress could be made.

Cost-effective implementation of the so-called process-control smart (Barney, 1985; Middelhoek et al., 1988) transmitters appeared in the early 1980s and required large-scale silicon circuits. A digital communication technique is required for configuring and interrogating the smart transmitters. The use of a digital modulation signal superimposed on the 4–20 mA analog signal enables a digital communication link to be established but it will be limited to use with the star connection topology. Maximum benefit from the use of the digital link will be obtained only when a serial connection of remote instrumentation can be achieved.

Outside of the process control industry a major driving force can be identifiedin the efforts being made to automate factory and office systems (Morgan, 1987).

This introduction has stressed the importance of the standards activity and briefly reviewed some of the key historical steps in the development of a practical serial data highway for measurement and control applications. Grimes (1982) (HBMS, vol.1, Chapter 13) describes the physical properties of connecting media. A more detailed chronology of the relevant standards and technological developments is shown in Table 42.1. Key dates in the development of integrated circuit technology have been included to provide an indication of the speed of development of one of the major driving forces for the creation of practical serial data highway systems.

The remainder of this chapter will discuss the OSI seven-layer model and its use in several proposed serial highway systems, including MAP. Several of the well-established bus systems will be described, namely MIL-STD-1553B (Military systems), ARINC 629 (Civil aviation), CAMAC (Nuclear instrumentation) and GPIB (Electronic instruments). This will be followed by a discussion on the emerging standards for serial highways, namely the field bus, the consumer electronic bus (CEbus), the control area network and the IEEE standard microcomputer serial bus. A concluding section will discuss future trends.

Table 42.1 Chronological development of measurement and control communication standards and related technologies. (The RS family of serial interface standards is not discussed in this chapter. For more information see Grimes (1982) and Freer (1987))

1951	Discrete transistors commercially available.
1955	4–20 mA process control instrumentation.
1960	Small-scale integration (3–100 components per chip).
1969	Large-scale integration (10^3–10^4 components per chip), RS-232 serial interface standard.
1970	BS 3586, 4–20 mA standard.
1975	Very large-scale integration (more than 10^4 components per chip), RS-422/423 serial interface standard.
1978	ISO seven-layer reference model for OSI.
1980	General Motors factory automation working group.
1981	Smart transmitters.
1983	ISO 7498-international standard for the basic reference model for OSI.
1984	MAP version 1.0
1985	IEC TC 65C/WG6 — field bus working group.
1986	UK DTI CIMAP 1553B demonstration field bus at the National Exhibition Centre, Birmingham, England.
1987	MAP version 3.0
1992	Field bus standard?

42.2 OSI AND MAP

42.2.1 OSI

The interconnection and inter-working of multi-vendor systems has been a major concern of engineers developing data processing and data collecting equipment. A need for a standardized communication architecture was recognized internationally in 1977 when the first attempts were made to produce a general description of the communication process. The Open System Interconnect (OSI) model comprehensively defines the tasks needed to establish an interconnection and subdivides these tasks into seven nested layers, with the bottom layer (layer 1) defining physical details and the top layer (layer 7) defining notation and operations specific to the particular application (Day and Zimmermann, 1983). It is important to note that the model does not define the protocols to be used by each layer. The first definition of the model appeared in 1978 (ISO 7498) but since then many additions have been made and these are now collectively referred to as the Open Systems Interconnect (MacKinnon *et al.*, 1990). The existence and acceptance of international standards specifying the connection of the lowest level measurement and control devices to the highest levels of data processing will have a considerable impact on domestic and industrial products.

The subdivision of the total communication operation into layers is arbitrary but it is a necessary requirement to achieve a flexible easily changed system. Each layer is designed to be independent of the layers above and below it so layers may be changed as technological capabilities change without affecting the operation of the overall system. Some of the specified seven layers will not be needed at all in some applications and some layers may be divided into sublayers.

Since multi-vendor operation is an objective of the standardization of the communication process it is clear that the data presented to the application layer should be in some standard, agreed format, and its scope and notation should be specified. Strictly, therefore, the OSI reference model should be an 8-layer model with the upper layer specifying the user field.

Layers 7 to 1 are called, respectively, the application, presentation, session, transport, network, data link and physical layers. Constructing a data frame for transmission involves each layer adding header bits to the user data, as shown in Figure 42.1. At the data link layer and the physical layer trailer bits are also added for error control and word recognition purposes. A data frame is reduced to user data by performing the reverse operation of successively stripping header and trailer bits as each layer acts on the frame and passes it on to the next layer. The function of each layer of the OSI model is very briefly listed below:

- *Application layer (7)*. Interfaces the user requirement to the seven-layer model. In MAP messaging services are provided by the manufacturing messaging standard, (MMS). MMS is essential for achieving vendor independent operability between field devices.

- *Presentation layer (6)*. This layer ensures translations between the way of sending and receiving systems construct data. It negotiates the selection of the transfer syntax and provides for conversions to and from it.

- *Session layer (5)*. The function of this layer is to set up, manage and synchronize dialogues between application processes.

- *Transport layer (4)*. This layer ensures reliable end-to-end data transfer. The transport layer provides a connection oriented service ie. it establishes the connection, exchanges the data as required and, arranges and controls the disconnection. A connectionless network (ie where data sent out carries all the route information needed to make a transfer) can provide faster transfers. A connectionless network layer has been devised to support the connection mode transport layer (Chapin, 1983; Dwyer and Ioannou, 1989).

- *Network layer (3)*. This layer enables the routing of messages between sub-networks and mixed media networks. This layer is not needed if communication across a number of networks is not required.

- *Data link layer (2)*. This layer carries out the functions necessary to ensure accurate transmission between nodes. The IEEE 802 specification (Madron, 1989) for the data link layer subdivides it into a logical link control (LLC)

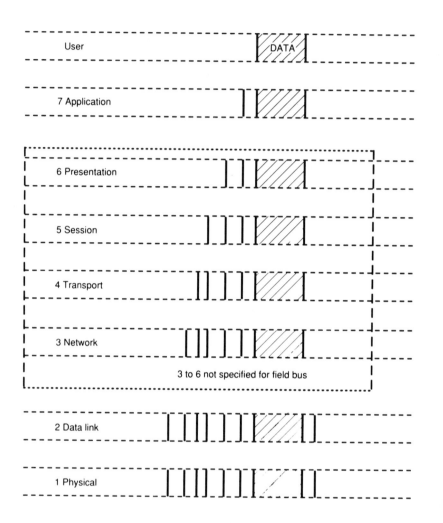

Figure 42.1 The ISO seven-layer model. Schematic diagram showing the use of leading and trailing bit patterns as data is prepared for transmission

sublayer (responsible for error control) and a media access control (MAC) sublayer.

- *Physical layer (1).* This layer defines the physical medium connecting the communication nodes.

Essentially the bottom four layers are concerned solely with the communication link while the top three layers are concerned with the user's application requirements.

42.2.2 MAP

In 1980 the American company, General Motors, established a group to develop an independent computer network capable of sustaining a multi-vendor equipment environment on the factory shop floor. This work has resulted in the creation of the Manufacturing Automation Protocol (MAP) and by 1984 the first specification (version 1.0) was published. An international MAP activity has been established and by 1987 version 3.0 of the MAP specification had been published (Dwyer and Ioannou, 1989). Technical and Office Protocol (TOP) is being devised for office and design applications and has the same methodological base as MAP. MAP and TOP are both based on the OSI seven-layer reference model (Morgan, 1987). At each of the layers of the model more than one standard can be used, with the result that care is still required to ensure that multi-supplier systems will operate correctly. The network, data link and physical layers are covered by the IEEE 802 family of standards (Madron, 1989).

The full MAP communication spine is a broadband 10 Mb/s system. It is multi-channel and access to the network is by the token-passing method. Carrier band MAP is single-channel, slower (5 Mb/s) and lower cost. Carrier band is

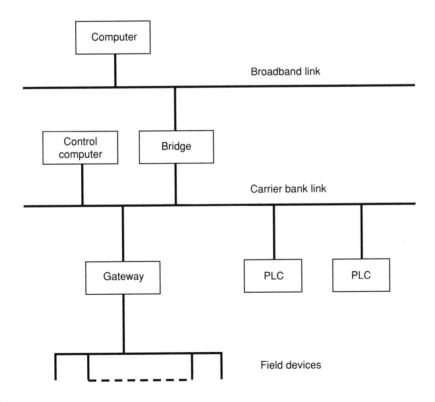

Figure 42.2   Manufacturing automation protocol (MAP). A typical topology

suitable for linking together clusters (cells) of instrumentation. The typical topology of a MAP system is shown in Figure 42.2. At the device level implementations of the full seven-layer capability is too complex, too slow and too expensive, and a method is required for sending short messages (with a modest data overhead) at frequent intervals. Several devices have been specified to enable communication networks to be interconnected. The most complex requirements arise when an OSI network is to be connected to a non-OSI network. The gateway architecture uses the full seven-layer stack to achieve this connection. Communication using a gateway is slow because every transaction is translated via its full seven layers. The bridge architecture links two compatible networks through a common data-link layer, hence networks linked must use the same addressing scheme and frame size. Since the bridge uses only layers 1 and 2 and they are transparent to the transmitting node the communication rate is reduced by its use. The bridge would be used, for example, to link a broadband MAP network with MAP carrier band subnetworks. A router links two or more networks with a common network layer protocol and enables different networks with a common network protocol to be linked. The router has to be addressed, it is not transparent to transmitting nodes. The existence of these inter-networking devices underlines the problems of establishing a universal standard approach to communication networks. Since user requirements are so varied it is likely that a need for these devices will always exist. The use of proprietary networks exacerbates this problem. It is clearly desirable that designers of future equipment should avoid specialized, non-standard networks.

The enhanced performance architecture (EPA) and the mini-MAP nodes provide a solution to the problem of providing MAP related communication for instrumentation within a cluster of equipment. An EPA node implements a seven layer link to full MAP and in addition allows an application to communicate directly with the data link layer. The mini-MAP node implements only the bottom two layers and requires to be linked to an EPA node to communicate with the full MAP system. Equipment manufacturers are now providing proprietory units for connecting field equipment to MAP systems. For example, the MAP Equalizer (Reflex Manufacturing Systems Ltd, Crawley, West Sussex, England) allows a large range of shop floor devices such as, PLC's, machine tools and robots which do not have MAP interfaces to be connected to a MAP network and communicated with via the MAP, MMS standard. It should be noted that MMS specifications cover numerical control machines, programmable controllers, robot controllers and process control systems.

42.3 ESTABLISHED BUS SYSTEMS

42.3.1 The MIL-STD-1553B bus

The MIL-STD-1553B bus is a time-division multiplexed, half duplex communications link with a 1 Mb/s data rate. A twisted pair cable links a maximum of 32 nodes

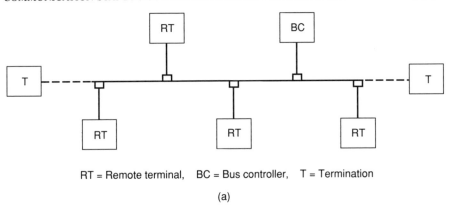

RT = Remote terminal,   BC = Bus controller,   T = Termination

(a)

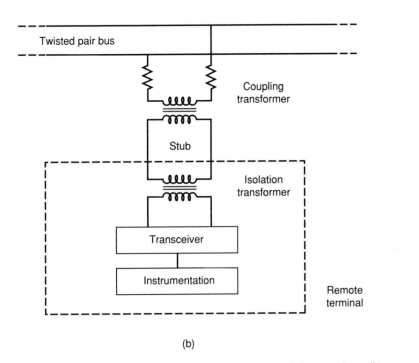

(b)

Figure 42.3   Physical features of the MIL-STD-I555B bus: (a) multi-drop topology; (b) transformer coupling

or instrumentation sites. Major physical features of this bus are shown in Figure 42.3. One of the nodes on the bus must act as a bus controller. Information is transferred in a command/response mode, with the bus controller having complete control over communications and no node speaks until spoken to. A 1553B node is called a remote terminal (RT) and it has a unique address on the bus.

Each RT has self-test facilities and fault tolerance can be enhanced by the use of an optional redundant bus.

There are three different word types used in the 1553B standard, defined as *command, status* and *data* words. A command word consists of a command sync followed by a five-bit RT address field. The following T/R bit determines whether the terminal is to transmit or receive data and the sub-address field specifies the address within the terminal to which the data will be sent or received. The final field specifies how many words are to be sent or received and this is followed by a single (odd) parity bit. The exception to this is when the sub-address field is 00000 or 11111, which indicates that the word count field contains a mode code. These codes are used to trigger internal events within the terminal such as reset, synchronize and self-test.

The data word is simply specified as a data sync followed by 16 data bits and a parity bit. The contents of the data field are user specific, but must be transmitted, most significant bit first.

The status word is defined as a command sync, followed by the address of the RT sending the word and the number of status flags. These flags are defined in more detail in MIL-STD-1553B.

A message is defined as a bus command followed by any relevant data and a status word. All messages start with a command from the bus controller. If an RT is to respond, then it should send its status word followed by any data words required. The number of data words to be transmitted or received is specified in the original command. The data must be continuous with no gaps between words and the RT must respond to the command within 4–12 μs with 1 μs corresponding to 1 bit time. There must always be a gap of at least 4 μs between each message.

Several organizations have already demonstrated practical use of the 1553 field bus in non-military applications. These include the European Nuclear Research Centre (CERN), the Rustronics Division of Ruston Gas Turbines and London Underground. ERA Technology Ltd. led a consortium of companies in the development of a field bus based on MIL-STD-1553B and they have demonstrated the practical feasibility of a field bus based on this standard.

Rausch (1989) has described the large system used to control the European particle accelerators. A token-ring network (conforming to the IEEE 802.5 standard) is used to interconnect process computers and workstations, and a multi-drop field bus (based on MIL-STD-1553B) is used for connecting accelerator instrumentation to local process computers. Figure 42.4 shows the main features of this system. A feature of this work is the use of low-cost commercially available industrial devices to implement the bus. In excess of 150 field bus systems have been linked, via bus controllers, to the token-ring network and it has been found to be a very reliable and adaptable system.

The Rustronic network (Clark, 1989) uses 1553 protocols, error checking and data validation procedures. These functions were implemented by using a remote terminal integrated circuit, the MA805 available from Marconi Electronic Devices Ltd. The 1553 bus was chosen to enable the high-speed data transfer required to monitor rapidly changing data obtained from a gas turbine.

The ERA project was particularly concerned with the adaptation of 1553B to suite the requirements of process control. The military standard could be used directly for factory automation since the 1553B 1 Mb/s data rate and the restricted few hundred metre maximum cable length was acceptable for many applications. Some process control applications require options for powering remote instrumentation via the bus cable. The ERA have demonstrated that the 1553B bus could be adapted to become an intrinsically safe, 2 km long bus, with power supplied (at a level to satisfy the requirements of existing process control devices) to three remote terminals via the bus cable (Stone, 1989). A slower bit rate of 62.5 kb/s was required for the intrinsically safe system. ERA have publicly demonstrated the extreme electromagnetic hardness of the 1553B bus and they

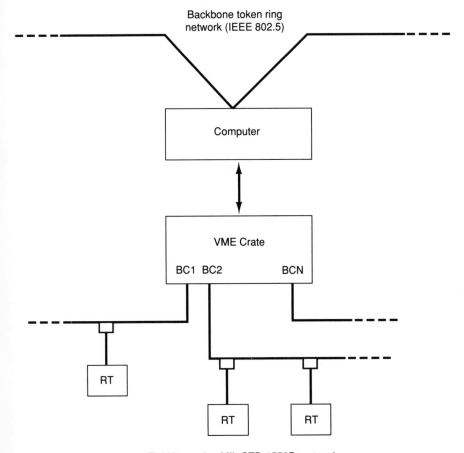

Field bus using MIL-STD-1553B protocol.

Figure 42.4   Schematic diagram showing the 1555B field bus used for accelerator control at CERN (From Rausch, 1989)

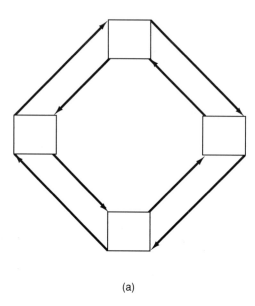

(a)

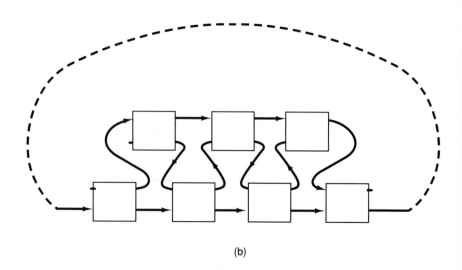

(b)

Figure 42.5 Fibre-optic rings with two-input/two-output optical transceivers: (a) use of two contra-rotating rings; (b) use of the skip connection method (Jordan *et al.*, 1988) for fault critical applications

have installed a trial demonstration at a water treatment plant. A field bus based on MIL-STD-1553B has many attractive features and, although the recently published IEC Draft Field Bus Standard documents precludes the use of of an unmodified 1553B bus, its continued use in non-military applications can be confidently predicted.

Fibre-optic implementations of 1553B are of increasing interest, particularly now that the cost of optical components has decreased significantly. Cruickshank and Kennett (1989) have reported the use of a local star topology implemented in a transmission format to link helicopter avionic systems. A derivative of frequency shift keying was used as the optical modulation technique for coding and decoding the electrical tri-level Manchester data signal. Progress has also been made with reflective star couplers. Graves (1989) has reported the use of fused fibre technology to produce a 32-port reflective star with a maximum loss of 19 dB. Clearly this is an area where technological developments can be expected to improve the capability and attractiveness of fibre-optic systems.

Fibre-optic rings, as shown in Figure 42.5, can be designed to operate at low power and tolerance to link failures and node failures can be achieved with a small increase in connection complexity (Jordan et al., 1988). Simple, low-cost LEDs and photodetectors can be used, and operation at 1 Mb/s is possible with power consumption as low as 3 mW. Lytollis et al. (1989) have reported a 1553 field bus with fibre-optic loop extensions. The optical loops will be transparent to the bus controller. The low power budget of the fibre-optic bus will help designers satisfy the requirements of intrinsically safe systems (Garside, 1988) and in addition, longer distance connections can be made with fibre-optic systems. This mixed media 1553 bus could form the basis for a field bus for all the applications discussed in this chapter.

42.3.2   ARINC 629

ARINC 629 defines a digital wire or fibre-optic link in which avionic subsystems may transmit and receive digital data using a standard communication protocol. The ARINC specification 429 (Mark 33 Digital Information Transfer System), adopted in 1977, has performed satisfactorily and with high reliability. The objective of ARINC 629 was to reduce airplane wiring and equipment interfaces, and simplify the implementation of centralised maintenance equipment where large and rapid data exchanges are required.

One of three media types may be selected for a given application, namely: unshielded or shielded twisted pair (current mode), shielded twisted pair (voltage mode) and fibre optic. With the current mode bus wires pass through a coupling transformer unbroken (ie no interconnections are required) and a clip on coupling arrangement is possible. A maximum of 120 terminals are specified on a linear bus (of maximum length 100 metres) with T-stubs of length up to 15 metres.

Carrier sense Multiple Access/Collision Avoidance (CSMA/CA) is used with Manchester II bi-phase coding. The transmission bit rate on the wire bus should be

2MHz. A dedicated bus controller is not required and bus access control is distributed among all of the participating terminals. A basic protocol allows terminals to communicate at constant intervals (periodic mode) during normal operation and, if the bus is overloaded, it will automatically switch to an aperiodic mode. A combined mode protocol provides for priority access for aperiodic data. The combined mode automatically limits the throughput of aperiodic data in a progressive fachion (from the lowest level upwards) according to the available times without affecting the periodicity of normally periodic data transmission. All seven layers of the open system interconnect are specified by ARINC 629.

42.3.3 CAMAC and the General Purpose Interface Bus(GPIB)

Computer Automated Measurement and Control (CAMAC) was developed in the mid-1960s for applications in high-energy nuclear physics and nuclear energy establishments. A 24-bit parallel bus (with a total of 86 connection wires) carries data and control signals around a unit (called a crate) into which 24 modules and a crate controller may be plugged. Up to seven crates may be connected in parallel via a 66-way bus extension. The CAMAC serial highway system uses a twisted pair connection to link up to 62 crates over longer distances up to 15 m. Much longer-distance transmissions are achieved with fibre-optic links. CAMAC is specified by IEEE, IEC and BSI standards (BS 5554 (1978), BS 5836 (1980), IEEE 583, 595, 596, 675, 683, 726, and 758).

The General Purpose Interface Bus (GPIB) was introduced by Hewlett Packard in the mid-1970s for use as a parallel interface for programmable electronic instruments. This parallel bus has 16 wires of which 8 are for data, 3 for handshaking and 5 for bus activity control. Transmission at 1 MHz is possible over 15 m. Units to serially extend (with fibre optic and coaxial cable) are available for communication over distances up to 1000 m with data rates of 50 Kb/s. This bus was originally specified by IEEE Std 488 (1975, revised 1978) (Loughry, 1978). It is now also specified by the following IEC standards: IEC Std 625 pt 1 (1979)—electrical, mechanical and functional details and IEC Std 625 pt 2(1980)—coding and formats. Careful initial design has led to the widespread, successful international use of the GPIB.

Devices connected to the bus must be able to perform one of three functions, namely:

(1) *listener*—a device capable of receiving data over the interface when addressed;

(2) *talker*—a device capable of transmitting data over the interface when addressed;

(3) *controller*—a device for management of communication by sending addresses and commands.

Devices able to talk and listen, and devices able to talk, listen and control are allowed but there can be only one active controller. A maximum of 15 devices may be connected in one contiguous bus. The GPIB is discussed in greater detail in Chapter 46.

These parallel bus systems were designed for the interconnection of units which are not widely separated and for rack-mounted equipment. CAMAC is relatively old but it is still actively supported by companies producing new equipment, for example, microcomputer based implementations of the standard. The IEEE 488 bus is currently the most important parallel bus for instrumentation systems. However, it should be noted that the well-established VMEbus (Freer, 1987) for digital systems has been extended to form what has been called the VXIbus (ERA, 1990). This new bus is compatible with and complementary to the IEEE 488 bus.

42.4   EMERGING BUS SYSTEMS

42.4.1   The Field Bus

The complete network hierarchy for measurement and control applications involves three main levels—plant, cell and field. MAP satisfies the higher-level requirements but it is not thought to be suitable for field devices where communication with sensors and actuators is required. The field bus is a digital, serial, multi-drop bus for communication with low-level devices such as sensor and actuators. Applications for the field bus will be found in products for process control (Basur, 1988; Caro, 1988), factory automation and intelligent buildings. Figure 42.6 shows the block diagram of a process control field bus system. Commonly quoted advantages of field bus systems are reduced material and installation costs, reduced weight of wiring, systems easy to change, improved accuracy of signal transmission and increased information flow.

Silicon technology will be required to minimize the cost of adding field bus capability to devices (some of which may be very simple, e.g. motor starters, relays and limit switches). In some process control applications it will be necessary to supply power to remote devices via the bus and when intrinsic safety is required for operation in flammable atmospheres design becomes a problem when cables are long. Factory automation generally requires a faster bit rate compared with process control, for example, for down loading large files to CNC machines. The fastest control loops in process control, e.g. flow control, typically requires a 0.1 s sample rate for sensor and actuator communication. At the present time fast control loops in manufacturing (e.g. in discrete parts assembly in the electronic industry) will normally not be connected directly to the field bus. It is clear that the intended wide application range of the field bus will lead to an equally wide range of requirements. While it is clearly desirable for a standard to be as comprehensive as possible there is a danger that documentation will become over complicated and difficult to use. It remains to be seen whether silicon circuits can be designed to implement field bus systems with an acceptably low cost overhead for the additional communications facility. An additional advantage of the silicon circuit is that the full complexity of the standard will be absorbed by the circuit designer and cast into the silicon.

The international field bus activity started in the mid-1980s with the formation of a field bus working group by the International Electrotechnical Commission (IEC). National groups (e.g. the SP50 committee of the Instrument Society of

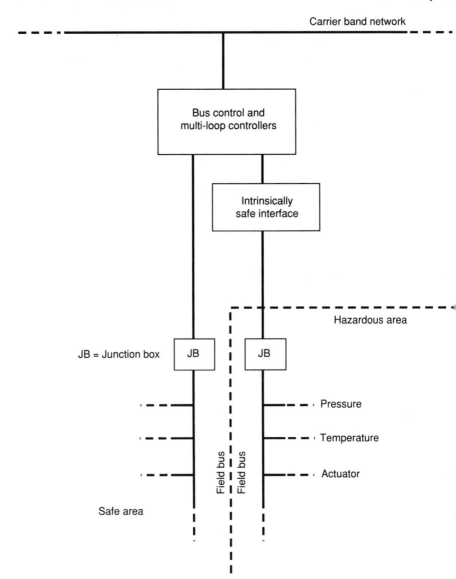

Figure 42.6   Block diagram of a process control field bus system. Note that communication and power will be carried by the bus.

America (ISA) and the AMT7 committee of the British Standards Institute (BSI)) have brought together a complete spectrum of industrial experience to work on the problem of specifying a serial data highway standard for measurement and control. Unfortunately a consensus view has been found to be difficult to obtain with the different philosophical basis of centralized media access control and token-passing access control being a particularly difficult area for agreement to be obtained (Wood, 1990; Phinney, 1988; Zielinski and Bergeson, 1988). In addition to the activities of the ISA in America and the BSI in Britain several other large groups are working on field bus problems. The Eureka Field Bus Project involves 15 companies from 7 European countries. A consortium of companies led by ERA Technology Ltd have investigated a field bus based on MIL-STD-1553B and they have successfully demonstrated the operation of a prototype system at a field trial site. National groups in France (the Factory Instrumentation Protocol (FIP) project) and in Germany (Profibus) have developed national standards for field bus systems with silicon circuits implementing these standards at an advanced stage of development. Fortunately all of these groups support the work of the IEC so the prospects for an international standard field bus are good.

Subcommittee 65C of the IEC group coordinating the field bus standard published a draft physical layer document in January 1990. It is not complete, sections on optical and radio links will be added when available. The main features of IEC Physical Layer specification are the use of multi-drop connections for up to 32 devices with twisted pair cables carrying asynchronous data transmission using Manchester coding and half-duplex communication. A voltage mode (with high-and low-speed options) and a current mode is specified. Including optical and radio links it is clear that a large number of options will be possible with this standard.

The bit rate of the slow voltage mode option is 31.25 Kb/s for cable lengths up to 1900 m and d.c. power for remote instrumentation may be carried by the cable in addition to the communication signal. For intrinsically safe applications at least five devices may be supplied with power with at least three of these devices located in an hazardous atmosphere. The data rate of the high speed option is 1 Mb/s for cable distances up to 750 m.

The current mode option features a 1 Mb/s data rate over distances up to 750 m. Clip-on transformer couplers will be used similar to those specified by ARINC. Power will be supplied over the bus by using a 14 kHz, 100 mA r.m.s. sinusoidal current source to drive the bus. It is claimed that an intrinsically safe field bus may be produced by this technique.

The message format (protocol data unit) for wire media is

Preamble, Start delimiter, Data, End delimiter

The data will be a sequence of bits determined by the data-link layer. Encoding rules are: logic 1—high to low transition during a bit time, logic 0—low to high transition during a bit time, N+ (non-data plus)—high during bit time and N–

—low during bit time. Transitions occur at mid-bit time. Using this notation the start delimiter word is 1, N+, N−, 1, 0, N−, N+, 0 and the end of frame delimiter word is 1, N+, N−, N+, N−, 1, To synchronize bit times a preamble is transmitted at the beginning of each sequence, namely: 1, 0, 1, 0, 1, 0, 1, 0. A minimum transmission gap of four bit times is specified.

The groups working on the data-link layer specification have experienced great difficulty in arriving at a document expressing the consensus view. One group supports the use of centralized media access which ensures that sampling periods are maintained (this is essential for good control loop performance) while another group supports the use of token-passing media access control based on IEEE 802. Token passing allows changes to be easily made (because each token holder can manage its own scanning program) but sample rate jitter can be experienced at high traffic loads. Timed token methods can be envisaged with bus idle times suffient to ensure that fixed scan times are achieved. A compromise document has been produced but it is complicated with many options and it is being reviewed to, for example, reduce the four token types and six address types specified.

The IEC field bus standard will specify only three of the seven layers of the ISO model. The physical layer and the data-link layer specifications will be joined by an application layer specification. The application layer will be a difficult specification to produce because the working groups have large differences in design philosophy and market sector interests. One important area where agreement is essential is the standardization of sense and actuator variables and notations (e.g. for pressure, temperature and current) to give a uniform user surface. This is almost a natural language interface requirement and it is sometimes useful to refer to it as an additional layer. General-purpose use of the hand-held communicator (the digital screwdriver of the field bus) requires the broadest possible consensus on the application layer that can be achieved in a reasonable time scale. Priority is being given to describing application layer services using the MMS terminology of MAP.

An EIA committee (Project 1393 (Hunt and Tomlinson, 1985)) was responsible for the initial work on the manufacturing message service specification. It was labelled RS 511 by the committee but it is now more commonly referred to as MMS. It is a connectionless protocol which enables frames of data to be assembled into complete messages and defines the message notation (i.e. defines the purpose, length and value of each element in a message).

MMS is now an international standard—ISO 9506. Hagar (1989) discusses the impact of MMS on future controllers and describes the mapping of an existing programmable controller into a software model that can be interfaced to the MMS. Warrior and Cobb (1990) discuss field bus messaging requirements.

Pimentel (1989) discusses the important question: how important is it for field bus networks to be OSI consistent? As currently specified the field bus has a reduced architecture and it is therefore not fully consistent with OSI. The physical, data-link and application layers are essential but the presentation and session layers are not needed. A four-layer architecture including the network layer is attractive because internetworking is facilitated. However, greater functionality could be achieved by replacing the network layer by the transport layer since network layer functionality can be supplied by bridges and routers. The trade-off between functionality, speed of response and the additional cost of implementing field bus facilities for field devices will determine what level of OSI conformance is appropriate.

Interest in radio telemetry for measurement and control applications has increased since the success of the digital techniques used in pagers and cellular radio telephones (Neve, 1990). Radio telemetry finds particular use when mobile or moving equipment is involved or hand-held instruments are required. Cost reductions can be expected where radio telemetry is used when installations are temporary and a short installation time is essential. Current spectrum regulations (which are Country specific) limit operating frequencies to 140–200 MHz and 400–500 MHz bands with 25 KHz and 12.5 KHz signal channels.

Neve (1990) defines three classes of radio telemetry for connecting field devices:

(1) low speed (less than 2400 b/s) for low-power operation with battery or solar power supplies;

(2) high speed (20 kb/s) for applications where power consumption is not critical e.g. pipe line automation;

(3) a high-speed system operating above 800 MHz for use in buildings with steel reinforcing.

Although radio telemetry is a well-defined area considerable research and development is required before an acceptable system is produced that will complement the function of the permanently connected field bus. In addition to the technical problems, severe legislative problems can be expected to be associated with spectrum allocation and safety certification.

42.4.2 Home Automation

The communication requirements of home automation are similar to that of industrial process control and factory automation so it is not surprising that several groups are working to produce a standard defining the serial interconnection of domestic equipment. The Electronic Industries Association (USA) formed its Consumer Electronics Bus (CEbus) standards committee in 1984 to create a standard for consumer equipment used in home automation systems. A parallel Japanese EIA group is working on a similar standard specification for what has been called the Home Automation Bus (Hamabe et al., 1986a, b; Hatari et al., 1986; Iida et al., 1986; Yashitoshi et al., 1986). In Europe several large companies (including Electrolux/Zanussi, GEC, Philips, Siemens, Thomson and Thorn-EMI) have formed a group to work on a home automation communication standard. All of these groups are reporting their results to an ISO/IEC joint committee and therefore a world standard is a long-term possibility.

Retrofit considerations are important for the field bus but they are particularly important for the home automation area. The building rate is low in Europe so retrofit is more likely to be the highest market for bus systems while in the USA a much higher building rate exists and therefore preparation of bus products for new houses may be more important. Factory built house techniques, pioneered in Sweden as a result of the 1973 energy crisis, offer great scope for a comprehensive implementation of bus and flexible power distribution schemes. Unfortunately this is still an expensive option so flexible (low cost) methods for installing domestic bus systems will still be required.

Mains-borne signalling is an attractive method for retrofitting a bus system in existing housing stock. Several commercial products are available and research papers describing improved security communication methods continue to appear. A British Standard for mains signalling (BS 6839) is available and a draft CENELEC standard has been produced. Amplitude shift keying (with extensive error checking and correction) is satisfactory for many applications but for the more stringent specification digital FM (using spread spectrum techniques) is used. Commercial energy management schemes have been designed to use a.c. power lines to carry control signals. The expense and time associated with installing traditional hardwiring for control signals is completely eliminated by the use of the power line.

A domestic communication bus is a very important enabling technology for home automation. Many product opportunities will arise once the bus standard is established because the domestic environment will become (or can be viewed as) one controllable engineering system. Interactive communication within the home will facilitate:

• management of security, safety and, energy and water consumption;

• control of heating, lighting, air conditioning, appliances and electronic entertainment systems.

External links (e.g. via the telephone network) will facilitate interactive communication with outside bodies (e.g. banking and information sources) and working from a home base will become a much more practical proposition. Telemedicine (health monitoring using remote hospital resources), telemetering (remote monitoring by gas, electricity and water authorities) and telecontrol (remote operation of domestic systems—cookers, heating and video recording and surveillance) will offer significant opportunities to the manufacturer of domestic equipment.

A multi-media network is likely to be specified by the home bus standard. The power line bus and the simple twisted pair will be used along with non-intrusive media, such as infra-red and radio links. Wider bandwidth media (e.g. fibre-optic and coaxial cables) will be required for video signals. Hence, unlike the previously described field bus, the home bus will require the network layer of the seven-layer OSI model to be included in the standard specification (Markwalter and Fitzpatrick, 1989). Routers used to link different media networks together

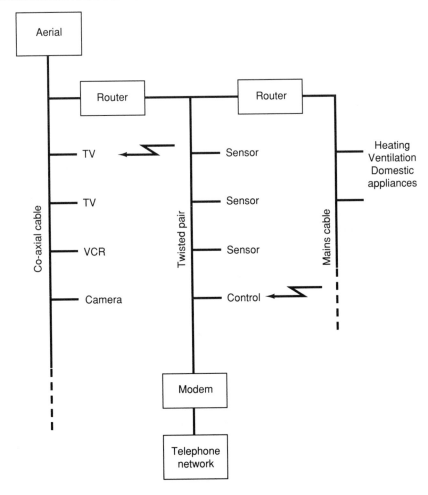

Figure 42.7   A multi-media network for home automation

will implement layers 1,2 and 3 while network nodes will implement layers 1,2,3 and, in addition, the application layer, layer 7. Figure 42.7 shows a typical block diagram of a multi-media network for home automation.

42.4.3   Controller Area Networks

Serial communication networks for vehicles (Page, 1987) have been discussed for many years. Protocols optimized for interrupt-driven, real-time functions are required to enable control modules in a vehicle to be linked by a single high-speed bus providing minimum latency and high-integrity data transfer for safety-critical driving operations. Three classes of data transfer can be recognized:

- *non-critical*: e.g. lights, mirrors, window control;

- *moderately critical*: e.g. connection of speed and temperature sensors to the instrument panel.

- *critical*: e.g. engine management and anti-lock brakes.

Bosch have developed an automotive serial bus (the controller area network—CAN) and they have granted licenses to Intel, Philips and Motorola to enable silicon circuits implementing the protocol to be developed. For example, the Intel 82526 is a VLSIC which, when connected to a microprocessor, implements the physical and data-link layer functions of the CAN protocol.

The CAN system will operate up to 1 Mb/s with a maximum latency of 150 μs for highest priority mesages at the maximum bit rate. It is a connectionless network, it will be easy to add or remove nodes and multi-master operation will allow any node to send data on the bus when it becomes free. Several error detection mechanisms are offered, including cyclic redundancy check (CRC) and coding rules. Repeat transmission is initiated when errors are detected and defective nodes are automatically located and disconnected. Bus contention is avoided by the use of dominant bit patterns which are related to the priority status of the message. Wide acceptance of the CAN protocol by the automotive industry and the availability of silicon implementations could lead to low-cost serial data highways for use in the complete range of measurement and control applications.

42.4.4   The Microcomputer Serial Bus

Parallel bus systems lead to cumbersome and expensive cabling, and they are generally not suitable for distributed, remote, sensor and actuator applications typically found in industrial automation and process control. Several serial bus systems have been proposed for what are essentially microcomputer systems and one of particular interest is the Intel BITBUS interconnect. This is an example of a silicon circuit manufacturer proposing an interconnect method and simultaneously introducing the silicon circuits that will lead to its low cost implementation. Another system product is the Philips I^2C bus, developed for consumer products but now widely used in industry.

An IEEE committee (Project P1118) is working to produce a standard for serial multi-drop communication between microcomputers. The standard is being aimed at the complete range of possible measurement and control applications of microcomputer based systems. Products conforming to the draft standard (for example, the Intel BITBUS system) are already available and in industrial use.

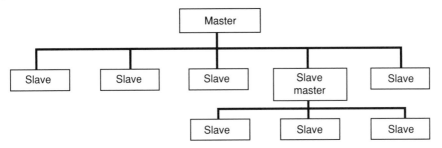

Figure 42.8   The BITBUS hierachy of bus systems. (BITBUS is a trade mark of the Intel Corporation)

The BITBUS physical medium is twisted pair and it uses the RS485 standard in a synchronous connection mode or a self-clocked connection mode. The synchronous mode uses two twisted pairs, one to carry the clock and the other to carry the data. The synchronous mode can interconnect up to 28 nodes over 30 metres at data rates between 500 kb/s and 2.4 Mb/s. If repeaters are not used the asynchronous, self-clocked connection mode requires only the data pair and communication over longer distances is possible.

The BITBUS uses central media access techniques and its data-link protocol is based on the IBM synchronous data-link control (SDLC) specification. The IEEE serial bus draft standard is based on HDLC (High-level data-link control), an ISO version (ISO 3309) of the bit-oriented synchronous protocol. At its simplest BITBUS can be configured as a single-level hierarchical network with a master controlling several slave units on a multi-drop bus. A multiple level hierarchy (Figure 42.8) is constructed by arranging for the slaves to have an associated sub-master which is used to control another multi-drop bus.

42.5   FUTURE TRENDS

The previous sections of this chapter have described emerging and established serial data highways for measurement and control applications. Four automation areas: industrial, vehicle, aircraft and domestic, can be identified, where significant commercial activity is leading to the production of standards documents. These areas have many similar requirements and in the future they may be unified under the OSI specification with one standard silicon circuit becoming available for all applications. The difficulties experienced by the various groups currently working on standards documents to obtain a consensus view suggests that this can only be a long-term proposition. However, the benefits of scale arising from the development of a generic serial data highway standard would be very large.

A feature of the current standards activity is that compromise proposals are avoided by allowing a number of options at each layer of the communication model. Hence different protocols will be available to perform the same function.

Whatever happens in the future this is likely to remain a problem because of the large installed base of communication equipment and because new protocols are developed as technology evolves. It is a fact of life that upward compatibility is often sacrificed for performance advantages so it can be expected that variants of a standard will appear soon after the standard is published. Techniques for protocol conversion are being investigated (Calvert and Lam, 1990), unfortunately they may not always provide a solution to the protocol mismatch problem.

Serial data highway applications in maintenance, condition monitoring, safety and security systems will require the sampling of data from high data rate sensors. Sensors monitoring vibration, strain, acoustic signals and images will be of particular interest. The communication rate on a field bus is likely to be 1 Mb/s with the access time per node being of the order of 1 ms in a 32-node system. Reducing the number of nodes to three gives a factor of ten improvement but the field bus would not then be used in a cost-effective way. Clearly the implied sample rate will be too low for many high-data-rate sensors. A data preprocessing system will be required before a high-data-rate sensor can be connected to a field bus.

In flexible manufacturing systems or in the so-called smart building a requirement exists to collect data from a wide range of sensors which are then 'fused' together to allow some form of optimal control to be achieved or to drive a decision-making algorithm. One idea here is that fusion of signals from several sensors of mediocre quality will give better results than could be obtained from a high-quality sensor (Holmbom et al., 1989; Luo and Kay, 1989). Clearly this approach will lead to an increased wiring requirement unless a field bus technique is adopted. Several of the sensors envisaged for building automation (e.g. vision and other security systems) will require signal preprocessing before being coupled to a bus and transmitted to a bus controller where fusion and decision-making algorithms are implemented. The collection of data for sensor fusion in flexible manufacturing systems will also benefit from field bus techniques.

Several attempts have already been made to introduce intelligent (or smart) instrumentation by superimposing a digital communication signal on the conventional 4–20 mA link. This work has had some success, for example, it has enabled the use of a hand-held communicator by field personnel to obtain calibration and maintenance information. However, intelligent devices connected by a serial data highway will significantly alter the way systems are operated and allow full advantage to be taken of smart features. It will be possible for the supervisory system to check device serial numbers and location, act on maintenance information obtained from an internal self-check and calibration procedure and, depending on the bus traffic, other information (e.g. vibration levels) related to the overall health of the plant could be supplied by the field bus devices. The silicon circuit overhead for the implementation of these additional features could be very low. Ultimately, peer-to-peer communication on the bus could be used to distribute control at the device level rather than at the controller level. Chapter 45 deals with smart systems.

Greater use of fibre-optic communication links can be expected in the future. Plastic fibre and visible LEDs will allow cost effective systems to be built for relatively short distance (a few hundred metres) links and silica fibre will be used for the longer-distance connections. Transmit and receive circuits are very simple so the overhead associated with adding a communication link to low-level field devices will be low. Low-power operation even at 1 Mb/s will help intrinsic safety requirements to be met. A separate power supply cable will be required to be run in parallel with the fibre-optic data link for process control applications that would normally use the bottom 4 mA of the 4–20 mA signalling method to supply power to remote nodes. The optical sensor/fibre optically linked (passive) approach to instrumentation is unlikely to progress beyond the research stage because it is difficult to build in the intelligent features (without using an electronic system) required by most users (Higham and Medlock, 1984). Various topologies for fibre-optic links have been proposed: passive stars, active stars, and parallel rings with skip connections for fault tolerance. The Society of Automotive Engineers are investigating a fibre-optic ring communication standard (Glass, 1989) for applications requiring sensor-to-processor communication at 100 MHz data rates. Standards for fibre-optic communication can be expected to make considerable progress.

The widely used 4–20 mA analog standard was published in a few sides of A4 paper. The field bus standard will be a multi-volume document with a total of hundreds of pages. This level of complexity is a very unattractive feature of the current standards activity; however, this complexity will be hidden from most users when a silicon implementation of the standard has been produced. Ultimately the first successful, widely accepted, silicon implementation of the standard will become the standard.

The standards discussed in this chapter will be successful if they allow multivendor systems to be easily constructed. It is essential that independent facilities are established for testing equipment for conformance to a standard (DTI, 1990). The European community have established a legal requirement for products sold in Europe to conform to specified standards (DTI, 1989). Unfortunately conformance to a standard will not guarantee the interoperability and interchangeability of equipment, especially where the higher layers of the communication model are involved with more than one applicable standard. It is generally agreed that users should play a much stronger role in standards development but it is difficult to see a clear commercial advantage arising from this increased involvement. Many of the advantages of standards-led product development will be lost if the commercial considerations of a few of the larger companies outweighs the socially desirable objective of optimizing standards to satisfy user requirements.

42.6   SOURCES OF FURTHER ADVICE

The reader is referred to the following textbooks and journals for further information on communication standards in measurement and control.

Books

Ciminiera, L., and Valenzano, A. (1991) *MAP and TOP* Addison-Wesley, New York.

Dwyer, J. and Ioannou, A. (1987) *MAP and TOP Advanced Manufacturing Communications* John Wiley, New York.

Giacomo, J. D. (1990) *Digital Bus Handbook* Mc-Graw-Hill, New York.

Green, P. E. (Ed.) (1989) *Network Interconnection and Protocol Conversion* IEEE Press, New York.

Rodd-Deravi, (1990) *Communication Systems for Industrial Automation* Prentice-Hall, Englewood Cliffs, NJ.

Journals

Control and Instrumentation Morgan-Grampian, UK.

IEEE Transactions on Consumer Electronics IEEE, USA.

Measurement and Control Institute of Measurement and Control, UK.

Measurement Science and Technology (formerly *Journal of Physics E: Scientific Instruments*) Institute of Physics, UK.

REFERENCES

Barney, G.C. (1985). *Intelligent Instrumentation*, Prentice Hall, London.

Basur, O.A. (1988). 'A control data framework with distributed intelligence', *Proc. ISA88*, Houston, Texas, 1153–69, ISA, Research Triangle Park, USA.

Calvert, K.L. and Lam, S.S. (1990). 'Formal methods for protocol conversion', *IEEE J. on Selected Areas in Communication*, **8** (1), 127–42.

Caro, R.H. (1988). 'The fifth generation process control architecture', *Proc. ISA88*, Houston, Texas, 659–67, ISA, Research Triangle Park, USA.

Chapin, A.L. (1983). 'Connections and connectionless data transmission', *Proc. IEEE*, **71** (12), 1365–71.

Clark, J.P., (1989). 'Applying 1553 technology to an industrial programmable electronic system', *Proc. ERA Conference MIL STD 1553B and the Next Generation*, London, 2.4.1–5.

Collins, G.B. (1968). 'A survey of digital instrumentation and computer interface methods and developments', *Proc. Conf. on Industrial Measurement Techniques for On-Line Computers*, IEE, Savoy Place, London, *IEE Conf. Pub. No. 43*, 1–8, 60–73.

Cruickshank, A.M. and Kennett, M.J. (1989). 'The development and flight demonstration of an optical MIL-STD-1553B data bus', *Proc. ERA Conference MIL STD 1553B and the Next Generation*, London, 3.1.1–9.

Day, J.D. and Zimmermann, H. (1983). 'The OSI reference model', *Proc. IEEE*, **71** (12), 1334–40.

DTI (1989). *The Single Market—New Approach to Technical Harmonisation and Standards*, Department of Trade and Industry, London, England.

DTI (1990). *The Single Market—Testing and Certification*, Department of Trade and Industry, London, England.

Dwyer, J. and Ioannou, A. (1989). *MAP and TOP: Advanced Manufacturing Communications*, Kogan Page, London, England.

ERA Report (1990). *The VXIbus: a Review of its Capabilities and Compatible Products*, ERA Report No. 90–0148. ERA Technology Ltd, Leatherhead, Surrey, England.

Freer, J. (1987). *System Design with Advanced Microprocessors*, Pitman, London, England.

Garside. R. (1988). *Intrinsically Safe Instrumentation: a Guide*, Safety Technology Ltd, Salisbury, England.

Glass, M. (1989). 'Overview of the SAE high speed ring bus', *Electronic Product Design*, May, 71–78.

Graves, R. (1989). 'In the light of experience', *Proc. ERA Conference MIL STD 1553B and the Next Generation*, London, 6.1.1–12.

Grimes, R.W. (1982). 'Transmission of data', in *Handbook of Measurement Science*, Vol. 1, Sydenham, P.H. ed. Wiley Chichester, 539–89.

Hagar, M.L. (1989). 'Impact of MMS on future controllers', *Texas Instrument Technical J.*, November–December, 48–55.

Hamabe, R., Murata, M. and Namekawa, T. (1986a). 'A revised proposal for standardisation of home bus systems for home automation', *IEEE Trans. Consumer Electronics*, **32** (1), 1–7.

Hamabe, R., Murata, M and Namekawa, T. (1986b). 'Home bus system (HBS) interface LSI and its standard protocol example', *IEEE Trans. Consumer Electronics*, **32** (1), 9–18.

Hatari. M., Mokuno, K., Iida, K., Ochiai R., and Harie, T. (1986). 'Home information and standardisation of the home bus', *IEEE Trans. Consumer Electronics*, **32** (3), 542–9.

Higham, E.H. and Medlock, R.S. (1984). 'The relevance of optical transducers to process industries', *Optics and Lasers in Engineering*, **5**, 193–209.

Holmbom, P., Pedersen, O., Sandell, B., and Lauber, A. (1989). 'Fusing sensor systems: promises and problems', *Sensor Review*, **9** (3), 143–52.

Hunt, G.G. and Tomlinson, J.R. (1985). 'EIA project 1393: a high level communication standard', *Robotics and computer integrated manufacturing*, **2** (3/4), 191–200.

Iida, K., H. Yahiro and Kubo, A. (1986). 'Housekeeping applications with bus line and telecommunications', *IEEE Trans. Consumer Electronics*, **32** (3), 558–63.

Jordan, J.R., Gater, C. and Mackie, R.D.L. (1988). 'Low power instrumentation with fibre optic links', *Proc. IMEKO XI Conf. Sensors*, 685–94.

Keefe, A.T., Moss, G. and Young, I.R. (1967). 'Data highway for process controlled plant', *Proc. IEE*, **114** (12), 1977–86.

Loughry, D. C. (1978). 'IEEE standard 488 and microprocessor synergism', *Proc. IEEE*, **66** (2), 162–72.

Luo, R.C. and Kay, M.G. (1989). 'Multisensor integration and fusion in intelligent systems', *IEEE Trans. Systems, Man and Cybernetics*, **19** (5), 901–31.

Lytollis, S., Jordan, J.R. and Kelly, R.G. (1989). 'A hybrid twisted pair/optical field bus based on MIL-STD-1553B', *Proc. ERA Conference MIL STD 1553B and the Next Generation*, London, 3.2.1–10.

MacKinnon, D.W., McCrum, W. and Sheppard, D. (1990). *An Introduction to Open Systems Interconnection*, Computer Science Press, New York.

Madron, T.W. (1989). *LANS—Applications of IEEE/ANSI 802 Standards*. Wiley, Chichester.

Markwalter, B.E. and Fitzpatrick, S.K. (1989). 'CEbus network layer description', *IEEE Trans. Consumer Electronics*, **35** (3), 571–6.

Mathews, P.R. (1986). 'Communications in process control', in *Instrumentation—A Reader* Loxton, R. and Pope, R. (eds.) 151–70, Open University Press.

Middelhoek, S., French, P.J., Huijsing, J.H. and Lian, W.J. (1988). 'Sensors with digital or frequency output', *Sensors and Actuators*, **15**, 119–33.

Morgan, E. (1987). *Through MAP to CIM*, Dept. of Trade and Industry, London, England.

Neve, B.D. (1990). 'Progress on radio field bus', *Measurement and Control*, **23** (2), 14–19.

Page, R.P. (1987). 'Automotive data buses—a review of architecture and media', *Proc. 6th Intl. Conf. on Automotive Electronics*, October, London. *IEE Conf. Pub.* No. 280, 209–12.

Phinney, T.L. (1988). 'An analysis of contending proposals for an ISA/IEC field instrument bus', *Proc. ISA88*, Houston, Texas, 675–82, ISA, Research Triangle Park, USA.

Pimentel, J.R. (1989). 'Communication architectures for field bus networks', *Control Engineering*, October, 74–8.

Rausch, R. (1989). 'Control of the large European LEP and SPS accelerators based on the 1553 field bus', *Proc. ERA Conference MIL STD 1553B and the Next Generation*, London, 2.1.1–12.

Stone, F.W.C. (1989). 'The 1553B field bus', *Proc. ERA Conference MIL STD 1553B and the Next Generation*, London, 2.2.1–13.

Warrior, J. and Cobb, J. (1990). 'Structure and flexibility for field bus messaging', *Measurement and Control*, **22** (1), 292–4.

Wood, G. (1990). 'Fieldbus standardisation in 1990', *Measurement and Control*, **23** (5), 135–9.

Yashitoshi, M., Ayugase, N., and Harada, S. (1986). 'Proposed interface specification for the home bus', *IEEE Trans. Consumer Electronics*, **32** (3), 550–7.

Zielinski, M. and Bergeson, D.W. (1988). 'Issues in the design of digital instrument networks', *Proc. ISA88*, Houston, Texas, 683–93.

APPENDIX 42.1: STANDARDS ORGANIZATIONS—COMMUNICATION SYSTEMS FOR MEASUREMENT AND CONTROL.

Airlines Electronic Engineering Committee (ARINC)
Aeronautical Radio Inc.
2551 Riva Road
Annapolis
Maryland 21401
USA

American National Standards Institute (ANSI)
1430 Broadway
New York
NY 10018
USA

Association Française de Normalisation (AFNOR)
Tour Europe—Cedex 7,
92080 Paris La Défense
FRANCE.

British Standards Institution
2 Park Street
London, W1A 2BS
UK

Deutsche Institut für Normun (DIN)
Burggrafenstrasse 6,
Postsach 1107,
D—1000 Berlin
GERMANY

Electronic Industries Association (EIA)
Standards Sales
2001 Eye Street
Washington
DC 20006
USA

Institute of Electrical and Electronic Engineers (IEEE)
Standard Sales—IEEE Service Centre
445 Hoes Lane
Piscataway
NS 08854
USA

International Electrotechnical Commission (IEC)
1 Rue de Varembe
Case Postale 56
CH—1211 Geneva 20
SWITZERLAND

International Organisation for Standardisation (ISO)
1 Rue de Varembe
Case Postale 56
CH—1211 Geneva 20
SWITZERLAND

Instrument Society of America (ISA)
Standards and Practices Working Group 50 (SP50)
PO Box 12277
Research Triangle Park
NC 27707
USA

Japanese Industrial Standards Committee
c/o Standards Department

Agency of Industrial Trade and Industry
Ministry of International Trade and Industry
1–3–1 Kasumigaseki, Chiyoda—Ku
Tokyo 100
JAPAN

National Electrical Manufacturers Association (NEMA)
Field Bus Working Group
2101L Street, N.W. Suite 300
Washington, DC 20037
USA

Chapter

43 W. E. DUCKWORTH, D. D. HARRIS AND P. H. SYDENHAM

Active and Passive Role of Materials in Measurement Systems

Editorial introduction

Materials are vital to the manufacture of effective sensors but they attract little teaching and research related to this purpose. Plastics and ceramics have made obvious inroads into sensor and instrument construction replacing traditional metals and natural materials. There appears to be an endless array of material choices. Their usefulness, however, is actually quite constrained because all materials have noticeable levels of activity to influence variables. Sensors, especially, demand levels of inactivity that are far smaller than needed in more common structural constructional requirements. This chapter provides an introduction to the use of *fine materials*, as they are being termed.

43.1 THE ROLE OF MATERIALS

A UK report Fellowship of Engineering (1983) 'Modern materials in the Manufacturing Industry', classified materials as having two main roles.

In their *active* (or *primary*) role they are 'hosts of potentially useful scientific phenomena, and always bring about a conversion of energy (or information) from one form to another'. They can provide this conversion using a gas (such as in a filled thermometer), fluid (liquid crystals in a display) or a solid (piezo-electric material in an accelerometer). The most prevalent active application area has been concerned with production of energy from fuels, not with sensor materials.

Now emerging is the far less developed sensor subclass of this family—those energy conversions concerned with provision of information about the physically

Handbook of Measurement Science, Volume 3
Edited by P. H. Sydenham and R. Thorn
© 1992 John Wiley & Sons Ltd

existing world. (The modulation of the energy link carries the required information from the system being studied to the observer.)

The importance of this class—usually referred to as *transducers* or *sensors*, is gradually being recognized in many ways. Jones (1990) and ERA (1989) are examples of recent overview studies exploring the market potential for sensing materials. The attendance at conferences on sensors is at an all-time high. The number of journals devoted to sensors is increasing. The number of installed sensors made from the 'new' sensor materials now exceeds billions.

In their *passive* (or *secondary*) role materials provide capability to connect, control, contain or display the primary sensing phenomenon provided by the active components.

Whilst extensive effort has been expended on the development and application of active materials for fuel purposes the same is not true for the sensor materials. Even in the material-producing industries, where the relationship is closest and the products very similar, sensing materials are rarely of interest. This is generally because sensing materials require different approaches to their production and the market has not, yet, been commercially attractive to bulk materials processors.

All passive materials are active to some degree in that they respond to one or more forms of external energy. They are not, however, regarded as transducer materials unless the transduction action is deliberately enhanced and controlled. For example, normal (passive) structural steel has mechanical hysteresis that varies far more than it does in the (active) instrument grades processed and specially alloyed with additive metals for making precision springs and load cells.

Another characteristic often, but not always, exhibited in a sensor material is *reversibility* of the transducer action. For example, piezoelectric material will produce electric charge when its shape is deformed; conversely when a electric charge is supplied to it the shape changes.

Another distinguishing parameter is that sensor materials ideally will provide a linear, or at least reproducible, energy transformation. Energy efficiency of the transformation is usually less important than its all-important fidelity.

The concern of technology for producing sensor materials is thus to enhance the efficiency of the information transformation action and to preserve reversibility where possible. A necessary target requirement is also that the transduction action remains constant regardless of time lapsed and the in-service level of influencing variables and operational signals.

Some sensors rely on loss of some mass to produce signals. Examples are many chemical sensors (where intimately related mass/energy transformations take place) and special, electrically resistive, tape used to monitor the abrasive level of magnetic recording tapes (its resistance rises as the tape is worn away). In these cases the design aim is to prolong the useful sensing life, usually by increasing the sensing sensitivity, thereby allowing lower mass consumption per unit time.

By the *traditional design* methodology a sensor is created from a mixture of different active and passive materials, chosen and set up in a manner that enhances the sensing effect whilst purposefully attenuating the unwanted internal and external influence effects. For example, a temperature sensor can be formed from an active material exhibiting significant thermal activity for its resistance. This active element is then mounted in a supportive containing structure, formed from passive materials, set up to ensure that the sensor is minimally effected by external influence variables and internal loss of heat through stray paths.

In sharp contrast the modern *integrated* form of sensor makes use of a common base material that is modified to enhance the sensing parameter(s) whilst adequately controlling the non-sensing ones. An example is a silicon temperature sensor integrated on a single silicon chip. In this *microelectronic* sensor form highly pure silicon is doped, in different ways with a range of materials, and at minute levels of impurity to make it sensitive to one of a vast range of measurands. By use of masks and other processing techniques the doping can be made to produce both active and passive areas in two-dimensional and three-dimensional geometry of a single chip. Thus, here the technology is required to not only produce active material but also to control, to extremes of positional precision, the passive properties. Another class of integrated sensor is the *fibre-optic* sensor where a basic element—the optical fibre—is modified by a range of methods to form many sensors.

Many transduction principles in use today were discovered as physical and chemical effects in the 19th century (Sydenham, 1979). What is new is the rate of expansion of this role in materials. This past century has been a period of enhancement of the passive and fuel uses of materials. Expansion of the information industries has begun to spur on effort in sensor materials.

A difficulty to be met in this new interest is that use of materials in sensing systems requires extensive interdisciplinary activity. Many observers, however, still regard sensing as a narrow, highly specialized, low demand activity of little economic importance.

This chapter is designed to provide an introduction to the use of materials in both active and passive roles and to provide awareness of the special skills needed from the many contributing disciplines. It provides this with respect to the three main material groups used—metals, ceramics and plastics.

43.2   DESIGN PARAMETERS

43.2.1   Properties of Materials

There exist three main groupings of material types that find application in measuring instruments—*metals*, *ceramics* and *plastics*. These variously form, depending on composition, processing or application arrangement, either active or passive components.

Brief mention is needed of the many natural materials that were once commonly used, such as wood, leather, bone, ivory and the like. In general, these are rapidly falling from favour because of their relatively high cost, problems of sufficient supply and lack of uniformity of properties.

Metals

This group comprises the common, relatively inexpensive, base metals and alloys such as steel, brass, aluminium and copper. Also used are several, less well-known, special metal alloys such as Invar, Ni-Span C, Nichrome and others with proprietary names.

Literature on the common metals is widely available, especially on those used for structural purposes. Information on properties of the specialist metals, however, must usually be obtained directly from the alloy makers. Sydenham (1986) gives an introduction to these.

The common metals have found time-honoured acceptance as passive engineering constructional materials due to their ease of working, wide temperature operating range, general availability, well-known history of performance and range of useful structural properties. They have, however, been developed primarily for purposes not associated with instrument manufacture, their use being by derivative adoption when and where appropriate. The engineering design of metals is covered extensively in terms of analysis of upper working limits with loads. Their use in instruments often, however, is more concerned with allowable deflection and stiffnesses under working loads.

Table 43.1 is a representative list of the metals used in instrument construction. Typical applications are given to indicate, not prescribe, which to use.

Those in the specialist subgroups of metals have been developed to suit certain performance needs such as provision of low thermal expansion rate, low mechanical hysteresis or enhanced sensor action.

Ceramics

The ceramic group of materials (glasses and semiconducting materials are included here but are not strictly defined as such) provide for such functions as thermal insulation or conduction, high-temperature mountings and can provide very useful transduction processes for use as both sensors and actuators. Whilst the use of ceramics is less common in instruments they can be vital because they augment metals and plastics by providing properties the latter two cannot. Table 43.2 lists representative ceramics used in instrument construction along with some typical applications. Silicon semiconductor material is the key to the advance of many miniature sensors. That application is dealt with Chapter 37 of this volume.

A useful introductory text on fine ceramic materials is Ichinose (1987) from which much of this explanation is extracted. More detailed data must be sought from the makers because ceramics, like plastics, are materials with very complex formulae, in which numerous controlled production factors decide the final performance.

Table 43.1 Metals commonly used in instrument construction

Metal	Application
Commonly available passive metals (often alloyed)	
Aluminium	Panels, mountings, leads, fittings
Brasses (copper–zinc alloys)	Diaphragms, mounts, contacts, springs
Bright steel (often hardened)	Frame, panels, shafts, gears, machined parts, pins, clips, springs, cams, knife edges, pulleys, chains, flexure strips
Bronzes	Cast frames, cast mechanisms, bearings, gears, cams
Copper	Leads, printed circuit connections, connectors contacts, heat sinks, heat conductors, shielding
Iron	Castings, temperature compensation
Nickel	Corrosion resistant needs, temperature match with glasses
Platinum, Silver and Gold	Contacts, electrodes, corrosion resistant layers
Solders (silver, tin–lead)	Connections, contact materials
Stainless steel	Springs, support frames, panels, shafts, flexure strips
Special passive alloy groups	
Beryllium–copper and phosphor–bronze (copper–tin)	Springs, sealing diaphragms
Copper–nickel, such as Monel	Corrosion resistance
Copper–nickel–chromium, such as Mu-Metal	Magnetic shielding
Nickel–iron, such as invar,	Temperature stable parts
such as Ni Span-C	Minimal mechanical hysteresis
Nickel–chromium–iron, such as Hastellow C	Diaphragms, springs
Nickel–chromium, such as Nichrome	Electrical heaters
Silicon–iron	Transformer laminations
Active metals for sensing applications	
Bimetal strips	Temperature actuators and detectors
Electrical resistance alloys	Stable resistors, strain gauge foils
Load cell materials	Diaphragms, load cell structures
Magnetic alloys	Thermomagnetic effects, magnetic detectors
Nickel alloys	Magnetostriction
Thermally sensitive electrical resistance alloys and thermocouple alloys	Temperature sensors

Some instrument ceramics are made with reasonably straightforward, specialized processes that convert natural materials such as sand and clays into the final form. Examples are the manufacture of fused silica (needing electric arc refining of special sands) and insulating porcelains (liquid pug is cast in moulds, followed by drying, followed by high-temperature firing).

The advanced materials—so-called hi-tech forms—are much more complex to manufacture. These *fine ceramics* require 'manufacture using highly refined raw materials, rigorously controlled composition and strictly regulated forming and sintering'. Production of ceramic powder, the starting material of many fine ceramics, is made with several processes, each being complex and needing expensive plant investment.

Table 43.2   Some ceramic materials used in instrument construction
(From Ichinose, 1987)

Passive constructional roles

Ceramic type	Application
Alumina	● Heat resistance support
Beryllia	● Heat conducting support
Fused silicas	● Low TEMPCO parts
Graphite	● Heat resistant support, colloidal lubricant
Hydroxyl apatite	● Artificial bones and tooth root repacement
Manganese oxide	● Insulators, IC substrates
Silicon carbide	● Cutting tools, low wear surfaces
Zirconia	● High temperature mechanisms

Active sensing roles

Alumina	● Sodium batteries
Barium titanate (a PZT)	● Piezo electric sensing and actuation
Ferrites	● Recording heads, RF transformer cores
Silicon dioxide	● Optical fibres
Stable zirconia	● Oxygen detection
Tin oxide	● Gas sensor
Zinc oxide—berylium oxide	● Voltage-dependent resistor

The powder is formed into the required shape by one of five main sintering processes—die casting, rubber mould pressing, extrusion moulding, slip casting and injection moulding. Next the item is *sintered* to bring about the minimum system energy through *densification* of the powder. These processes require high temperatures (1400–1800°C) to be present whilst the pressing action is in place. The final mechanical properties are decided by some twenty major parameters, each needing strict control of several parameters.

To illustrate the complex production involved consider making silicon 'chip' sensors. This is a long, many-stage process that requires plant so expensive that many countries cannot afford to have such facilities. The process starts by *pulling* a silicon crystal followed by zone refining with heat. This is vital to obtain the highest purity material. The crystal is then sliced and polished to form wafers of pure silicon. These are then subjected to several similar cycles of masking, doping and developing to form the miniature electronic and mechanical components within the silicon. Vacuum deposition is then used to add the metal connections. The silicon *die* is then tested and mounted on a support package, electrical leads are bonded from the die to the connection pins and the product is packaged and labelled. Most silicon microelectronic manufacturing plants are built to make electronic circuits and, as such, are not always capable of providing the processes needed for certain sensor manufacturing needs—micromachining, for example, requires special additional processes.

Once a ceramic item is made it cannot be easily remanufactured so these components must be ready formed for specific applications. The high plant costs must be supported by large-volume production so ceramics are not easily modified to suit each application; the designer of small-run sensors has to make use of stock proprietary forms.

Plastics

Plastics have steadily displaced metals in many instrument application areas, such as frames, cases and even precision mechanisms. Caren (1988) has summarized their place well.

Plastics permit a greater amount of structural freedom than any other material. Plastic parts can be large or small, simple or complex, rigid or flexible, solid or hollow, tough or brittle, transparent, opaque or virtually any colour, chemical resistant or biodegradable, and materials can be blended to achieve any desired property, or combination of properties.

However, success in their use is very much

affected by the part design and processing. The designer's knowledge of all of these variables can profoundly affect the ultimate success or failure of a consumer or industrial product.

Plastics fall into two major groupings:

- *thermoplastics*, which can be melted and re-used; and

- *thermosets*, which are chemically changed by the manufacturing process and cannot be re-melted.

Typically, thermoplastics are shaped by melting, forcing the (often highly viscous) liquid into the desired shape and allowing it to freeze. Injection moulding (for bulk material) and thermoforming (sheet material) are processes of this type. Thermosets are compressed into a mould and heated, the heating process changing the chemical structure and forming the moulding shape. However, these divisions are becoming blurred, and injection-mouldable thermosets are in increasing use. The heating required to set off the thermoset is also often aided by catalysts (as in dough moulding) and in fact, the chemical change can be brought about by catalysis alone with no external heat as in composite lay-up (for example, fibreglass) construction.

Crystalline thermoplastics are generally stronger, more chemically resistant and often self-lubricating. Examples are polyethylene, nylon, polyprophylene and acetal. Amorphous thermoplastics such as polystyrene, vinyl (PVC), acrylic and polycarbonate are more easily formed, decorated and fastened and can be transparent.

Many plastic materials can be formed as *foams* with closed or open cell structure. Polystyrene foam is familiar as a rigid, yet light and strong, packaging material. Polyurethane foams can be rigid through to soft for applications like cushioning material in instrument carry cases. Materials can also be made to foam during moulding, forming a solid external skin with a foam core. This provides structural benefits (high strength to weight) and material savings as well as moulding benefits of reduced cycle time and surface sinks.

Individual plastic materials are supplied with a wide range of properties affected by the *molecular weight* of the bulk material, *additives* for modification of production performance and *final product properties*, and *colouring agents*. Suppliers have ranges of product to suit different processes and provide specialty blends for specific requirements.

An important subset of thermoplastics is the *elastomer* group. These are plastics with rubber-like, elastic properties. SBR (styrene butadiene rubber) is used for protective boots, shock-absorbing mounts and grommets. Polyurethane (a thermoset) can be made in hardnesses ranging from glass hard to soft and rubbery. It is prepared by mixing two liquids and finds applications in cast-in-place elastic elements. Silicone rubbers provide rubbery consistency with excellent chemical resistance, although if used in contact with sensitive items (such as foil strain gauges) a grade should be chosen which does not emit acetic acid on setting.

Some combinations of polymers can also be mixed together to form *copolymers* (where the chains of the different materials cross-link) or *alloys* where the ingredients remain discrete. ABS is a typical terpolymer (a copolymer of three ingredients) while PC/PVC is an alloy.

While the range of plastic materials available presents a large spectrum of opportunity for the designer, the possibilities added by *additives, fillers* and *reinforcers* increases this enormously. These can modify finished part strength, impact resistance, density and colour. The performance of the material in production can also be affected. Materials can be added to plastics in bulk, during moulding or by hand lay-up.

In general, the instrument designer need only be aware of the possibilities arising from the use of additives, leaving details to the plastics supplier and manufacturer. Such additives as biocides, fungicides, heat and light stabilizers and anti-static agents are quite vital to the success of a product, especially those that relate to dimensional stability and to life.

Fillers are mainly inorganic materials such as talc, kaolinite, feldspar and glass microspheres. They can improve processability, reduce shrinkage, increase stiffness and reduce cost.

Reinforcers are fibrous materials including various glass fibres, aramid, carbon and more exotic materials. Their major use is increasing strength and stiffness, often anisotropically, providing finished products which, on a strength-to-weight basis, are often comparable with high-strength steels.

All additives can affect surface finish and subsequent decorative processes, those added for physical property improvement frequently adversely affecting finish.

The selection of plastics is a highly skilled task. Plastics manufacturers, however, provide extensive data and design services including electronic forms of adviser, EPOS (ICI, 1988) being an example. Kroschwitz (1990), Rubin (1988) and its various chapters such as Jones (1988), and Plastics Industry Yearbooks provide greater and more generic detail.

Table 43.3 is provided as a first guide to the type of plastics commonly used for the construction of various instrument parts.

Table 43.3   Plastics used in instrument construction
(Compiled from industry catalogues–a guide only–manufacturers must be consulted in making detailed choice)

Common plastic name	Application
Passive constructional roles	
ABS	● Telephone handsets, high-quality cabinets, PCB plugs, transparent covers—is platable
Acetals	● Gears, pawls, links, cams, cranks—is platable.
Acrylics	● Optical lenses and covers
Cellulosics	● Blister packaging, poor durability
Epoxys	● Circuit boards, hard surfaces
Fluoroplastics (includes PTFE)	● Electrical insulation, bearings, valves, linings, implants
Nylons	● Gears, snap-on parts, hinges, guide rolls, cams, plug connectors, brush holders, print wheels, hot-water valve bodies, switches, coil formers press buttons, bearings
Polycarbonates	● Tough, transparent structures, lenses, safety enclosures, cases
Phonelics	● Low-cost moulded parts
Polyimids	● High-temperature, low-creep, instrument parts
Polyesters	● Resin for glass, carbon and aramid fibre-based composites, containers, films
Polyethylenes	● Nozzles, containers, cable clamps, cases, cable insulation
Polystyrenes	● Lamp shades, diffusers, instrument frames, panels, video cassette parts, foam thermal insulation and packaging
Polypropylenes	● Appliance housings, cable sleeves, sterilizable parts
Polyurethane	● Cushion soft to glass hard needs, fascias, tubing, cords

(continued overleaf)

Table 43.3 (*continued*)

Common plastic name	Application
Poly (vinyl chloride)	● Cable ducts, pipes insulation on cables swith covers, medical parts
Silicones	● Flexible seals, protective covers, moulded seals
	Active sensing roles
Polyvinylidene flouride (PVDF)	● Piezo electric
Polyacetylene (and others)	● Electrical conduction

Influence effect on materials

Table 43.4 is provided to complete this section. It lists often-met transduction effects that may give unwanted activity in a material normally regarded as passive. Due attention to such, often unexpected, factors is needed to reach a satisfactory design outcome. Most of the effects are reversible. Generally only one or two effects would be expected to be significant in a chosen material but as the system sensitivity is increased more effects will become significant and need attention.

Table 43.4   Some physical effects giving rise to unexpected activity in materials

Term describing effect of	*Input variable*	on	*Output variable*
	(most are reversible)		
Gas law effect (at constant volume)	Pressure		Temperature
Electrochemical effect	Electric current		Material decomposition
Electromagnetic effect	Electric current		Force
Electrostatic effect	Electric voltage		Force
Faraday effect	Magnetic field		Polarization angle
Hall effect	Magnetic field		Current/voltage
Magnetoresistive effect	Magnetic field		Electrical resistance
Magnetostrictive effect	Magnetic field		Length
Photoelastic effect	Strain		Optical polarization
Piezoelectric effect	Force		Charge
Piezoresistive effect	Force		Electrical resistance
Poisson effect	Longitudinal load		Transverse length
Pyroelectric effect	Heat		Charge
Rectifying effect	Electric current		Flows only one way
Resistance–effect	Strain		Electrical resistance
Self-heating effect	Electrical power		Temperature
Shear modulus effect	Shear load		Length
Temperature velocity effect	Temperature		Velocity of sound
Thermoelastic effect	Temperature		Elasticity
Thermoelectric effect	Temperature		Voltage
Thermoresistive effect	Temperature		Electrical resistance
Young's modulus effect	Tensile load		Length

43.2.2 Selecting a Material

General considerations

Metal and ceramic materials tend now to be used in instruments to provide design solutions where plastics are unsuited. For instance, where high temperature and high mechanical stability are needed.

Another factor that dominates the choice of material is the volume of the production run. For small volume runs tooling costs associated with the, often preferred, use of plastics are usually too high compared with the cost of machining metal parts. Ceramics, usually requiring very elaborate manufacturing processes to obtain the forms needed, are often used in stock shapes with minimal machining. Their use is somewhat inhibited by their manufacturing difficulties posed for small volume production.

It might be thought that as materials have been developing since earliest times of man that the designer always has to hand materials with the right combination of properties to suit any task. Unfortunately this is far from the case because the number of well-defined and controlled material properties required is great and each material application will usually involve need for several particularly stringent properties at the same time.

To provide an appreciation of the material selection problem consider the abridged list, Table 43.4, of the physical effects that might be involved when selecting a material for making a specific part.

Measuring instruments are made in production volumes ranging from one-off to many thousands. Plastic materials, and the production processes required to convert them into products, are generally more suited to high-volume production. For example, injection moulding is one of the most widely used processes, but the simplest moulding dies can cost around $10000 with more complex dies costing very much more. Injection moulding is an elegant and versatile process but production runs of less than several thousand inevitably carry a high tool amortization cost per part.

Die-cast metals can also be used but these are falling from favour except where plastics cannot cope because of lack of a key property such as needing to have metal shielding. Even there, however, special electrically resistive, sprayed, coatings and plated coatings can be used to shield the contents of plastic cases. For very high temperatures metal castings are still needed. Metal castings can also be more suitable where high connection strength is needed for joined parts.

Many plastics can be machined and joined by mechanical fastenings or adhesives. However, the cost of such reworked plastic components may well be higher than metal materials. In very low volume production plastic materials are normally selected for their physical properties rather than for low cost or appearance.

A rationale for selecting passive materials

The first step towards material selection for the passive part of a sensor is to set up the system architecture that will satisfy the operational requirement for the measurement system. This means deciding the functions needed and how they are to be provided as organic modules.

The module connecting the observed process to the subsequent signal processing and the output actuating device both need an active sensing material, whereas the mountings and following signal processing assembly are made from sufficiently passive materials.

These building blocks, in turn, are made from a combination of parts. This reticulation process eventually identifies the basic components needed, calling for ready-made or new parts to be made from appropriate materials.

Crucial in this design process is the simultaneous identification of which salient physical effects, called *influence parameters*, will arise that need to be allowed for. For example, the steel to be used as the active load cell component of a weighing system used in electroplating out aluminium in a refinery must have very low magnetostrictive properties lest it be influenced by the high magnetic fields existing in such locations. This could lead to apparent load existing as d.c. electrolysis currents vary.

The passive role of materials in instruments can be found by considering the generalized role of the passive components.

Reuleaux, in 1875, defined a machine as

a combination of resistant forces so arranged that by their means the mechanical forces of nature can be compelled to do work accompanied by certain determined motions.

Considered in a more general context the concepts of *resistant forces* and *determined motions* are applicable to any form of structure, whether it be mechanical, electrical, optical, etc.

Materials must be chosen to ensure their interacting properties are in accordance with the requirement. As instrument design is highly interdisciplinary there are usually many different such resistant forces and determined motions to be considered.

To illustrate the generalized design path to material selection for the passive requirements consider the best-known design regime—mechanical.

In mechanical design the main system requirement aims to allow movements, in a three-dimensional space, of as many of the six possible degrees of freedom as are needed. These freedoms, for an extended object in space, can occur as three translations and three rotations. A *framework* has to hold all six stable so that parts mounted on it are constrained to the needed dimensional limits. A *slideway* allows one degree of translation, a *pivot* one degree of rotation and so forth, up to systems with many degrees of freedom such as measuring machines and robots. (The concept of *degrees of freedom* applies to all energy regimes.)

Two approaches can be used to allow the necessary freedoms in a mechanical system. The first is to make use of materials, selected in shape and type, to form a structure that is effectively non-elastic. These are formed to allow the needed freedoms. An example is a ball bearing where the rollers and tracks are made to exhibit as little deformation as possible. This method is known as *kinematic* mechanical design.

The alternative is to purposefully provide certain parts with elasticity so that movements can occur at intended places. An example is a flexure strip hinge used to allow limited rotation. This is known as *elastic* mechanical design.

In general kinematic design has the widest applicability but when relatively small movements are needed its options often cannot compete with the cost effectiveness of the elastic method.

Once the overall constraint of freedoms strategy is decided a suitable physical framework is evolved to support the various parts needed to provide the functions. A collection of basic parts set up to provide a particular function is termed a *mechanism*. This, in turn, is formed of *members, linkages, joiners, bearings*, and parts that provide *transfer* motion (Sydenham, 1986). An overview of the design of such parts for fine instruments is covered in Trylinski (1983).

Parts and roles identified by this top-down design process then allow the choice of material. As can be seen from the abridged list of material properties given in Table 43.4 the choice of a suitable material is often complex and often impossible to adequately satisfy.

Altering the level of passive or active behaviour

An often met difficulty is that the material that must be used is not passive enough to some external influences. Conversely it may be desirable to increase the activity level. Several methods can be used.

First, these material effects can be reduced by placing the active part, or the whole unit, in a suitably controlled environment.

Second, some form of compensation can be incorporated. For example, to render a mechanical length formed from a metal bar less variant with temperature the bar can be mounted to another bar that expands in the opposite direction. By use of materials with different TEMPCOs and different lengths it is possible to set up a system that has finite length. Opposing action is needed because very few materials have negative TEMPCOs.

Another compensation method is to measure the influencing variable and from knowledge of the systematic nature of the activity change a value in the system at some appropriate place.

The principle of these two compensation methods can be applied to systems in any energy regime. Often transducers are used to convert energy forms in order to make the comparison and compensatory actuation. An example of this type is use of a second resistance strain gauge placed in one arm of a Wheatstone bridge interrogation circuit such that its temperature-induced changes cancel out those of the sensing gauge. This second gauge is mounted such that it is not subjected to any strain but is subjected to the same temperature as the sensing gauge.

There does sometimes arise need for increasing the effective activity of a material. A commonly used method is to place more than one element in an arrangement that adds or multiplies the effect. An example is the use of a set of series-connected thermocouples to form a thermopile. Another example is where the cold junction of a Peltier cooling cell is used to mount a second unit, the temperature difference then being more than one unit alone can provide. These methods add the effects. A mechanical lever system can be used to increase a small deflection by multiplication.

43.2.3 Material Selection—an Example

As an example of the necessary interplay between the active and passive roles of materials forming even a simple component consider the material needed to form a spring for a weighing balance. The measurement purpose of the spring element (the clearly defined active component) is to produce a reproducible elongation proportional to the mass hung on its end.

If the material from which the spring is made changes length with temperature it will exhibit a displacement error for a fixed load as its ambient temperature changes. This suggests, on first considerations (but see later), that the spring should be made from a material that has a zero thermal coefficient of expansion (TEMPCO). Figure 43.1 shows how this coefficient varies in typical instrument materials, showing that virtually zero coefficient materials are, indeed, available.

However, measurement of spring displacement is made by *comparing* the length of the spring with a reference support framework (that should at first sight be made of a passive material—but again see later). What is actually required, however, is that the framework and the spring have the same thermal expansion coefficient so that their changes in length track each other as temperature changes. As it is expedient to make a frame from steel it does appear feasible to use a steel spring, despite its being quite active to temperature effects. However, there is another aspect still to be covered.

A vital physical property needed of the measuring spring material is that its *modulus of elasticity* remains adequately constant with variation of operating temperature—this is expressed as its *thermoelastic coefficient* (TEC).

Study of the properties of the steel family shows that this is relatively large. A steel spring is prone to temperature error in that its weighing sensitivity significantly increases as the temperature rises. For precision weighing springs steel is not a suitable material.

The material needed has to have a thermal coefficient of expansion suited to structural materials yet have a low TEC. No entirely suitable material has yet been developed—a common situation to be faced in instrument design. The nickel–iron–chromium alloys, however, provide some scope. Figure 43.2 shows how the TEC varies with alloy composition for this family.

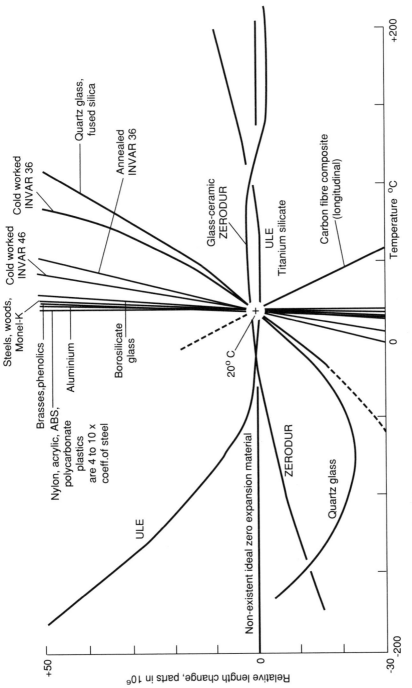

Figure 43.1   Thermal expansion of metals, plastics and ceramic materials

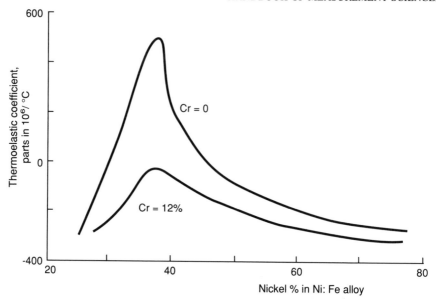

Figure 43.2   Thermoelastic coefficient variation versus alloying composition for the Ni-Fe-Cr alloys. (Copyright INCO EUROPE Ltd. Used with permission)

Even if the above needs are met there is still another most important parameter to be considered—the spring should return to the same displacement position when a mass to be weighed is removed. This is its *mechanical hysteresis.* Steels are quite poor in this regard. Special alloys, such as Ni-Span C, have been developed to exhibit insignificant mechanical hysteresis, but their TEMPCO and TEC are not at the preferred values.

To make material selection still more complex the properties mentioned above change with history of use, with temperature of operation and sometimes with level of present and past history of stress. Additionally they are usually time-variant, often in non-systematic ways.

The same considerations apply for the support frame design as that provides the positional reference framework. They are, however, usually of less significance.

In practice precision springs can be made to only certain performance limits. Use of some form of active or passive compensation can be used to squeeze a little more sensing performance from the spring (Sydenham, 1986). These material limits were met in the design of spring–mass clock movements finally being replaced by better defined elastic properties of the piezoelectric quartz crystal now commonly used in electronic watches.

The design and operation of highly precise and stable springs is a matter of setting up a system that recognizes the sources of likely error arising from material activity and somehow operates with all of the material restraints in manner that optimizes the operation. In the spring case this may mean tightly controlling the temperature to keep the spring at its optimum operation condition—Hugill (1984) is an example of the lengths that are needed to obtain a top-performance geophysical gravity meter—which is, in effect, an ultra-sensitive spring balance.

Thus it is seen the design of a sensor is not just a procedure of selecting an active material, setting it up within a system of passive supporting and containing materials. To get the best performance sensor design invariably needs complex interaction of material properties.

The problem of interaction between material properties will be found in any design regime. For example, in optical lens design obtaining a certain magnification power for a given spectral bandwidth will be a matter of juggling between availability of a few materials having limited ranges of refractive index and varying wavelength dependent transmission losses.

43.3   STRUCTURAL ELEMENTS AND THE MATERIALS USED TO MAKE THEM

43.3.1   General Remarks

Structural parts are mostly made from metals or plastics with small use of ceramics (Ichinose, 1987) for special support situations—for example, in electronic hybrid forms of sensor or for high-temperature support.

The main structural elements of instruments include support casings, frames, panels and covers. Metals and ceramics are mostly used by post-processing basic stock forms such as sheets, bars, shafts, tubes, and more complex extruded cross-sectional shapes.

The more complex support structures are usually made by metal sheet fabrication, by moulding the shape with metals and plastics, or by machining the shape from, usually, solid metal stock or from roughly shaped castings. Ceramics are the most difficult material to shape (Ichinose, 1987) and thus find little use as complex structural parts, especially if the size is not small and the production insufficient to cover the specialized tooling costs.

Commonly needed instrument parts include the casing for the sensor, the housing for the signal-processing electronics or a carrying case for the complete instrument. The structure often is required to hold mechanical components in a strict working relationship, such as the optical parts of an interferometer. In this case the designer must be aware that plastic materials often cannot provide the dimensional stability of metals or ceramics.

The structural use of materials is well ordered and documented but generally lacks instrument design emphasis. It has developed on a reasonably theoretical basis supported by good quality experimental data about material properties. Its literature is found under the name *strength of materials*. Such works, however, emphasize strength and failure performance rather than the deflection and non-linearity aspects important in measurement systems. Good understanding of elastic properties is essential, a summary being provided in Sydenham (1986).

Compared with that of plastics the behaviour of metals and ceramics is well defined and understood—at least until practical limits are met in very precise sensing situations where non-linearity, history of processing and material property data become uncertain. Plastics, however, are revolutionizing instrument design but are still less well understood. This section, therefore, concentrates on plastics.

43.3.2   Using Plastics to Form Structural Parts

Because the structural properties of plastics are so different from metals, on which much of engineering practice is based, and because manufacturing processes and material selection have such significant effects, the instrumentation designer is cautioned to obtain advice early in the design stage if plastics are to be used. This section concentrates on the use of plastics. Important factors to consider are:

- Stress–strain relationships for plastics are similar to metals only to a limited extent. The time of application of stress is an important modifier and operating temperature has a large influence.

- Young's modulus of elasticity for plastics is typically around 1% of that for steel (1800–2400 MPa vs 200 000 MPa). Plastic products are, therefore, much more compliant.

- Yield stress is typically in the range 20–50 MPa (compared with 230 MPa for structural steel). However, this is for short-term stress application and varies widely with temperature.

- If stress is applied for long periods, most plastic materials will creep. For low stresses (as in the typical supporting role of instrument housings) this may be unimportant but if significant loads are applied (such as maintaining spring tension) and especially if elevated temperatures are encountered, creep can be highly significant and possibly lead to failure.

- Many plastic materials are highly susceptible to environmental conditions. The UV in bright sunlight rapidly degrades many plastics, polypropylene, polystyrene and PVC being particularly susceptible unless protected with appropriate colour (preferably black) and/or anti-UV additives. Nylon absorbs atmospheric moisture, expanding in the process.

- Plastics, like metals, exhibit fatigue.

● Joining plastics requires care. Adhesives will not work with all plastics while with others (such as polystyrene) the solvents associated with adhesives can cause surface cracking. Creep must be considered when using mechanical fasteners. On the other hand, thread-forming screws and other fasteners taking advantage of the softness of plastic materials are very effective. Ultrasonic welding will often join difficult-to-join materials.

If production volume is sufficient to support highly productive processes such as injection moulding, ultrasonic welding, reaction injection moulding (RIM) or compression moulding, supported by sophisticated material selection and formulation, then cost and throughput will probably be sufficient reasons for the decision to use a plastic. Given the volume to cover the considerable tooling and set-up costs, additional benefits will include:

● Freedom of shape. Almost any shape can be moulded, allowing *customized* support for internal components, ergonomic shape, good appearance and freedom of choice in colour and decoration.

● Light weight combined with rigidity. Foamed materials can be used with benefits.

● Electrical insulation.

● Resistance to shock and damage in use or abuse.

● Location of moulded-in features such as internal supports, hinges, cable channels, snap fastenings, electronic board retaining slots, protective surrounds and access openings. Once in the tool, these features introduce no added part costs.

Some of these features can be obtained with low volume methods such as fibreglass lay-up techniques. Moulds are low pressure, and, therefore, low cost. Female moulds are generally used, forming the outer surface of the item. The outer plastic layer is a gel-coat of unreinforced material (usually epoxy or polyester) backed up with further layers of material reinforced with glass fibre cloth. It is rare in this type of application to use the more exotic reinforcing materials as their high performance is not required. Parts are generally hand made within the mould.

Sheet-moulding and dough-moulding processes form thermoset materials, filled with short-fibre reinforcement, in relatively low-pressure moulds, using heat to cure the product. These are moderate to high-volume processes, generally used for relatively large items of simple shape (such as large instrument carry cases.).

The good surface finish, fine-dimensional tolerances and self-colouring available with plastics frequently allow the plastic product to be used in the as-moulded condition. However, other finishes are available, decorative or performance-enhancing, depending upon the material used. They include:

- Painting. Generally, however, paint cannot be satisfactorily applied to those plastics having low-friction surfaces, such as acrylic, polyethylene or nylon.

- Hot stamping. This allows metallic lettering and related effects to be used. Where this decoration is moulded into the part, crisper hot-stamped effects are possible.

- Vacuum metallizing provides a thin surface of aluminium on parts. The surface is fragile and should be protected. Some materials can be chromium plated, giving a surface which is wear resistant as well as decorative.

- Anti-static and anti-EMI coatings. These conductive coatings are normally applied to the inner surface of instrument housings to minimize interference from external electromagnetic radiation.

Figure 43.3 shows an electronic ignition timer module made from plastic moulded around the necessary metal parts. Note the use of a ceramic substrate supporting the hybrid electronic circuitry and various electronic connection points.

43.4 MATERIALS USED TO MAKE MECHANISMS

To obtain space-varying mechanical functions, such as a linear to rotary motion conversion, it is necessary to use a collection of basic components to form a class known as *mechanisms*. Typical mechanism-building components are *fasteners*, *pivots*, *links*, *gears*, *wheels*, *cams*, *tension* and *compression elements*, *screws* and *springs*.

Traditionally mechanisms were made from hard wearing, easily machined metals such as brasses, bronzes and steels. Today there is increasing use of plastics in this role as most elements of mechanism can now be formed in plastics. Ceramics are also finding more uses here—one application (Ichinose, 1987) is as biomedical replacement parts.

If production volumes allow for tooling and precise material formulation then plastic elements can be low in cost as well as having desirable material properties. In low-volume production the parts can be machined from a wide range of plastics in bar, block, tube or sheet form.

As previously noted, plastics cannot provide the dimensional stability of metals. By the use of composites and special formulations plastics can, however, often challenge the metal equivalent. Where dimensional stability is not critical they provide many benefits over metal parts. These include:

- light weight;

- self-lubricating properties, without need for added lubricants;

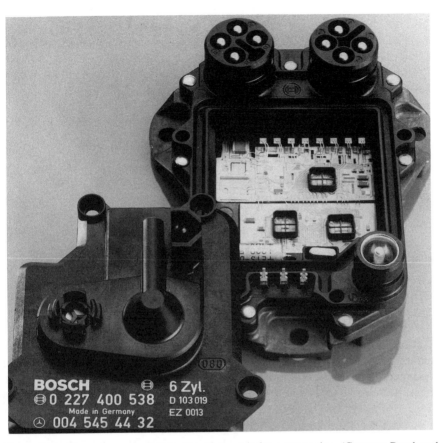

Figure 43.3   Computer-controlled ignition timer unit for a car engine. (Courtesy Bosch and BASF)

- electrical insulation;

- soft, quiet operation, with inherent shock absorption.

A widely used plastic for mechanism parts is acetal. It is strong and stiff, with good resistance to abrasion, heat and chemicals and has low creep and good fatigue properties. It also has a low surface coefficient of friction, making it ideal for sliding parts.

It is often possible to design some elements of a mechanism into an injection-moulded part. Bearings, cam surfaces and springs can be formed in this way. A servicing disadvantage is that they can then only be repaired by replacement of the complete moulding.

Examples of the now extensive use of plastics in instruments are to be seen in electronic consumer products such as audio, video and compact disc players. These use very few metal parts. At the more highly stressed level of use of plastics in mechanism design are power hand tool examples which also now have only minimal metal in their construction.

A limit switch for use with a hoist transmission system is shown in Figure 43.4. Gears, cams, toggles and the case are all made from plastic. The only metal parts are the support shafts (which require stiffness, strength and bearing surface) and metal contacts and electrical connections (which require high electrical conductivity.)

43.5 MATERIALS USED FOR SENSING

43.5.1 General Remarks

As has already been pointed out, materials that exhibit a marked degree of activity of energy transformation are used to form the sensing part of a sensor.

The number of such effects is truly vast and to date no comprehensive classification structure has been developed. There do exist, however, several schemes that at least assist an understanding of the variety of possible sensor principles. That by Middelhoek and Noorlag—see Sydenham *et al.* (1989) for a summary and its application to structured sensor design—is based on the energy forms of the input, output and excitation ports of the sensor block. Another addresses the potential sensing functions that exist for the silicon sensor, Middelhoek and Hoogerwert (1987). The problem of classification with respect to the materials of sensors is addressed in Jones (1990).

Metals and ceramics can be used variously to form the basis of a range of sensors. Obvious examples are the metal platinum resistor used to sense temperature and the optical glass fibre used to sense numerous variables. In each case the material is usually prepared especially for the application to ensure it exhibits good stability, fidelity and is in a form that can be fashioned into a sensor unit.

The nature of plastic materials is such that they have few properties which are predictable enough for use as sensing elements, and where such properties exist they are usually heavily dependent upon formulation, manufacture and working environment. Examples do exist (such as the piezo-electric effect of some conductive plastics) but these are rather specialized.

43.5.2 Materials Involved in Commonly Needed Sensors

This section now presents an overview of how a range of common measurands can be sensed, also discussing what materials are involved. This is achieved by

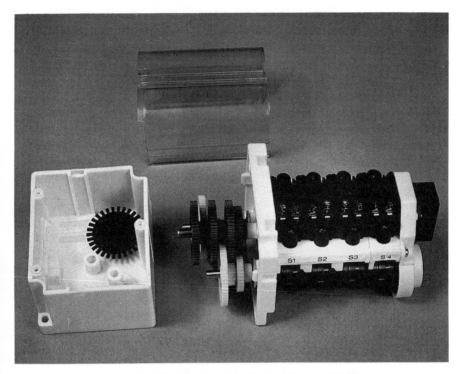

Figure 43.4   Electrical limit switch for use with a hoist. (Courtesy BASF)

working through a collection of sensors, given as Table 43.5, providing brief comment of the state of each. The list can only be seen as an introduction—the full list would take many volumes to describe.

Light measurement

The need here is for a sensitivity of measurement to match the resolution of the human eye. Finer and finer patterns are being etched on silicon charge-coupled devices in the visible spectrum and on cadmium mercury telluride photodiodes in the infrared. The size limitation imposed by single-crystal technology has been overcome by the development of amorphous silicon.

The digital and discrete nature of the photon response requires ever-improved systems electronics to match the continuous gradations which the human eye can detect. Reversible chemical detectors are currently too insensitive and are unlikely to respond to development. With the human retina as an example, however, there should be substantial scope for biological sensors when the technology of extracting usable information from cellular material has developed sufficiently.

Table 43.5　Sensor types classified according to the measurand being sensed

Effect Measured	Transformation			Transduction		
	Physical	Chemical	Biological	Physical	Chemical	Biological
Light	–	1	2	3	4	5
Heat	6	7	8	9	10	11
Sound	12	13	14	15	–	16
Pressure	17	–	18	19	20	21
Displacement	22	–	–	23	–	24
Time	25	26	27	28	–	29
Electric field	–	–	–	30	–	–
Magnetic field	31	–	32	33	–	–
Particle flux	34	35	36	37	–	–
Concentration	–	38	39	40	41	42
State of Matter	43	44	–	45	46	47

A Detection of Light
  1　Silver iodine transformation
  2　Photosynthesis
  3　Photodiode
  4　Photochromic glass
  5　Human eye
B Detection of Heat
  6　Permanent deformation
  7　Permanent colour change
  8　Cell death
  9　Expansion
  10　Reversible colour change
  11　Nervous system
C Detection of Sound Frequency
  12　Fracture
  13　Reaction Kinetics
  14　Cell death
  15　Microphone
  16　Nervous system
D Detection of Pressure
  17　Permanent deformation
  18　Cell death
  19　Piezo electricity
  20　Chemical reactions
  21　Nervous system
E Measurement of Displacement
  22　Marker Pen
  23　Radar
  24　Nervous system
F Time
  25　Stalactite

26　Candle
27　Tree rings
28　Clock
29　Biorhythms
G Measurement of Electric Field
  30　Cathode ray tubes
H Measurement of Magnetic Field
  31　Magnetic tape
  32　Electric eels
  33　SQUIDS
I Measurement of Particle Flux
  34　Colour change
  35　Dye bleaching
  36　Cell death
  37　Geiger counter
J Measurement of Chemical and
  Biological concentration
  38　Permanent colour change
  39　Antibody reaction
  40　pH
  41　Reversible colour change
  42　Noses
K State of Matter including chemical
  composition, integrity, state of
  division.
  43　Permanent resistance change
  44　Chemical reactions
  45　Optical observation
  46　Catalysis
  47　Human eye

Heat

Increasing use is now being made of pyroelectric materials and until their potential has been more fully exploited there is unlikely to be a demand for even newer materials. Organic sensors such as PVdF are being studied because of advantages of convenience, performance and economy.

Liquid crystals have proved popular in aspects of temperature measurement not requiring high accuracy but do not offer advantages in control situations.

Chemical and biological materials have a readily detectable response to heat but again do not meet the criteria for sensitivity and ease of information handling. In this area, therefore, physical sensor materials will always be dominant.

Sound frequencies

The electrets introduced some ten years ago represented the first major advance in sound-detecting materials for some 100 years. There is now some challenge from the polymeric piezo-electrics, such as PVdF, which offer advantages in special situations such as the random-noise cancelling microphones because only one membrane is involved. If sufficient of these special situations develop then the reliability and cost of the piezoelectric polymers may match those of the electrets. The broadband response of the polymers does make them more suitable for high-frequency ultrasound than ceramic piezoelectric materials which tend to 'ring'.

Sound sensor technology has tended to lag behind other technical developments in the sound-transmission regime because of the tolerance of the human ear to distortion. If the commercial demand for speech and other sound recognition facilities increases then the pressure for materials able to translate vibrations into digital signals will increase. There appears to be ample scope in the development of improved piezoelectric materials on which limited work is already taking place.

Once again the relative sound insensitivity of chemical and biological materials offers little scope for development.

Pressure

There is a great demand for more sensitive and convenient pressure sensors for on-line process and engine control, load cells, altimeters etc. Where a single crystal of silicon can be used, problems of fatigue are greatly reduced. The search is on for strong, tough, single-crystal membranes and particularly for materials which themselves are strain sensitive so they can be used in a strain-gauge mode and involve no other post-processing transduction element.

The relative slow rate of advance in this area illustrates the need for much interdisciplinary cooperation between chemists and physicists. The biologist should also be involved since muscle and nerve tissues are themselves pressure sensitive and generate convenient voltage pulses.

Displacement

Sensor materials have little to contribute to further advances in displacement measurement. Optical means necessarily hold sway and their sensitivity is limited by the wavelength of the radiation used and by the ability of the information-handling system to interpolate wavelength. Linear analog electrical sensors can detect to nanometer noise levels.

Lapsed time

Materials development has been of prime importance because of the key role time measurement plays in so many scientific and industrial determinations. With the development of the quartz crystal for industrial use and the atomic clock for scientific purposes further improvement would seem marginal and as with displacement, the future emphasis is on systems development for improved information handling.

Electric field

Cathode ray tube phosphors have undergone intensive development. Further improvements would only have a marginal affect on performance. Once again systems development is paramount.

Magnetic field

Considerable attention has been paid to the development of materials for recording of signals on tape and disk. Further development of the storage media may yet lead to a magnetic field sensor. Very small fields are detected by superconducting Josephson junctions and superconducting quantum interference device (SQUID) transducers may be useful for medical purposes and for submarine detection. Recent developments in high-temperature superconductors will possibly produce SQUIDS which are economic to use. This is an example of an advance in materials technology producing lower-cost sensors without involving a new phenomenon.

Particle flux

Higher-sensitivity ionizing radiation-detecting materials are required for safety and scientific reasons. The higher the atomic number of a material the greater its sensitivity to particle flux, hence the present emphasis on materials such as mercuric iodide and calcium tungstate.

Measurement of chemical and biological concentration

This is one of the major areas of sensor material advance because of the increasing need for on-line and rapid means of measuring concentration and concentration changes during industrial processes, in medical patient monitoring and diagnosis and in control of pollution of the environment.

Considerable effort is being directed to the development of sensors specific to particular constituents or activities of a chemical or biological analyte able to work in direct contact.

The term *crosensors* is used to describe chemical sensors which detect biological activity and biologically based sensors which detect chemical activity. There is thus scope for much confusion. Here the former are referred to as chemical sensors.

Potentiometric measurement has been the most fruitful area of advance to date and this is now epitomized in the silicon field effect transistor or *chemFET*. The major problem with silicon is that it is reactive to moisture as are most, if not all, of the transistor materials. There is need for sensor materials able to produce the components of an integrated potentiometric element in a moist chemical and biological environment.

The use of enzymes as a coating on a chemFET, or in an associated gel or membrane, increases the selectivity of these devices and serves to blur the distinction between chemical and biological sensors. Selectivity is a major problem of potentiometric based chemical sensors, others are detection limits and dynamic response times.

To overcome these problems and produce 'true' biosensors by the definition used here amperometric reactions are being developed. In these a current proportional to the specific binding reaction is generated rather than a voltage where the relationship is logarithmic. The sensitivity is thus improved. Redox enzymes are the present favourites for these reactions.

Other true biosensors are now being developed using ionophores in a membrane which generate a potential difference when exposed to specific ion concentrations.

The detection of antigens using monochlonal antibodies in immunosensors is another area where the binding of the antibody to the antigen can produce a potential difference in a membrane. In other types of immunosensor, an addition agent such as an enzyme is added to determine the extent of binding of the antigen. These fluoresce or luminesce and are detected by fibre-optic means or by producing a sensed colour change. Other reactions involve the generation of heat.

Biosensors provide an interesting example where two sensor materials are combined. One involving a transformation—the biological element, the other a transduction—the physical element.

The major problem with the commercialization of many biologically allied chemical sensors and the biosensors themselves is that reproducible manufacturing methods have yet to be developed. Calibration problems also arise because the performance of a device changes during its shelf life due to changes in the structure or decay of the biological element.

Measurement of concentration in gases is an easier problem because the moisture content is not usually high. Light absorption is the most convenient means, employing photodiode materials which are well developed and understood.

Once again, systems development is the key to accuracy and performance.

Two very important gases, hydrogen and oxygen, are not, however, detectable by light absorption.

There is thus a need for materials which exhibit a simple transduction phenomenon susceptible to the presence of a hydrogen or oxygen atom. One example of interest is the electrical conductivity of tin oxide and like materials.

Biosensors are covered in more depth in Chapter 40.

State of matter

In this category are included many parameters about the condition of materials that are important in industrial and other contexts. These cover:

● state of division;

● cell or crystal size;

● internal or external integrity;

● impurity content;

● atomic or molecular structure.

The main investigative techniques for these complex quantities are optical, electronic, or acoustic microscopy; X-ray, electronic and acoustic examination or detection. The sensor materials are components of the examination system and therefore have, in the main, been considered earlier. Systems development is, once again, the major requirement.

An increasing area of activity is condition monitoring where the relevant state of matter is being interrogated in service. Here the search is for on-line monitoring systems such as acoustic emission, which can continually detect and analyse signals related to the matter state or for sensor materials which can either simulate or detect the performance of the materials in service. (See chapter 35 for further details of conditioning monitoring.)

One example of this type of sensor is a wear debris monitor (developed by Fulmer Ltd, UK) in which a resistance element is exposed to a fluid circuit. Abrasive matter in the fluid causes a reduction in the thickness of the resistance element and hence a rise in its resistance which is related to the presence of wear debris and hence gives an indication of the state of matter in the fluid circuit.

This is a rare example of a permanent transformation being used for monitoring purposes but is perhaps a guide to the type of behaviour needed in continuous state of matter monitoring.

43.5.3 Conclusions about Sensor Materials

Adequate, if not ideal, sensor materials exist for most of the information-sensing requirements. The main needs for improved materials, and hence for improved added value in performance, are for the sensing of light, sound frequencies, pressure and chemical concentration. These areas would open up enormously if better sensors could be developed (Jones, 1990).

Most development follows the traditional path of physical transduction because this is the area of most advanced information-handling technology. To make most progress the skills of the physicist, and increasingly of the mathematical modeller, must be allied with those of the chemist and metallurgist to make maximum use of available phenomena and to improve such factors as consistency of performance and ease of manufacture. This is especially the case with piezo-and pyro-electric materials and with membranes for pressure sensing. Semiconductor materials able to operate in moist chemical and biological environments would advance process control in these environments. To combine disciplines in an effective manner the role of the device physicist has been heavily involved in many major industrial companies concerned with instrumentation and control.

Chemical phenomena alone seem to offer little scope in sensor materials except for recording purposes where some, such as the light exposure behaviour of silver iodide, will remain dominant.

Biological reactions offer great potential for many applications particularly in the measurement of concentration and activity but await further advances in the conversion of signals into electronic form and in the consistency of behaviour of the biological structures.

The development of device chemists and device biologists would greatly accelerate progress as well as realization of the expected added-value multiplier.

43.6 SPECIAL PARTS NEEDED FOR INSTRUMENTS

Modern plastic formulations have made it much easier to provide a range of devices needed in instruments that were, until recently, comparatively difficult to manufacture in metal. Some are now discussed.

Connections in the various energy regimes

Electrical connections normally use off-the-shelf plugs and sockets. These make major use of plastics and frequently include features minimizing the effects of pulling and twisting of the cable in service. The role of the instrument designer is normally limited to selection from available proprietary items.

Mechanical connections, such as connecting rods, in tension or compression are typically made from acetal. They can include snap fastening to mating parts. The designer should be aware of their high (relative to metal) thermal expansion.

Fluid connections frequently include plastic tubing. For non-critical, aqueous or air connections, off-the-shelf irrigation fittings and PVC tubing may be adequate. In the field of bio-engineering, PTFE (e.g. Teflon) tubing is almost universally used and a number of standard types of fittings are available to accept the range of tubing available. Rigid PVC tubing is also available that may be applied for large-volume, non-critical applications.

Coupling protection is frequently provided by elastomeric boots. These can be used to protect sensitive moving parts from contamination by dust and water splash. With more careful design they can protect against more severe environments such as immersion. When volume permits, injection-moulded thermoplastic elastomers may be used. For small numbers, polyurethane rubbers can be cast in moulds, but skill is required in both material selection and moulding to provide a durable result. These boots and special shapes are used to protect sensors where they are subject to water and oils, such as in machine tool readout optical sensors.

Coatings

Plastic coatings are applied to (typically) metal components to provide protection against damage and corrosion as well as to enhance appearance.

Paints, consisting of inorganic and organic pigments and fillers in an organic carrier, are plastic coatings. A full treatment of paint finishes is well beyond the scope of this text, but the instrument designer should be aware of the very wide range of paint finishes, offering various levels of environmental protection and decorative finishes. For good results, paint application processes and surface preparation are at least as important as the paint material itself, with chemical cleanliness being a critical factor.

The designer should be aware of the availability of a range of pre-coated metals, in particular galvanized steel sheet, which can be extensively deformed without damage to the coating. Some, but not all, plastic materials can be painted.

Powder coating of plastics provides thicker coatings and greater control over materials. Application of the powder can be by dry spray, in a process similar to spray painting, or by dipping hot components into a fluidized bed of powder. Coating thickness can be quite high, providing high levels of corrosion and impact protection.

An important coating plastic is PTFE. This is frequently applied to metals to provide a surface which is:

- *Non-stick*. Very few materials will adhere to PTFE and it is able to withstand quite high temperatures (up to 200°C). Two uses are on industrial heater strips and the well-known frypan.

- *Self-lubricating*. PTFE has a very low coefficient of friction in combination with a number of other materials. PTFE is used in bulk, as a coating and as a filler in lubricant-free bearings. One instrument use is as contacting strips used to keep two closely placed optical parts from touching in their sensitive areas.

● *Non-contaminating.* PTFE is non-toxic and may be used to coat sensors in contact with food.

Seals

Standard engineering sealing elements, such as gaskets, lip seals and O-rings are frequently made from a range of plastic materials for different operating environments.

For low-volume applications, polyurethane rubbers may be cast to form sealing elements. A common technique is to use one mating part as part of the mould, the sealing element adhering to, and becoming part of, this element.

For sealing penetrating elements (where, for example, a lead penetrates a housing) an elastomeric grommet or boot is normally chosen. For low-volume applications, silicone rubber can provide a solution. This is available as a gel which sets on exposure to air. The standard material emits corrosive acetic acid on setting.

Fasteners

A wide range of standard fasteners made from plastic materials is available. These cater for such needs as cabling, snap mounting of circuit boards to panels and snap-in bearings. Equivalents of standard machine screws and nuts, generally in nylon, are also available. They are normally used when electrical insulation or corrosion resistance are important. Another important application of plastic threaded fasteners is when vibration resistance is needed. Nylon wing nuts can have quite high thread interference, making them resistant to vibration loosening. Other fasteners (e.g. Nyloc nuts) use plastic inserts to achieve a similar result.

The range of special-purpose metal parts used in instrument assemblies is large. They include thread inserts, locking tabs and washers, solder terminals and circlips.

The range of metal, plastic and ceramic ancillary items can be seen by inspection of the annual supply catalogues published by electronic component supply companies.

43.7 SOURCES OF FURTHER ADVICE

The reader is referred to the following textbooks and journals for further information on materials for measurement systems.

Books

Crawford, R. J. (1987) *Plastics Engineering* Pergamon, Oxford.

Farag, M. (1989) *Selection of Materials and Manufacturing Processes for Engineering Design* Prentice Hall, Englewood Cliffs, NJ.

Flinn, R. A. and Trojan, P. K. (1990) *Engineering Materials and Their Applications* Houghton Mifflin, MA.

Nunez, C. E., Nunez, R. A. and Mann, A. C. (1990) *CenBASE—Materials in Print* John Wiley, New York.

Crane, F. A. A. and Charles, J. A. (1984) *Selection and Use of Engineering Materials* Butterworths, London.

Journals

Journal of Materials Science.

Measurement Science and Technology (formerly *Journal of Physics E: Scientific Instruments*) Institute of Physics. UK.

Measurement IMEKO, UK.

Sensors and Actuators A—Physical B—Chemical Elsevier Sequoia SA, Lausanne.

REFERENCES

Caren, S. (1988), 'Product design, basic parameters, basic requirements', in *Handbook of Plastic Materials and Technology*, Rubin I.I. (ed.) Wiley, New York.

ERA (1989), *Silicon Sensors in Europe*, ERA Report 89-0644, ERA Technology, Leatherhead, UK.

Fellowship of Engineering, (1983) *Modern Materials in Manufacturing Industry*, Fellowship of Engineering, UK.

Hugill, A.L. (1984). *The Design of a Gravimeter with Automatic Readout*, PhD Thesis, Flinders University of South Australia.

Ichinose, N. (1987). *Introduction to Fine Ceramics*, Wiley, Chichester.

ICI (1988). *EPOS—Plastics Selection System*, Electronic diskette, ICI Plastics, UK.

Jones, A.J. (1990). *Sensor Technology—Part 1: Materials and Devices*, Department of Industry, Technology and Commerce (DITAC), Canberra.

Jones, J.A. (1988). 'Product design—structural', in *Handbook of Plastic Materials and Technology*, Rubin I.I. (ed.), Wiley, New York.

Kroschwitz, J.I. (1990). *Concise Encyclopedia of Polymer Science and Engineering*, Wiley, Chichester.

Middelhoek, S. and Hoogerwerf, A.C. (1987). 'Classifying solid- state sensors: the "sensor cube" ', in *Sensors and Actuators*, Middelhoek, S. and Van der Spiegel, J. (eds.), Elsevier, Amsterdam.

Rubin, I.I. (1988), *Handbook of Plastic Materials and Technology*, Wiley, New York.

Sydenham P.H. (1979). *Measuring Instruments: Tools of Knowledge and Control*, Peter Peregrinus, Stevenage.

Sydenham, P.H. (1986). *Mechanical Design of Instruments*, ISA, Research Triangle Park, North Carolina, USA.

Sydenham, P.H., Hancock, N.H. and Thorn, R. (1989). *Introduction to Measurement Science and Engineering*, Wiley, Chichester.

Trylinski, W. (1983). 'Mechanical regime of measuring instruments', in Sydenham, P.H. (ed.), *Handbook of Measurement Science* Vol. 2, Wiley, Chichester.

Chapter

44 D. HOFMANN

Quality Control through Measurement

Editorial introduction

Survival in product manufacture has become a case of who can manufacture items at the least cost in the shortest time, whilst also improving the product's quality. This has led to the manufacturing systems now involving prolific numbers of sensors located throughout these processes in order to provide timely and accurate systems management data. Quality measurement systems exemplify the multi-sensor systems design task, how it is put into practice and how it is influencing the move toward international standardization.

44.1 INTRODUCTION

Increasing competition in the international market has, in many cases, led to qualitative structuring of industrial production processes in order to squeeze more productivity from capital expenditure. As a result, the question 'What is quality?', has received ever-increasing attention.

Quality is the totality of features and characteristics of a product or service that influence its ability to satisfy stated or implied needs (see references ISO 8402 and ISO 9000 through 9004).

The term 'quality', as used here, is not about expressing the degree of excellence in a comparative sense, nor is it used in a quantitative sense for technical evaluations. Basic issues involved are:

Handbook of Measurement Science, Volume 3
Edited by P. H. Sydenham and R. Thorn
© 1992 John Wiley & Sons Ltd

- quality measures have to be defined;

- quality of today is not the quality of tomorrow;

- *quality control* (QC) is about the operational techniques and activities that are used to fulfil given requirements for obtaining quality (ISO 9000);

- *quality assurance* (QA) is about the planned and systematic actions necessary to provide adequate confidence that a product or a service will satisfy given requirements for quality (ISO 9000).

Modern concepts of quality control and quality assurance for production not only concentrate on the manufacturing process but also include all elements of the quality cycle.

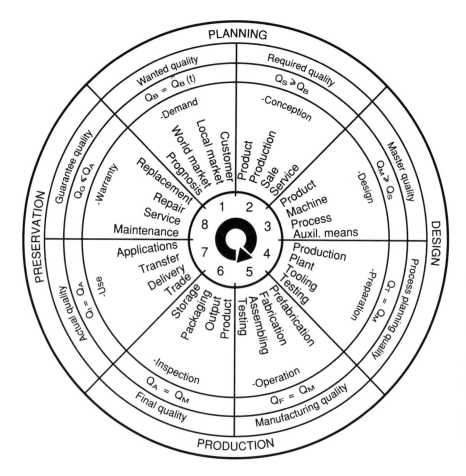

Figure 44.1   The total quality cycle

The *quality cycle* is a conceptual model of interacting activities ranging from planning through design, production and to the preservation of a product or service—see Figure 44.1 (Hofmann, 1988, 1989).

Consideration of the total quality cycle is a matter of overall economy. For example, in a production cycle the design of a product costs about 5% of the total costs but it pre-decides about 70% of the total expenditure (Figure 44.2).

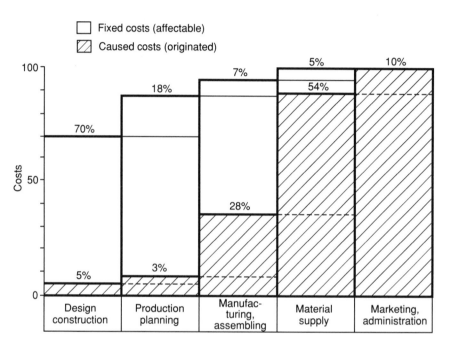

Figure 44.2   Sources and distribution of quality costs

Another reason for using strict quality control is the so- called 'factor 10 rule of failure costs'—see Figure 44.3. The later the failure is detected the higher the costs of the flow-on effects of that failure; and these are then independent of the failure source. For instance, the total recall of a product to rectify a safety related item that only needs, say, a resistor to be replaced can consume all of the product's profit and in some cases even result in litigation, the costs of which can cause an enterprise to fold completely.

To suppress systematic and random errors and failures, which are a fact of engineering and production unless carefully controlled, different methods are practised. In planning and design the *failure mode and effects analysis* (FMEA) is increasingly being adopted. In manufacturing and assembly *statistical process control* (SPC) has found wide application. Failures in application normally are noticed by the customer, maintenance, service (CMS), see Figure 44.3 (Kamiske, 1989).

Modern philosophies of failure prevention are heavily based on *computer integrated manufacturing* (CIM) and *computer-aided quality control* (CAQ), Figure 44.4.

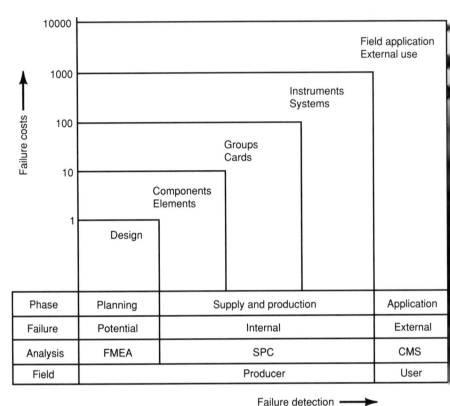

Figure 44.3   Factor of 10 rule of failure costs throughout the product life cycle

44.2   THE IMPORTANCE OF QC AND QA.

Quality control (QC) and quality assurance (QA) are the technological allies of evolutionary and innovative changes taking place in our social and individual lives (ICQC'87, 1987); Kohoutek, 1988). This new life style is characterized by the following facts.

● The qualitative new type of consumer demands are centred on reliable and low-cost products with a lifetime determined by a morally reasoned expectation of an adequate wear-out term for goods like clothes and shoes, laundering machines and refrigerators, radio and television sets, cars and motorcycles,

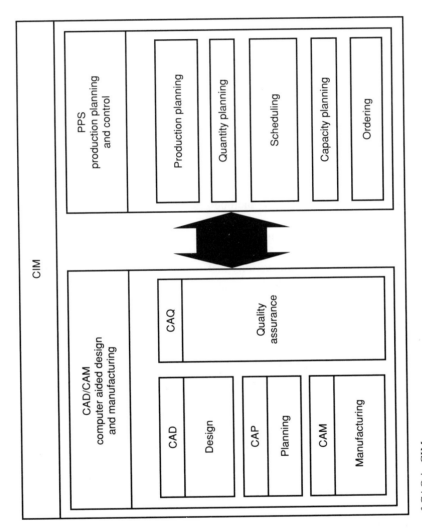

Figure 44.4   Position of CAQ in CIM

air conditioners and copiers, personal computers and personal word processors and more.

- The qualitative new global market which is over-supplied with redundant high-technology products and heightened competition between producers favoured by the consumers from within the more industrialized countries.

- The qualitative new social life-style bringing with it rapid growth of matter, energy and information consumption, social and individual transport dependencies and world-wide communication systems demanding high performance, reliability and security of nuclear power plants, bullet trains, air buses and communication satellites as well as more of the former needs such as thermal power plants, local trains, cars and telephones.

- The qualitative new organization of industrial production processes such as industrialized agriculture and farming, computer-integrated manufacturing and automation of plant and office work, all having a stepwise increase in complexity and converging methodologies (Figure 44.4).

Quality issues have well and truly entered the legal domain. The purchase of innovative products, or the use of innovative services from wherever they are found, now has legal ramifications that must be addressed if a producer is to remain socially and individually competitive (see Chapter 34).

To assure quality the old concept of simply using just quality testing of an outgoing product has now to be replaced by the modern philosophy of *total quality control* (TQC), (Ishikawa and Lu, 1985). It is applied to all stages of the life cycle of a product with emphasis not only on the outgoing product, or the productivity of the manufacturing process, but also to the entire quality cycle from planning and design through production right up to preservation of the product or service (Figure 44.1).

44.3 RELATIONSHIP OF MEASUREMENT TO QC AND QA

Quality control (QC) and quality assurance (QA) belong to all of the swelling flows of resources that are transforming matter and energy using information (Figure 44.5).

Qualitative measures of quality are needed at all stages, from country to country, company to company, plant to plant and workplace to workplace.

Quality is, however, a fuzzy term. The content, objectivity, accuracy, reproducibility and compatibility of quality descriptions are not trivial as they involve extensive crossing of discipline boundaries.

Due to the existence of extensively developed world-wide structures and globally recognized services related to legal and plant metrology, quality parameters

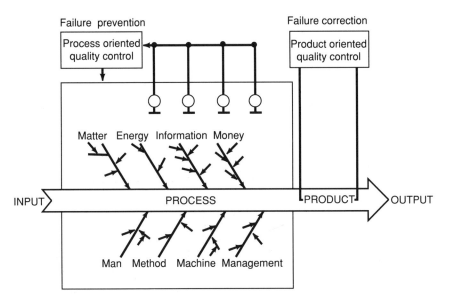

Figure 44.5 Cause-and-effect diagram in quality control

must be well defined and analysed by physically measurable quantities or their measurable representatives, (Sydenham, 1982, 1983; Kemeny and Havrilla, 1986).

Legal metrology has the task to ensure that there exists a unified system of measures and an adequate level of correctness of measurements made (Hofmann, 1983, 1986).

Plant metrology has the task of providing the measurements needed for production control and quality assurance. It must also make available quality management information at the right time and at the right place (Figure 44.6), (Hofmann, 1983, 1986). As can be seen from the figure, this means an extensive measurement system has to be created.

The *INQUAMESS* (intelligent quality measurements) philosophy states:

Objective quality information acquisition and comparable quality audit can only be accomplished if enough, appropriate, measurements are made.

To illustrate the very basis for quality control consider, for example, the gearbox housing shown in Figure 44.7. The quality of the gear box in Figure 44.7a is unknown as there exists no metrological data or declared qualtitative requirement. Its function may appear to be obvious but how well it can perform its function is still to be found.

From an understanding of the functional purposes of the part the quality variates can be decided and declared as shown in Figure 44.7(b).

The symbols used have the following meanings:

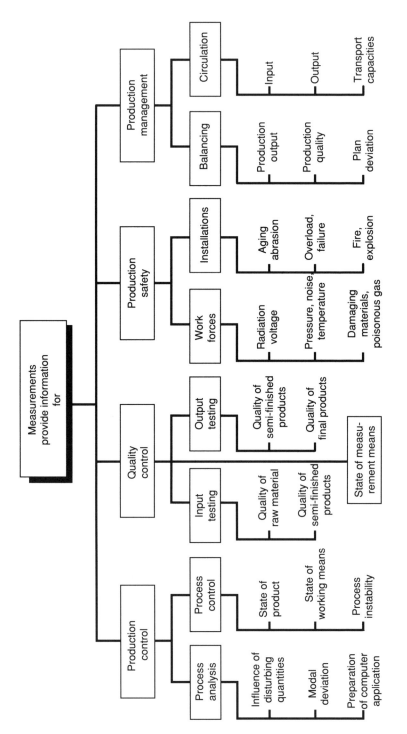

Figure 44.6   Tasks of measurement in production

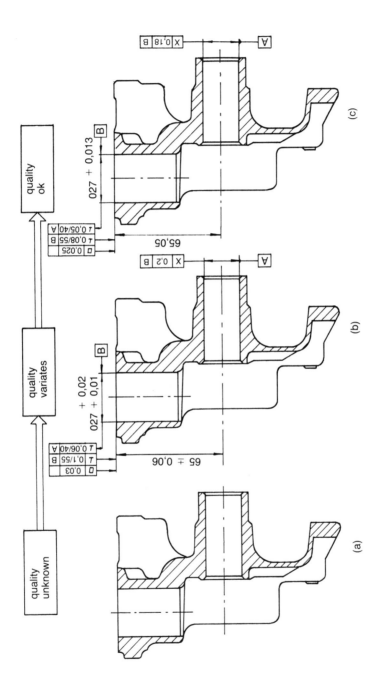

Figure 44.7 Definition and testing of quality variates: an example—a gear housing

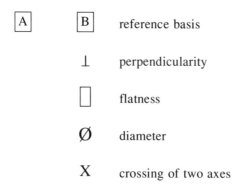

| A | B | reference basis |

⊥   perpendicularity

☐   flatness

Ø   diameter

X   crossing of two axes

Only after measurement has taken place of the necessary Q-variates, and after these have been compared against the desired values to ascertain if they fall within acceptable values, is it known that the tested part is satisfactory. The measured values of the part must lie within the declared tolerances. The part in Figure 44.7(c) is a *GO* part. If any measured part failed to meet the criteria set then it is a *NO GO* part and in that state represents a less valuable component that may even be valued as scrap. Obviously, if a NO GO part is allowed to enter the next step of its production cycle then from that point it contributes negative productivity to the overall process. Without a satisfactorily detailed measurement system the further wasted work on NO GO parts cannot be detected and avoided.

44.4 STRUCTURE OF QUALITY MEASUREMENT SYSTEMS

Approaches to quality control and quality assurance are shifting away from the older idea in which ex-process sorting took place in so-called, *big quality control loops*, the purpose of which was to weed out NO GO parts at the final product, or perhaps a sub-assembly stages.

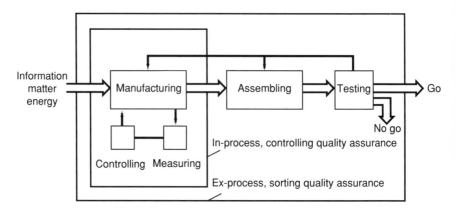

Figure 44.8 Small and big quality loops

It is now realized that more intensive *in-process* methods are required to assist control quality assurance in *small quality control loops* (Figure 44.8).

The main reason for this shift is the realization that quality has to be manufactured into the product and cannot be forced by testing—final type measurements can only tell what parts are NO GO. It gives little scope for timely rectification. Final selection between good parts and scrap work in an uncontrolled production process is now too expensive to be competitive.

The fundamental structure of modern quality measurement systems is as shown in Figure 44.9.

Measurement data acquisition can be accomplished using a great variety of sensors available on market for the measurement of electrical quantities (e.g.

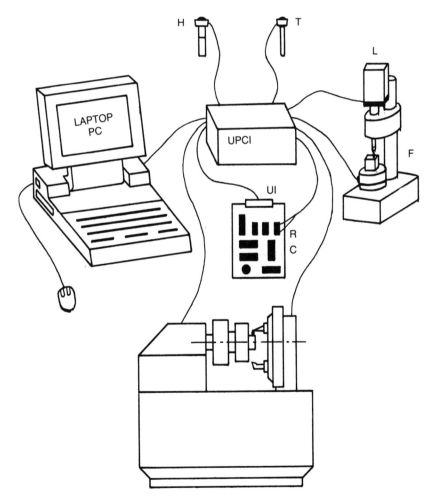

Figure 44.9 Structure, at the basic level, of a modern quality measurement system

voltage and capacitance); for mechanical quantities (e.g. length and force); for thermal quantities such as temperature; for analytical quantities like humidity and for many others. Further details can be found in Sydenham (1982, 1983), and Hofmann (1983, 1986).

For measurement signal interfacing and formatting the universal personal-computer-based instruments (UPCI) are used. The low, consumer based, price of the PC machine and the availability of numerous suitable programming languages and software packages usually makes this option more attractive than the development of a dedicated, customized, microprocessor-based computing system.

Measurement data processing will be generally carried out using portable personal computers of the laptop, briefcase or palm type. They will also use intelligent methods. The data base becomes the interface between different models of program packages for quality control (Figure 44.10).

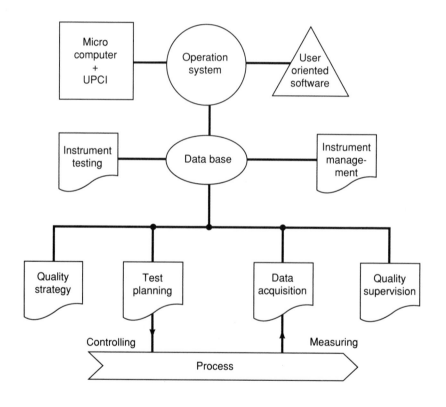

Figure 44.10   Position of data bases in CAQ systems

The physical aspect of these systems will always possess:

● a computer with its keyboard and screen;

- a data base;

- means to conveniently interface, to set standards, the necessary sensors and actuators to the computer, this usually meaning use of a digital data bus system;

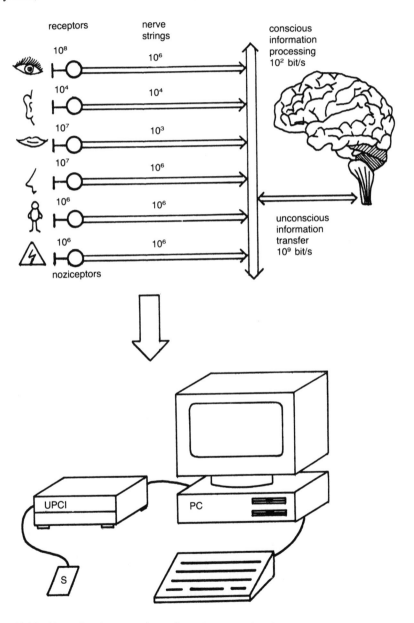

Figure 44.11 Natural and man-made quality measurement systems

- software to organize collection of the measurement data and drive output devices and to make the necessary decisions by comparing processes measurement data with data base values;

- means to update programs and add sensors as the process control is improved.

44.5 MAIN METROLOGICAL METHODS IN USE

Quality measurement is the experiential comparison of a measurable quantity with a known quality (measurement) standard by a quality measuring system.

Subjective information (today termed *heuristic* knowledge that is increasingly finding more relevance now that computing can make effective use of it) about quality is gained by perception by the human through information acquisition and processing by a person using a powerful set of natural biological sensors (eyes, ears, mouth, nose and skin), a central processing unit (brain), algorithms (experience) and information storage for reference standards (natural memory) (Figure 44.11).

Objective information on quality is gained by intelligent type measurements using information acquisition and processing obtained with technical equipment possessing many (over 200 000 proprietary items are offered for sale) artificial technological analog and digital sensors, central processing provided by the microcomputer, algorithms embedded in the software and information storage for reference standards held in the data bases artificial memory (Figure 44.11).

Conventional quality measurements are often organized as mixed procedures of subjective and objective quality information acquisition. Computer-aided intelligent quality measurements demand, and render possible, objective quality information acquisition based on formalized principles.

A central problem is the definition and practical realization of quality measurement standards. The official definition of a measurement standard in the vocabulary of an international official nomenclature (OIML, 1984), is:

A measurement standard is a material measure, measuring instrument or system intended to define, realize, conserve or reproduce a unit, or one or more known values of a quantity, in order to transmit them to other measuring instruments by comparison.

This definition reflects the situation for considering conventional measurements. To enhance the productivity and reliability of the quality measurement information processes required in industry, the number of objective information acquisition and processing procedures made by measurements must be enlarged.

Quality measurement standards should be understood as an integrated set of metrological, particular and virtual standards (Figure 44.12). More developed definitions are needed to suit the needs of quality measurements.

A *metrological standard* is a measuring instrument intended to define, to represent physically, to conserve or to reproduce the unit of measurement of a

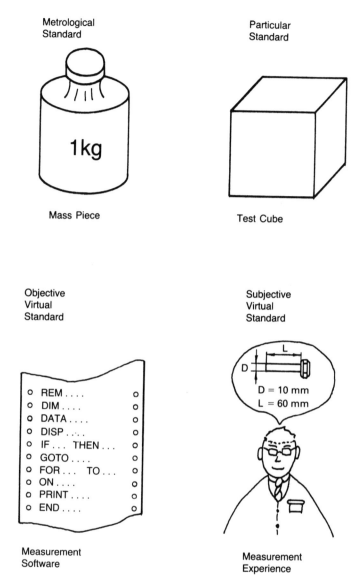

Figure 44.12   Standards for intelligent quality measurements

quantity, or to transmit it to other measuring instruments by comparison. Examples are a linear scale (mechanically or optically produced) used on coordinate measuring instruments or a laser etalon.

A *particular standard* is an objective pattern intended to represent physically, to conserve or to reproduce the identification marks of an object or an event in order to transmit it to other measuring instruments by comparison. Examples are the test cube used to calibrate a multi-coordinate measuring machine or chemical or physical reference probes.

An *objective virtual standard* is a software program or data base intended to define, to represent numerically, verbally or graphically, to conserve or to reproduce the identification signs of an object or an event in order to transmit it to other measuring instruments by comparison. Examples here are the diagnostical patterns of machine tool performance obtained under changing working conditions or the construction patterns of different products during design procedures.

Particular and virtual standards are not yet arranged in a hierarchy. They are not yet *traceable* to a national primary standard. They are agreements made by cooperating firms, laboratories, groups or individuals. Few standardized methods of evaluation and comparison exist. Their accuracy is not yet defined in an accurate manner. Information processing with particular and virtual standards is only at a rough classification stage sitting within the deeper understanding of metrology as a discipline.

The guide in Table 44.1 gives the distribution of measured quantities found to be adequate by industry in order to obtain an assured quality of *batch production* (BP) or *flow production* (FP).

Table 44.1

	BP (%)	FP (%)
Analytical quantities	2	4
Electrical quantities	8	3
Force, pressure	4	10
Length	25	1
Level	1	6
Mass, volume	5	5
Mass flow, volume flow	4	15
Number of parts	25	1
Temperature	8	50
Time	15	4
Others	3	1

For the measurement of these quantities a great variety of sensors are in use. Detailed information is given in the three volumes of this Handbook (Sydenham, 1982, 1983 and here).

44.6 UNIVERSAL PERSONAL COMPUTER INSTRUMENTATION

To obtain quality control it is necessary to use many, varied, sensors whose signals are fused to provide the necessary actions. Although the sensors needed

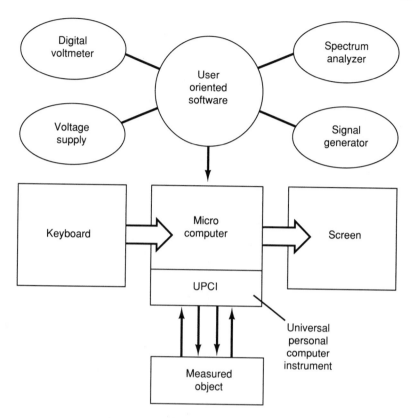

Figure 44.13 Structure of an 'add-in' universal personal computer instrument

must be tailored to each specific application the method of interfacing them to a central computer can be standardized. Several proprietary systems are offered for this purpose. The concept of a *universal personal computer instrument* (UPCI) is relevant here.

The universal personal computer instrument can be either implemented as an add-in board in a free slot within a personal computer (Figure 44.13) or as an add-on box in addition to a personal computer (Figure 44.14).

The add-in technology is characterized by, high-flexibility, high-signal–noise ratio for sensing input circuits, a great number of free slots for system expansion and high capacity of an individual power supply. The disadvantage is the extra cost compared with the plug-in board alternative. For laboratory use and for multidimensional multifunctional quality measurements in industry the add-on technology has convincing advantages (Figure 44.15).

A great number of different sensors for spatial, chronological, mechanical, electrical, thermal, optical, analytical measurements can be easily connected to the external microcomputer.

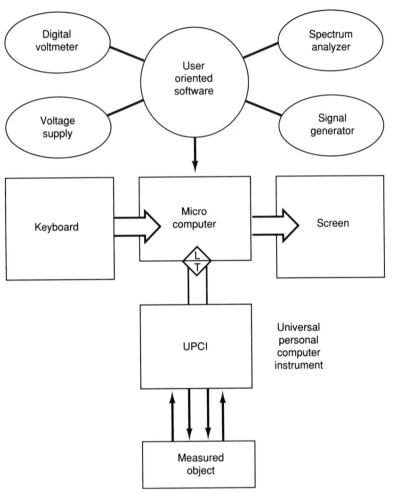

Figure 44.14   Structure of an 'add-on' universal personal computer instrument

As an example, Figure 44.16 shows an internal single-chip microcomputer (INT 2.0) that controls the signal flow from the sensors to the standardized interfaces (RS-232C or IEC 625) organizing the preprocessing of quality measurement data from inductive sensors through an INDAS 2.0 interface, and from incremental sensors through an INKRAS 3.0 interface—and so on as needed.

Increasing availability of 'build as you go' data acquisition systems is making the task of setting up multi-sensor systems ever more one that can be satisfactorily undertaken by the system user with little help from instrument experts. The advantages of expert-system based, decision support, systems and icon programming methods are rapidly becoming recognized. What is not properly understood is how to select, apply and verify the correct sensors needed, for these need

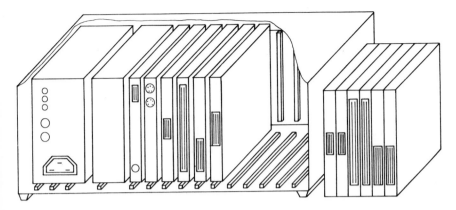

Figure 44.15 General appearance of a universal personal computer instrument

careful choice and installation to obtain the quality measurements so vital to quality control.

44.7 MODERN ORGANIZATION IN FACTORY OPERATION

Modern organization in factory operations is mostly based on the model of a closed quality cycle (see Figures 44.1 and 44.4). In this organizational model the

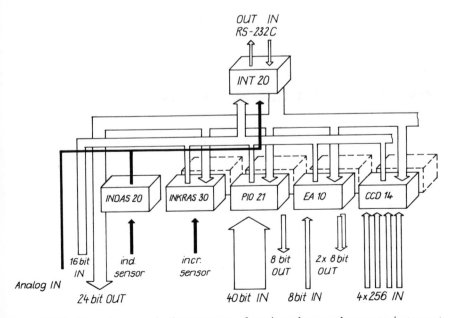

Figure 44.16 Internal communications structure of a universal personal computer instrument

flows of information between office and plant automation are not used to the fullest because the various data are flowing in different streams (Figure 44.17). Truly effective quality control requires the information flow between the different segments of the quality cycle to be integrated and streamlined into a coherent system of both technical and manpower management information.

Problems impeding this necessary integration are that the type and quantity of data produced around a plant and the product management practices are rapidly increasing making integration more difficult. Obviously the task of reorganizing a large, already on-stream, manufacturing plant to give it more efficient quality operations is a costly venture that can also have negative impact on output during the conversion period. Favourable parameters are that computer power per unit cost is increasing rapidly, computers are becoming easier to program and use and the communication standards needed are emerging—such as the Manufacturing Automation Protocol (MAP) and the Technical Office Protocol (TOP), covered in Chapter 42.

The task is not limited by the technological availability of sensors, data collection networking systems, computing power or user interfaces—but there is always room for improvement in each of these aspects. It is limited much

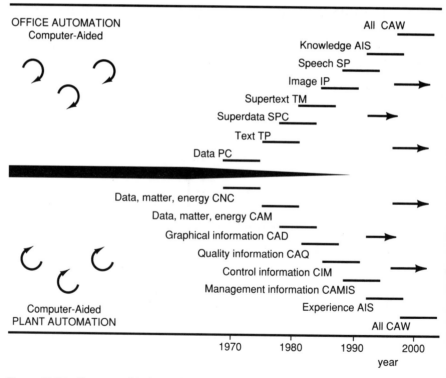

Figure 44.17 Changes, with time, of the constituent information streams in computer-aided automation

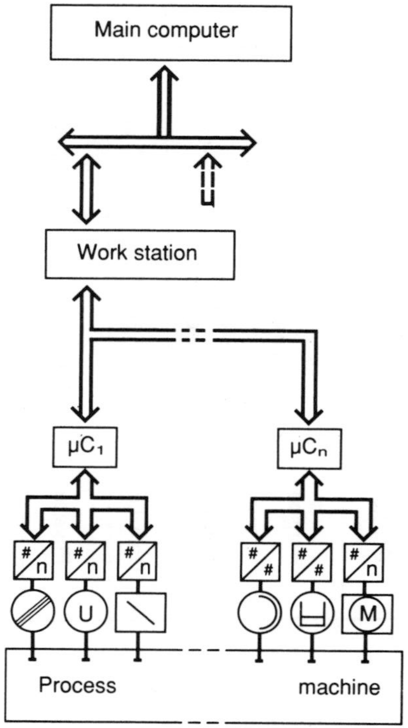

instrumentation	function	level	room	distance	location	distur- bance	environ- ment	data rate	data transfer
	quality manage- ment	factory, enter- price	main office and computing centre	1...5 km	central	small	excellent	M Byte	h
	quality informa- tion pro- cessing	plant	quality control and measuring centre	100... 1000 m	centra- lized	medium	good	K Byte	min
	quality informa- tion reduction	cell	cell	to 200 m	decentral	great	standard	Byte	s
	quality informa- tion acquisi- tion	sensor process, machine tool	working place	to 10 m	scattered	hostile	bad	bit	ms

Figure 44.18 Hierarchy and features of computer-aided quality control information

more by systems aspects such as agreement on data network standards, on field bus standards for connecting sensors, on terminology and on how to set up the complete system. Demonstrations of extensively integrated systems have been given in the USA (under ISA) and in the EEC (under an Esprit programme) so it can be expected that integration will gradually be solved as more experience is achieved.

Returning to the measurement aspects in more detail quality information is transferred, by communication networks, from the sensor, process and machine tool levels streaming through the cell and plant levels upward to the factory or enterprise level. Figure 44.18 summarizes this process and gives the overall characteristics of the information involved at each step. The sensor data are at a low flow rate but many of them pass into the system via a field-bus aggregating to pass through a microcontroller into a more major computer ranging from a PC to a workstation. From there it is networked into a central computer or cluster of such. The communication bearers need to have increasingly larger data rate as the information stream grows. Thus connections begin with shielded cables passing through to coaxial and optical fibre cables.

Interfacing of the different levels is most commonly, but not exclusively, done according to the MAP convention for which Figure 44.19 gives information on the various layers of the system. (See also the more formal definition provided in Chapter 42 on communications standards in this volume.)

OSI-Model	MAP	Mini MAP
Application	ISO Case/MMFS FTAM	MMFS
Presentation	Null	Null
Session	ISO Session Kernel	Null
Transport	ISO/NBS Class IV Transport	Null
Network	ISO Internet Connections	Null
Data Link	IEEE 802.2	IEEE 802.2
Physical	IEEE 802.4	IEEE 802.4

Figure 44.19   Simplified explanation of the Manufacturing Automation Protocol (MAP) layers

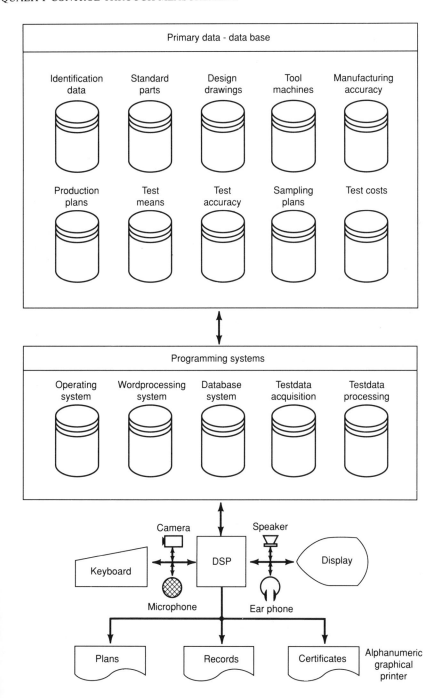

Figure 44.20 Component parts of a CAQ computer workstation

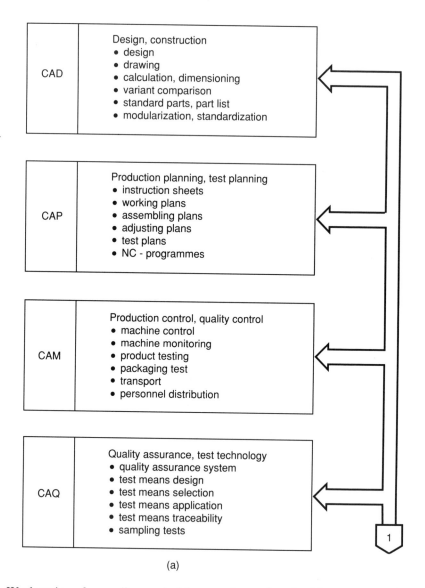

(a)

Workstations for quality control have to be equipped with a great number of data bases and programming systems. Figure 44.20 shows the extent of such systems capability and the complexity that is needed to be successful.

Different levels of the production process work with specific data bases but also need to use the information from other data bases. Typical examples are given for computer-aided design (CAD), planning (CAP), manufacturing (CAM), quality control (CAQ), administration (CAA), computer-based education (CBE) and computer-integrated manufacturing (CIM) (see Figure 44.21).

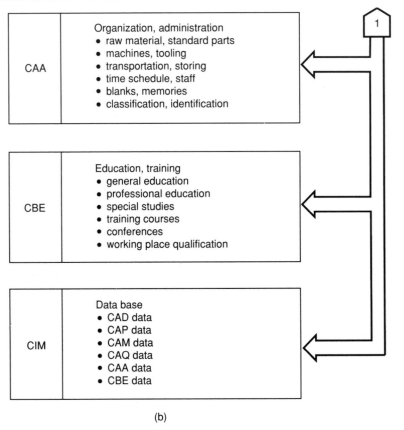

(b)

Figure 44.21   Segments needed in a CIM data base: (a) CAD, CAP, CAM and CAQ parts; (b) CAA, CBE and CIM parts

To assure the quality in a production process a number of methodical aids for factory operation are available (Kamiske, 1989). Figure 44.22 lists these and shows where they are relevant to the production cycle.

It should be appreciated that a combination of several methods is required to the positive results needed in quality control. The methods mentioned have often only been developed for specific applications; adoption to other uses can be a resource-thirsty operation.

44.8   THE QUALITY MANUAL AND STANDARDIZATION

The International Standards Organisation (ISO) has published a vocabulary for quality (ISO 8402) and a number of international standards dealing with models for quality assurance in design/development, production, installation and servicing (ISO 9000–9004).

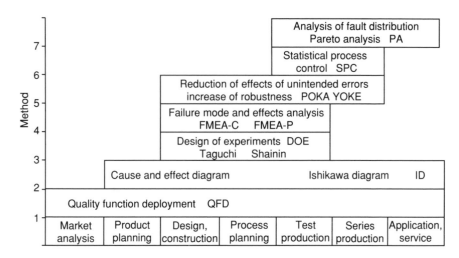

Figure 44.22   Methods and roles for intelligent quality control in the production process

It is said that a prime concern of any company or organization who wish to remain competitive must be the quality of their products and services. In order to be successful a company must offer products or services that:

● meet a well-defined need, use or purpose;

● satisfy the customer's expectations;

● comply with applicable standards and specifications;

● comply with statutory requirements of society;

● are made available at competitive prices;

● are provided at a cost which will yield a profit.

An important part of the documentation used to design and implement a quality system is the 'quality manual'.

A cross-reference list of the standards documentation clauses needed in a quality system (based on ISO 9000) is given in Figure 44.23. This provides rapid recognition of the standardization requirements.

For the production of complex parts the typical contents of the quality manual are the required quality characteristics for the machine tool and for the process. Those issued in Ford Q-101 (1985) are given in Figure 44.24.

In order to achieve maximum effectiveness and to satisfy customer expectations the quality management system should be appropriate to the product or service being offered. The quality manual should be developed by the manufacturer and must be confirmed by the customer as appropriate and acceptable.

Clause No. in ISO 9004	Title	Corresponding clause Nos. in		
		ISO 9001	ISO 9002	ISO 9003
4	Management responsibility	4.1 ●	4.1 ◐	4.1 ○
5	Quality system principles	4.2 ●	4.2 ●	4.2 ◐
5.4	Auditing the quality system (internal)	4.17 ●	4.16 ◐	–
6	Economics - Quality - related cost considerations	–	–	–
7	Quality in marketing (Contract review)	4.3 ●	4.3 ●	–
8	Quality in specification and design (Design control)	4.4 ●	–	–
9	Quality in procurement (Purchasing)	4.6 ●	4.5 ●	–
10	Quality in production (Process control)	4.9 ●	4.8 ●	–
11	Control of production	4.9 ●	4.8 ●	–
11.2	Material control and traceability (Product identification and traceability)	4.8 ●	4.7 ●	4.4 ◐
11.7	Control of verification status (Inspection and test status)	4.12 ●	4.11 ●	4.7 ◐
12	Product verification (Inspection and testing)	4.10 ●	4.9 ●	4.5 ◐
13	Control of measuring and test equipment (Inspection, measuring and test equipment)	4.11 ●	4.10 ●	4.6 ◐
14	Nonconformity (Control of nonconforming product)	4.13 ●	4.12 ●	4,8 ◐
15	Corrective action	4.14 ●	4.13 ●	–
16	Handling and post-production (Handling, storage, packaging and delivery)	4.15 ●	4.14 ●	4.9 ◐
16.2	After-sales servicing	4.19 ●	–	–
17	Quality documentation and records (Document control)	4.5 ●	4.4 ●	4.3 ◐
17.3	Quality records	4.16 ●	4.15 ●	4.10 ◐
18	Personnel (Training)	4.18 ●	4.17 ◐	4.11 ○
19	Product safety and liability	–	–	–
20	Use of statistical methods (Statistical techniques)	4.20 ●	4.18 ●	4.12 ◐
–	Purchaser supplied product	4.7 ●	4.6 ●	–

● Full requirement
◐ Less stringent than ISO 9001
○ Less stringent than ISO 9002
– Element not present

Figure 44.23 Cross-reference list of quality system elements in the ISO quality standard

44.9 FUTURE TRENDS

An obvious first trend exists towards:

● replacement, of serial, off-line, single-channel, one-dimensional, single function, quality measuring instruments and systems having hard keys, fixed functions and without facilities allowing guidance by the operator,

with systems that offer

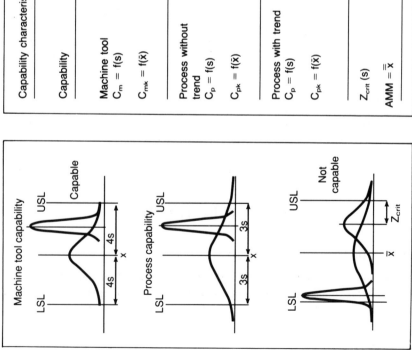

Capability characteristics

Capability	Calculation	Minimal demands	Condition
Machine tool			
$C_m = f(s)$	$C_m = \dfrac{USL - LSL}{6s}$	$C_m \geqq 1.33$	Standard distribution
$C_{mk} = f(\bar{x})$	$C_{mk} = \dfrac{Z_{crit}(s)}{3s}$	$C_{mk} \geqq 1.33$	Standard distribution
Process without trend			
$C_p = f(s)$	$C_p = \dfrac{USL - LSL}{6s}$	$C_p \geqq 1.0$	Standard distribution
$C_{pk} = f(\bar{x})$	$C_{pk} = \dfrac{Z_{crit}(s)}{3s}$	$C_{pk} \geqq 1.0$	Standard distribution
Process with trend			
$C_p = f(s)$	$C_p = \dfrac{USL - LSL - AMM}{6s}$	$C_p \geqq 1.0$	Standard distribution
$C_{pk} = f(\bar{x})$	$C_{pk} = \dfrac{Z_{crit}(s) - \dfrac{AMM}{2}}{3s}$	$C_{pk} \geqq 1.0$	Standard distribution

$Z_{crit}(s)$ = Critical difference between arithmetic mean $\bar{x}$ concerning the specification limit SL in units of standard deviation s

$AMM = \bar{\bar{x}}$ = Average moving mean

Figure 44.24 Machine tool and process capability in the Ford Q-101 specification

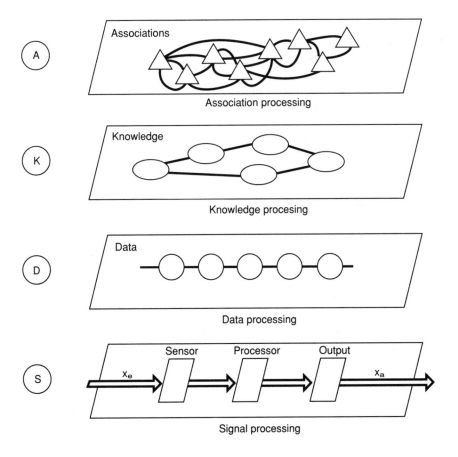

Figure 44.25 Levels of information processing, from basic data gathering through to associative learning

- parallel, on-line, multi-channel, multi-dimensional, multi-functional measuring instruments and systems that also include data management and networking technology offering real-time operation and having advanced operator interfacing facilities such as use of sound and display screens with 'window' style operation.

The main driving force behind this trend is the economic benefits that ensue, for with such systems the manufacturer is better able to be more competitive in cost, delivery time and product quality.

An important component of the system must be the quality of the measurement information itself as it is the 'feedstock' of the information system on which success depends so much. The technico-economic relationship for assessment of the quality of the measurement information Q is given by:

$$Q = \text{lb } m/(t_E \ E) \quad \text{which is expressed as bits/(s \$)} \quad (44.1)$$

where lb—binary logarithm
     m—maximum metrological discrimination capability in bits
     t_E—response time in seconds
     E—expense in monetary units
     bit—binary digit of information transmitted
     s—second
     \$—monetary unit.

From this it can be seen that the higher the quality of measurement information, the higher is the technological value and commercial competitiveness of the quality measuring system.

A second clear trend is that instrumentation for quality measurement applications is emerging that has improved software having better portability, standardization and metrological characteristics. This is largely due to the supply of more highly standardized software and firmware at the international level, which in turn is due to vendors moving to supply systems suitable for open systems interconnection (OSI).

There are increasing needs for better system reliability and compatibility in the general computing arena, correctness of quality measurements and the possibility of obtaining metrological verification and validation of the means by which quality measurements were set up.

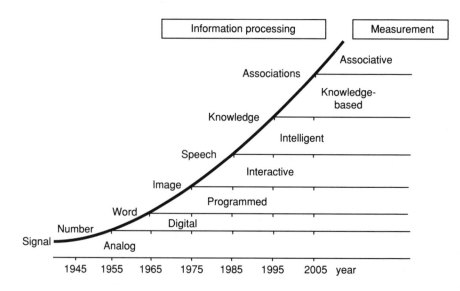

Figure 44.26 Trend, over time, of the form of quality measurements

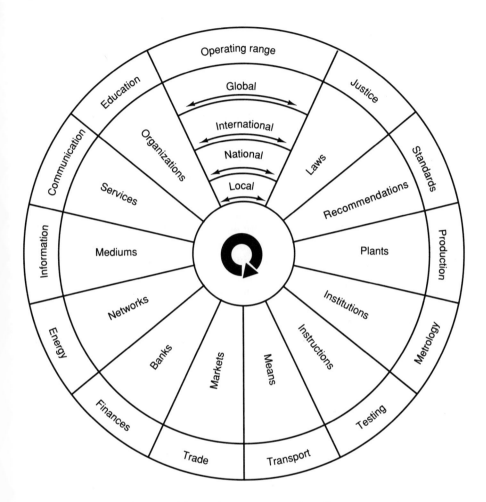

Figure 44.27 Globalization and interactive components of quality management

A problem with quality management systems to date is that they often produce different quality decisions for the same process or product in cases where the quality measurement data is acquired and processed by different means. This is not so surprising given that the general level of understanding about sensors and sensor systems is not always as advanced as it is for other system aspects. Take for example the problem of being able to assess whether a sensor system is obtaining the knowledge needed. This is an emerging research area with much still to be done (see Chapter 41 on the CAD of instrumentation).

A third trend is toward the 'intellectualization' of quality measurements by adding in an intelligence component. The main reason for this addition is that it greatly improves the efficiency of the system's purpose.

The main driving force for this trend is the introduction of higher levels of quality data processing, such as computer-aided processing of knowledge and associations in quality assurance systems have now emerged from the research stage to provide useful tools (Figure 44.25), after Sztipanovits (1989).

Quality measurements have developed from their original relatively unsophisticated analog forms through to digital methods where it is possible now to make use of still relatively crude intelligence features. They will increasingly become part of knowledge-based and associative learning systems in future. Figure 44.26 shows a time chart of these changes.

A fourth trend in this field is brought about by the necessary constituents of quality assurance systems and the globalization of quality assurance demands that have gone well beyond the boundaries of individual countries. These demands have their sources in laws, standards, production, metrology, testing, transport and so on—see Figure 44.27—and more recently also in environmental pollution, energy conservation, traffic jams, nuclear accidents, hunger in the world and symptoms of increasing stress of Man's living and working conditions.

A way to overcome these problems is the wide spread adoption of quality assurance programmes possessing intelligent quality measurements (INQUAMESS), standardized operating systems and software which, by being available at affordable prices and in use in numerous applications, becomes an internationally accepted world standard.

44.10   SOURCES OF FURTHER ADVICE

The reader is referred to the following textbooks and journals for further information on quality control.

Books

IMEKO Secretariat (1989) *Proceedings of International Symposium on Metrology for Quality Control in Production* IMEKO, Budapest.

Ishikawa, K. and Lu. D. J. (1985) *What is Total Quality Control? The Japanese Way* Prentice Hall, Englewood Cliffs. NJ.

Krishnaiah, P. R. and Rao, C. R. (1988) *Quality Control and Reliability* Elsevier, Amsterdam.

Pyzdek, T. (1988) *What Every Engineer Should Know About Quality Control* Marcel Dekker, New York.

Robinson, S. L. and Miller, R. K. (1986) *Automated Inspection and Quality Assurance* Marcel Dekker, New York.

Journals

IEEE Transactions on Reliability IEEE, USA.

Measurement IMEKO, UK.

Quality and Reliability Engineering International John Wiley, UK.

REFERENCES

Ford, Q-101 (1985). *Statistische Prozeßregelung, Maschinenund Prozeßfähigkeit (Statistical Process Control, Machine Tool and Process Capability)*, Ford Company, Koeln, Germany.

Hofmann, D. (1983, 1986). *Handbuch Meßtechnik und Qualitäts-sicherung (Handbook Measurement Engineering and Quality Control)*, Friedr. Vieweg und Sohn, Braunschweig/Wiesbaden; VEB Verlag Technik, Berlin.

Hofmann, D. (1988, 1989). *Rechnergestützte Qualitätssicherung (Computer-aided Quality Assurance)*, Verlag Technik Berlin; Dr Alfred Hüthig Verlag, Heidelberg.

ICQC'87 (1987). *Proceedings of the International Conference on Quality Control*. Union of Japanese Scientists and Engineers, Tokyo.

Ishikawa, K. and Lu, D.J., (1985). *What is Total Quality Control? The Japanese Way*, Prentice-Hall, Englewood Cliffs, N.J.

ISO 8402. *Quality vocabulary*, ISO, Geneva.

ISO 9000. *Quality Management and Quality Assurance Standards—Guidelines for Selection and Use*, ISO, Geneva.

ISO 9001. *Quality Systems—Model for Quality Assurance in Design/Development, Production, Installation and Servicing*, ISO, Geneva.

ISO 9002. *Quality Systems—Model for Quality Assurance in Production and Installation*, ISO, Geneva.

ISO 9003. *Quality systems—Model for Quality Assurance in Final Inspection and Test*, ISO, Geneva.

ISO 9004 (1989). *Quality Management and Quality Systems—Guidelines, ISO*, Geneva.

Kamiske, G.F. (1989). *Proc. Die Hohe Schule der Qualitätstechnik (The High School of Quality Technology)*, Technical University (West), Berlin.

Kemeny, T. and Havrilla, K. (eds.). (1986). 'Intelligent measurement', *Proc. 5th TC 7 Symposium*, Jena, Nova Science Publishers, Commack, N.Y.

Kohoutek, H.J. (1988). 'Intelligent instrumentation: a quality challenge', *Proc. IMEKO XI World Congress*, Vol. 1, 257–65, Instrument Society of America, Research Triangle Park N.C.

OIML (1984). *International Vocabulary of Basic and General Terms in Metrology*, BIPM, CEI, ISO, OIML, Geneva.

Sydenham, P.H. (1982). *Handbook of Measurement Science*, Vol. 1, *Theoretical Fundamentals*, Wiley, Chichester.

Sydenham, P.H. (1983). *Handbook of Measurement Science*, Vol. 2, *Practical Fundamentals*, Wiley, Chichester.

Sztipanovits, J. (1989). 'Intelligent instruments', *Measurement*, 7(3), 98–108.

Chapter

45 S. HOWELL and S. HAMILTON

Intelligent Instruments

Editorial introduction

Today's multi-sensor, highly distributed, sensing systems cannot be maintained at affordable costs without being given the capability to be at least a little clever. Smart sensors are now common place and the parameters of intelligence are generally agreed. The technological availability of affordable single-chip computers and CASE tools for generating the software for large electronic systems have found their way into sensing systems. This chapter covers how intelligence is placed into sensor systems using these latest technologies.

45.1 INTRODUCTION

Instrumentation for data acquisition and control forms a vital component of many diverse systems, ranging in cost, complexity and performance from simple instrumentation to large-scale process control and manufacturing applications, where a wide range of complex instrumentation is often required. The demands placed on such instrumentation are continually escalating as the criteria and compromises between cost, performance, complexity and characteristics evolve to meet the nature of ever-demanding applications. This natural progression has resulted in significant improvements to the basic instrumentation, ranging from the facilities provided by the simplest analog meters, to 'state of the art', high-performance personal workstations and 'intelligent' or 'smart' instruments, which include a dedicated processing element to control and manage the operation of that instrument. Between these two extremes is a vast spectrum of

Handbook of Measurement Science, Volume 3
Edited by P. H. Sydenham and R. Thorn
© 1992 John Wiley & Sons Ltd

instrumentation with widely differing characteristics. Some are simple *stand-alone* units with a simple analog meter being the only mechanism by which information is presented to the user. Others, characterized by the terms *intelligent* or *smart* instruments, are completely autonomous units with an embedded computing ability; they are able to independently manage the real-time acquisition of data or assume responsibility for the autonomous control of a specific process or operation.

The historical development of such intelligent instrumentation is itself of interest, beginning in the 1960s with the development of 4-and 8-bit microprocessors which were embedded in simple instrumentation that required the design of custom hardware interfaces, together with the development of control software implemented in assembly language. This combination clearly resulted in instrumentation that was difficult to maintain and upgrade, with the flexibility of the device being largely determined by the hardware. In addition, the lack of standardization, both in the hardware and the supporting software, ensured there was little or no compatibility between instruments developed by different manufacturers and users.

The subsequent development of the General Purpose Instrument Bus or GPIB, later ratified as IEEE Standard 488.1, did much to alleviate this problem by defining a standard interface between a central 'host' computer and distributed instrumentation, together with a standardized command set that could be implemented by any manufacturer. (See Chapter 46 for further details.) The success of this bus was also greatly enhanced by the development of the BASIC programming language in the 1970s, which provided a convenient and easy mechanism for programming GPIB instruments.

In the early 1980s instrumentation was revolutionized once again by the appearance of the IBM PC and subsequent compatibles, which provided the opportunity to add specific I/O boards to the computer's backplane, whose number and function could be selected for the specific application. Coupled with this flexibility was the ability to program these boards via an ever-increasing number of languages and software packages, which in most cases allow the programmer to define the function of the instrument, rather than the hardware. At the same time, the facilities of the PC permit the convenient and powerful manipulation of data acquired by the system, providing facilities not readily available to the previous generation of stand-alone instruments, including off-line storage and retrieval of data and the implementation of complex analysis and manipulation functions.

Numerous software packages were developed to take advantage of this new flexibility, many of them offering *integrated* facilities that provide a complete data acquisition, management and analysis environment. Perhaps the most exciting of these is the concept of the virtual instrument, initially developed by National Instruments, which allows the screen of the host computer to emulate the physical layout of the conventional instrument. The operation and programming of the instrument is performed by conventional input devices including keyboard and mouse, and the user is able to rapidly and efficiently acquire and manipulate information that is of direct concern, without resorting to complex hardware and software development.

Although the development of PC-based instrumentation has been a significant milestone, emulated by the manufacturers of other popular computer systems, there is an obvious bottleneck in that many sources of information eventually rely on the limited computing and display capacity of the host computer. To alleviate this problem, recent years have witnessed the development of intelligent instrumentation, which effectively implement a *self-managing* instrument that is capable of autonomously acquiring, manipulating and analysing data, relying on the host computer only for the display and analysis of the most condensed form of that information. In addition, the host will usually assume a *master* status, in that it can determine and override the operating mode of the remote instrument.

This chapter examines the implementation and capabilities of such 'intelligent' instrumentation, and describes the hardware and software facilities that are available to support the design and development of intelligent boards, systems and instruments.

45.2 DESIGN OBJECTIVES

45.2.1 Introduction

The internal function of an instrument may be represented as three inter-related tasks, implemented both in hardware and software, and defined as follows (Figure 45.1).

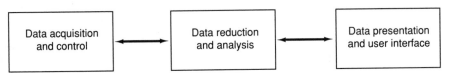

Figure 45.1   Internal function of a typical instrument represented as three inter-related tasks

● *Data acquisition and control.* Acquisition of data or the generation of stimuli that may control the acquisition or define the physical state of the process of interest is readily accomplished using a wide variety of hardware platforms. These range from simple instrumentation incorporating an RS232 or RS485 serial interface, to VME/VXI boards and custom designs. The principal functions of such hardware are summarized in Figure 45.2.

● *Data analysis.* Subsequent manipulation of data may include simple tasks such as the calculation of a mean and standard deviation of a set of measure-

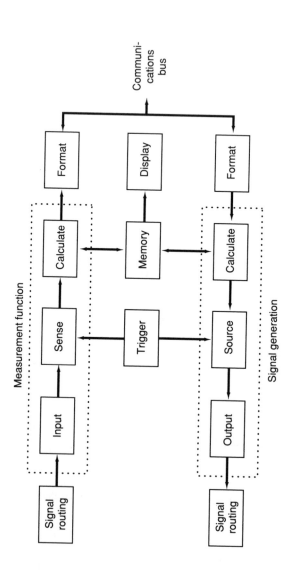

Figure 45.2   Generic model defining the internal functions of a typical instrument, which is acquiring data (measurement function) and generating stimuli (signal generation), which define the physical state of a specific process

ments, compensated for any relevant offsets and calibration constants. More complex applications may include array operations, smoothing, curve-fitting and convolution.

● *Data presentation.* The mechanism by which the user interacts with the data is of great importance, and includes such crucial factors as report generation, database management, file I/O, provisions for hardcopy and high-level graphical facilities that will provide a convenient and appropriate summary of the acquired data.

The inclusion of local intelligence within an instrument typically has a direct influence on the first two of these areas, in that a local processing capability should produce a clear improvement in the data acquisition and control facilities provided by the instrument, together with its ability to pre-process the acquired data, before transmitting the resulting information to a remote host computer. Only in the most complex and expensive instruments will the intelligence be applied to the subsequent presentation of the data, particularly as this will usually necessitate the inclusion of a local keypad or keyboard and monitor/LCD.

45.2.2 Distributed Intelligence

The distribution of intelligence around a complex data acquisition and control system can result in significant improvements to the performance of that system, whilst also providing facilities that are difficult or impossible to attain using conventional techniques. This is particularly important in complex experiments and large-scale process control applications, which often require the acquisition and processing of large volumes of data and the intimate control of many parameters, without undue degradation in the performance of the host. The availability of distributed intelligence ensures that the latter is then available for more important tasks, particularly real-time display and data analysis, rather than being relegated to the routine, time-consuming operations associated with the acquisition of this information.

The advantages of using distributed intelligence to control and acquire information from an experiment or industrial process have been well reviewed by

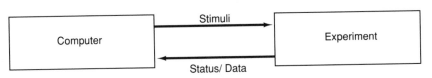

Figure 45.3   Conventional approach for interfacing to a specific application. The computer has sole responsibility for generating all stimuli and interpreting status information and experimental data

many authors see, for example Brignell (1984, 1985), Brignell and Dorey (1983), Favennec (1987), Westbrook (1989), and Zhu (1989). This account deals with the less described implementation aspect. In a typical system in which distributed intelligence is not employed, a computer is used to generate one or more stimuli that will define the physical state of a process—common examples being voltages, currents, angular and linear motion and temperature (Figure 45.3). To assess the effect of these stimuli, the computer should also be able to obtain status information from appropriate sensors, which will provide an indication of the present condition of the process being controlled. This usually entails the use of a dedicated computer which must routinely process raw data from such sensors and continually generate information to provide new stimuli. As a result, not only is the computer being used very inefficiently as it is performing relatively trivial operations at frequent intervals, but it will also usually be unavailable for other desirable tasks, notably the manipulation and analysis of data.

An alternative regime allows for the distribution of intelligent units between the sensors or controllers and the host computer, which will provide independent control and interpretation of a small component of a specific process (Figure 45.4). Consequently the host may be relieved of many mundane and repetitive tasks associated with data acquisition, allowing it to provide more useful and powerful facilities for the user. In a typical example a standard transducer may be interfaced to an intelligent unit and provided with a pre-defined protocol for communicating with a host computer. At any appropriate time, the latter will issue a terse command to perform a specific task, which may, for example, require the acquisition of 1000 readings from an ADC. On reception of such a command the local intelligence will assume responsibility for acquiring the data and pre-processing it before the host is interrupted and provided with a summary of this data. Such a summary may simply be the mean and standard deviation of 1000 measurements, automatically corrected for any sensor defects such as offsets, drift and non-linearities.

Figure 45.4   Incorporating 'intelligent units' between a process and the host computer allows independent management and interpretation of one or more interfaces

Intelligent units used in this way must clearly be both inexpensive and powerful as significant numbers may be required, and each one must be able to meet the demands of the appropriate sensor or controller. It is also important that the hardware and software that constitute the intelligence be easy to develop and apply.

45.2.3   The Use of Intelligence

To incorporate local intelligence within an instrument or system, the designer must be convinced that the additional complexity that necessarily arises is justified (Figure 45.5). This will require the identification of processes that incur extensive or frequent interaction with a central computer, or the realization that the capabilities of an instrument or subsystem could be enhanced through the inclusion of local intelligence. The penalties for such capabilities must also be considered, since the task of designing, constructing, debugging and maintaining intelligent systems is often compounded by the increased sophistication of the hardware and controlling software, both for management of the instrument and for the protocols that allow interprocessor communications.

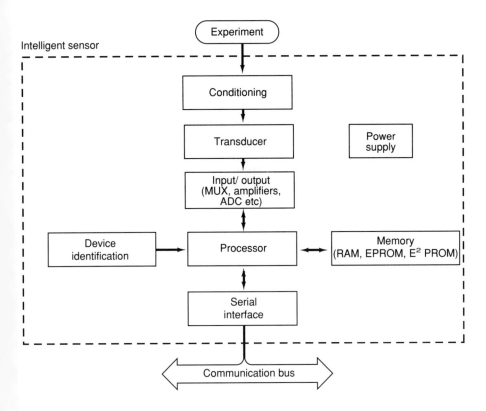

Figure 45.5 Schematic diagram of a typical intelligent instrument. Conventional sensors would simply require analog conditioning and a suitable transducer

Nevertheless there are significant advantages associated with the use of local intelligence, the most important of which are identified below:

- Autonomous supervision of mundane and repetitive tasks ensures that the host is free to perform more important functions, in particular real-time display of data acquired from many sources and the ability to perform complex data analysis operations, unhindered by routine data acquisition. Similarly, real-time events requiring immediate service may be allocated intelligent managers that handle the event and temporarily store large blocks of resulting data in local RAM, before intermittent transfer of this information to the host.

- Intimate control of local hardware, with the ability to provide complex status and diagnostic information about an instrument. This may include auto-configuration and verification that the hardware is operating correctly following internal self-checks, which is important for maintaining the integrity of large systems of instrumentation.

- Automatic measurement and correction of defects inherent in the use of sensors, including offsets, time and temperature drifts and non-linearities. An intelligent instrument may periodically recalibrate itself by acquiring the appropriate calibration constants and storing the resultant information in local RAM or EEPROM.

- The compilation of a summary of the acquired data and the application of relevant correction factors, together with the transmission of this data in a convenient format, which will usually be in the appropriate engineering or scientific units. In complex systems, the information transmitted would usually include an identifier to specify the unit that is supplying the data, together with any relevant status information.

- *Dynamic reconfiguration* of an instrument, in which different control programs are downloaded from the host and executed by the local processor, allowing the instrumentation to adapt to different situations.

- As the performance of the host computer will remain largely independent of the number of intelligent units interfaced to it, it is possible to configure large systems of intelligent controllers and sensors, each with individual addresses, which are able to communicate both with other units and the host over one or more common data buses. In this case the use of a popular serial bus (e.g. RS232C or RS485) ensures that units may be interconnected with great ease and minimum expense, and allows controllers and sensors to be located at considerable distance from the host. This is often an important consideration in industrial applications and for the acquisition of data in hostile or electrically noisy environments.

- The versatility of intelligent instrumentation has significant advantages for industrial applications. Independent verification of instrument operation and the announcement of alarm conditions is clearly advantageous for the reliability and servicing of process control instrumentation, whilst greatly reduced wiring costs and the possibility of using less powerful central computers results in considerable financial savings. In addition, a general-purpose intelligent 'block' can be interfaced to a wide range of sensors or instruments, providing a standardized, modular approach that greatly simplifies the servicing of large industrial systems (Favennec, 1987). In simple configurations the overall effect is to increase both the reliability and credibility of data acquisition, whilst minimizing down-time and human interaction. More complex instrumentation may take further advantage of its intelligence to provide fault tolerance and recovery.

The most significant advantage of incorporating local intelligence within an instrument is that it becomes self-managing, and no longer involves the simple transmission of analog signals which must then be interpreted by a central computer. Complex communication networks and protocols may be constructed (Atkinson, 1987) in which a host computer determines the mode of operation of an instrument by specifying such parameters as its range, number of measurements to be acquired and acquisition trigger conditions. In addition, the host should be able to interrogate an instrument at any time to obtain diagnostic and status information or intermediate data, and the instrument itself may independently instigate communications with the host or another module, for example to flag alarm or error conditions. Consequently, the intelligent instrument possesses a communications facility that is considerably more powerful than the crude alternative offered by unidirectional transfer of analog data to a remote host computer.

45.3 IMPLEMENTING INTELLIGENCE

45.3.1 'Smart' Integrated Circuits

Once suitable applications for distributed intelligence have been identified the designer must determine the most appropriate method for implementing the intelligence. These range from purpose-built digital control systems based on SSI, MSI and LSI technology, to commercially available microcomputers, single-board computers and microcontrollers. Each approach may have distinct advantages in specific applications and a compromise often has to be made between cost, development facilities, physical size and capabilities, in terms of processing power, memory space, I/O provisions and support devices, including storage media if appropriate. A number of different approaches for implementing intelligence are now described, representing a fairly broad spectrum of technologies that range from individual chips to complete systems. The first to be considered is the smart integrated circuit (IC).

A number of semiconductor manufacturers have incorporated limited intelligence within conventional integrated circuits, with the simple aim of managing the operation of that specific component. A typical device is the CSC5012 12 bit self-calibrating analog-to-digital converter, which is part of the *SMART Analog* family produced by Crystal Semiconductors. The ADC may be configured, controlled, calibrated and monitored by the microcontroller at power up/reset or on software intervention; alternatively, the device may be operated independently of its on-board intelligence. As a result the accuracy and linearity of the converter can be maintained over any period of time and across a wide range of temperatures.

This approach of combining digital and analog architectures within a single device clearly has significant advantages, and it is anticipated that this technique will be applied to a wide range of integrated circuits used for the routine acquisition or transfer of information.

45.3.2 Microcontrollers

Microcontrollers, also known as single-chip computers, offer a particularly attractive means of providing local intelligence because of the wide range of hardware features that are provided within a typical device, which removes the costly and time-consuming process associated with the design and debugging of standard hardware. This versatility, coupled with their small physical size and the ease with which application hardware and software may be developed, has resulted in the wide-scale use of these devices in both small and large-scale control applications. Most major semiconductor manufacturers produce one or more families of microcontrollers, including those marketed by Intel (MCS 48, 51 and 96 families), Zilog (Z8 and ZS8), Mitsubishi (MELPS 720/760 and 780), National Semiconductors (COPS 400, 880 and HPC), Motorola (68HC11), Rockwell (R6500), Philips (93C110) and NEC (PCOM-75, 87 and 78).

The heart of any standard microcontroller is a 4-, 8-, 16-or 32-bit CPU, together with limited on-chip memory which usually includes both ROM and RAM for application code. Typical devices also usually provide a UART with a baud rate generator, support for DMA and interrupts, one or more 8/16-bit counter/timers with corresponding pre-scalers, and digital I/O, usually implemented as several bidirectional parallel ports that may be programmed at the bit, nibble or byte level (Peatman, 1988).

A microcontroller family is composed of many distinct members, each of which incorporate additional on-board hardware for specific applications. This may include enhanced counter/timers, parallel I/O and high-speed digital interfaces, together with multiplexed ADCs with their associated sample and hold amplifiers, watchdog timers, pulse outputs and display drivers. Full support is also provided for interfacing to additional external hardware, should this be necessary. A summary of the features provided by four typical microcontrollers is presented in Table 45.1.

Table 45.1 A comparison of typical 4, 8, 16 and 32-bit microcontrollers

	M50760	i8051AH	Microcontroller HPC46083	PC93C110
Technology	4 bit	8-bit	16-bit	32/16-bit
Implementation	CMOS	HMOS	CMOS	CMOS
Clock	400 kHz	12 MHz	30MHz	15 MHz
Number of instructions	37	111	55	56 types
Internal RAM	48×4	128×8	256×8	512×8
Internal ROM	$1K \times 8$	$4K \times 8$	$8K \times 8$	$34K \times 8$
Total external memory	—	64K	64K	2M
Interrupts	1	5,2 priorities	8, prioritized (vectored)	8, plus I/O bits (7 levels)
Low-power modes	—	—	IDLE HALT	IDLE HALT
Counter/timers	1,7-bit	2,16-bit	8,16-bit	3,16-bit
Parallel I/O	14 bits	32 bits	52 bits	40 bits
Serial I/O	—	1, full duplex	1, full duplex	1, full duplex
Watchdog	—	—	Supported	Supported
Notes	—	—	(i) PWM outputs (ii) 4-input capture registers (iii) MICROWIRE-PLUS serial interface	(i) I^2C bus (ii) 256×8 EEPROM (iii) External I/O select (iv) 68000 and 80C51 buses

It is often possible to construct complex control systems through the addition of only a small number of external components. These usually include TTL/RS232 translators for the serial port to allow interaction with a remote host for the development of application code and the transmission of data, together with additional ROM or RAM and a single latch to demultiplex the address/data bus. A typical system based on an 8-bit microcontroller with 4K EPROM and 2K RAM is illustrated in Figure 45.6. It clearly demonstrates the ease with which simple controllers may be implemented. Comprehensive development tools are widely available for most microcontroller families, and the development cycle then becomes both fast and inexpensive, providing a convenient solution that can be tailored to specific applications. This should be compared with dedicated microcomputer systems and single-board computers, which usually entail considerable expense and redundancy, particularly in large networks of distributed intelligence.

Interfacing to microcontrollers has recently been made significantly easier following the advent of devices incorporating a simple bus interface, intended

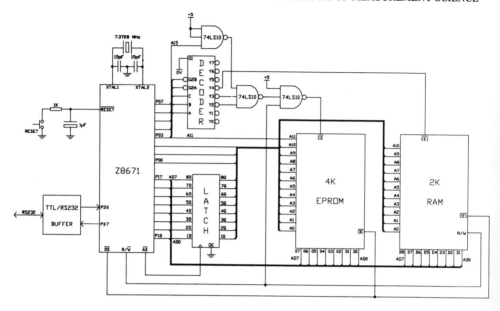

Figure 45.6 Simple yet powerful controller. The Z8671 single-chip computer also provides an on-chip BASIC interpreter, 16-bidirectional digital I/O lines, two 8-bit counter/timers, six interrupts and a UART for serial I/O

for local communications over short distances. A good example of this is the Inter IC (I^2C) bus, which provides a standard two-wire synchronous interface that relies on a clock and bidirectional serial data. Many microcontrollers now include an I^2C bus interface, and a large number of common devices are now manufactured in support of the bus, including RAM, EEPROM, real-time clocks, digital I/O and ADCs/DACs. The advantages for the systems designer on basing instrumentation on such hardware are considerable, as it allows complex instrumentation to be constructed extremely rapidly with minimal hardware and software interfacing, particularly as low-level drivers are often available commercially. As a result, a suitable microcontroller or single-board computer can be rapidly and conveniently expanded to include specific I/O devices, and standard software drivers employed to interact with them.

Although the I^2C bus has found widespread application, particularly in consumer markets, other alternatives are also available, including an increasing number of standard I/O devices that are now being provided with serial interfaces (e.g. Trim DACs and ADCs). In the simplest case such devices only require connection to power and the communications interface (four wires in total!), together with the specification of the device address within a standard software driver before useful communications may commence. It should, however, be noted that although this approach can offer significant savings in hardware and software development, the problems associated with fault-finding and debugging a serial interface that often relies on a complex communication protocol can be significant.

An exciting feature provided by a small number of microcontrollers is the inclusion of *software on silicon*, in which internal ROM is factory-programmed with an operating system or high-level language, ensuring the rapid and convenient development of application software. Tiny-BASIC, for example, is implemented on the Z8671 8-bit microcontroller, whilst the i8052 provides an enhanced and highly structured version of the same language (MCS BASIC-52); this is a fast tokenized interpreter that supports 14 commands, 50 operators and over 40 statements. As an alternative, FORTH is also available (e.g. RSC-FORTH on the R65F11/12 and ChipFORTH).

In the commercial world, microcontrollers have found widespread use in the automotive and process control industries, together with a vast range of applications in domestic appliances, ranging from burglar alarms and stereo systems to calculators, *smart* cards and washing machines. Despite this widespread commercial impact, there have been fewer applications in the research environment. Khan has described a multi-channel analyser based on an i8751 microcontroller (Khan, 1987), whilst Howell and Hamilton have implemented several boards and systems in which microcontrollers are employed to considerable benefit in providing distributed intelligence within complex instrumentation (Howell *et al.*, 1989, 1990). Recently, a number of manufacturers have also announced industrial sensors and signal conditioning units in which an embedded controller is used to provide the communication and computational heart of the unit.

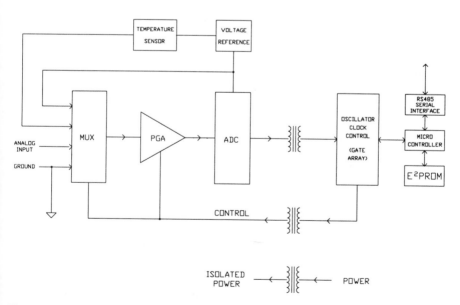

Figure 45.7   Block diagram of an Analog Devices 6B module, which provides an intelligent sensor based on a 8/16-bit microcontroller

A typical example is the Analog Devices 6B series of industrial I/O units, in which an i8052 microcontroller, memory, ADC and serial interface are integrated within a single module (Figure 45.7). The embedded controller manages the operation of the signal conditioning front-end and maintains communications with a remote host computer. It also provides many other important benefits, including autocalibration and ranging, linearization and compensation for changes in ambient temperature and internal reference voltages. Consequently, as all calibrations and settings can be periodically updated and stored in local EEPROM, the accuracy and stability of the unit is considerably enhanced.

The multiplexer MUX (Figure 45.7) selects an analog voltage generated by the appropriate sensor and routes it through a programmable gain amplifier (PGA) to an A/D converter. The multiplexer can also select a compensation signal, the sensor ground, the A/D converter reference and a voltage proportional to the local temperature, allowing offsets and calibration data to be calculated and stored. Transformer coupling is used to isolate all power and digital signals, and a gate array interfaced to the microcontroller manages the acquisition of data.

There are clear advantages associated with the use of microcontrollers over conventional microprocessors and support devices, and this will become increasingly apparent as the microcontroller market continues to expand and mature. Eight-bit microcontrollers based on a RISC architecture with 5 MIP performance are now available, together with 32-bit controllers with awesome capabilities. In addition, microcontrollers are now available that are based on a digital signal processor (DSP) core such as the TMS32020, but also incorporating I/O capabilities that allow a quick and convenient interface with external devices. Consequently it becomes possible to implement a *single-chip* DSP system based on microcontrollers such as the DSC320C14 and TMS320C25 for applications ranging from correlation to digital filtering and smoothing.

45.3.3 Single-board Computers

Single-board computers (SBCs) are widely used in industrial environments to provide autonomous and distributed instrumentation systems, as they offer a quick and convenient 'off the shelf' solution to many problems. A typical SBC provides an 8-, 16- or 32-bit CPU together with memory and several I/O devices on a board of pre-defined size, which usually includes an interface to a standard bus such as STE or VME. A typical SBC is given in Figure 45.8. The board measures 100 mm by 115 mm and includes 128 kbytes of Static RAM, 512 kbytes of EPROM space, a 40×4 LCD port and an interface to an RS485 multi-drop network. Additional I/O facilities are implemented both on the microcontroller and the Philips Inter IC (I^2C) bus, which provides access for up to 128 peripheral ICs with a maximum transfer speed of 100 Kbits per second.

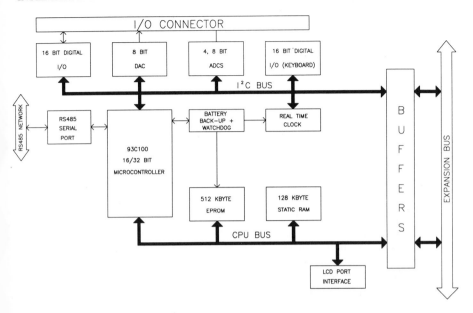

Figure 45.8 Facilities offered by a typical single-board computer, in this example, an 8 MHz 68000 compatible 93C100 16/32-bit CMOS microcontroller (PSI Systems Ltd. Mini Module)

Full development facilities are provided, including a subset of Modula-2, which provides a high-level, multi-tasking language that allows rapid and convenient development of application software.

An SBC usually includes parallel and serial I/O, counter/timers, significant RAM space for application software and one or more ROMs that may contain monitors, assemblers/disassemblers and high-level languages such as BASIC or FORTH.

A large number of SBCs are available commercially for specific buses, together with a wide choice of I/O cards that may be used to enhance the capabilities of the original computer. These cards may also incorporate local intelligence if they are associated with processor-intensive tasks, and it is now common practice to include high-performance microprocessors on disk controllers and serial interfaces. To increase system performance further, this approach has recently been applied to data acquisition boards that include DAC, ADC and stepping motor interfaces, and specific languages have been developed to manage and program this distribution of intelligence within a system. (An example is the CLiP specification, which describes a Control Language for intelligent Peripherals developed jointly by Arcom Control Systems Ltd and British Telecom Microprocessor Systems (CLiP, 1989).) Although the convenience of commercial availability, ready-made hardware and ease of software development are of great importance, there are significant disadvantages associated with SBCs, particularly in large-scale applications. They are relatively expensive, both in terms of the board and the supporting hardware, which may include additional cards, backplanes and power supplies. In addition, many features of the standard SBC are often redundant or simply inappropriate for a specific application.

45.3.4 Personal Computers

The use of commercially available computer systems to provide distributed intelligence appears to be unattractive at first sight, on grounds of both cost and physical size, and for many years this has been totally justified. However, the recent advent of inexpensive PC clones has removed these restrictions, making it feasible to construct single instruments or systems around a dedicated computer. The advantages of this are clear: the design and construction of crates, backplanes and power supplies becomes unnecessary as the PC will provide all these facilities, together with a full development system that includes a keyboard, monitor and disk drives, coupled with a staggering choice of standard languages that may be used to develop application code. Once this has been written and debugged it can be transferred to EPROM or loaded automatically from the PC's AUTOEXEC.BAT file at power up or reset. Some situations may also allow the removal of the keyboard, monitor or disk drives, but quite often these will provide an invaluable mechanism for interaction between the user and the instrument (see, for example, Hamilton *et al.*, 1990).

Communication between the central computer and the PC-based instrument is also greatly simplified, as standard serial interfaces are widely available, together with diagnostic communications software that can be invaluable for debugging large systems of distributed intelligence that communicate over shared buses.

In applications where there are constraints on the physical size of the PC, the development of surface mount technology and the recent advent of PC chip sets has allowed designers to integrate a PC onto a single board. For example, the Megatel Quark PC II includes an 80386 CPU, 4 Mbytes of DRAM, an SCSI interface, floppy disk controller, VGA compatible video/LCD interface, a parallel port, 2 serial ports, a real-time clock and legal AT-compatible basic input/output software (BIOS) all on a single board measuring 100×150 mm. There is also a provision for expanding this processor core through an external PC backplane for incorporating additional hardware, should this be necessary for a specific application. Thus not only is the hardware both inexpensive and widely available, but the inclusion of a legal BIOS allows any version of PC or MS DOS to be booted, providing access to an enormous range of software packages and application languages for developing a dedicated, embedded controller. Even more remarkably, Chips and Technologies have recently announced an implementation of a PC on a motherboard the size of a credit card, providing a CPU for industrial and scientific applications where weight or physical size are the primary concerns.

45.3.5 Programmable Logic Controllers

Programmable logic controllers (PLCs) are widely used in the manufacturing and process control industries, where they provide inexpensive, reliable and flexible control and monitoring of information in what is often a very hostile and demanding environment. They offer compact, efficient and economical replacements for a wide range of hard-wired relay circuits, timers and sequencers, which have previously incurred significant costs in design, development, installation and subsequent upgrade and servicing.

The PLC was developed during the early 1970s to provide intelligent management of specific applications, rather than for general-purpose development work. The heart of the PLC is a microprocessor or controller, often with limited I/O facilities, around which a large number of standard modules may be added to provide the capabilities required for specific applications. A typical PLC family, of which there are many (see, for example, the Mitsubishi F1/F2 series and the Siemens Simatic S5 range) will thus include one or more central processing modules, to which a vast array of extension modules may be attached, ranging from high-density digital I/O to sophisticated analog and counter/timing facilities. These are usually provided with particularly convenient connectors for interfacing directly to the real world.

The PLC is readily and conveniently programmed using simple high-level commands by temporarily attaching a suitable programming keyboard and screen which can then be removed once the code has been tested and subsequently committed to E^2PROM or EPROM. With this flexibility they are conveniently and rapidly reprogrammed for subsequent modifications or changes to the control application. In large-scale systems, the advantage of using a single, well- documented programming language and standardized hardware which can nevertheless be tailored to meet specific and individual demands, are numerous, and the continued use of PLCs both in simple and complex industrial control applications is clearly guaranteed.

45.4 SOFTWARE

45.4.1 Introduction

The importance of software and associated development facilities is often overlooked during the design of intelligent instrumentation, and this can have a drastic effect on the ease with which such instruments can be implemented and subsequently modified, and may even determine the eventual complexity of the software that can be developed for a specific application. Although the availability of hardware support is clearly desirable, particularly in the provision of suitable emulators for microprocessors and single-chip computers, it is probably the software support that is so readily overlooked, and clearly the convenience with which suitable software may be created, ported to the target, debugged, modified and supported is of critical importance.

The key factors involved in the performance and flexibility of the software are related to the choice of the operating system, the application language and the provision of development facilities.

45.4.2 Real-time Operating Systems

Real-time systems for intelligent instrumentation impose demanding requirements on the operating system, which must be able to respond to critical real-time events with sufficient speed and reliability to ensure an appropriate response within a specified time. As the size of the operating system is often an important consideration also, the choice of a suitable system is clearly vital to the performance of the instrumentation. In selecting suitable candidates, the following points should be considered:

● The provisions available for real-time multi-tasking operation, which is often an important facility for intelligent instrumentation. In particular, task scheduling and latency times should be compatible with the proposed application.

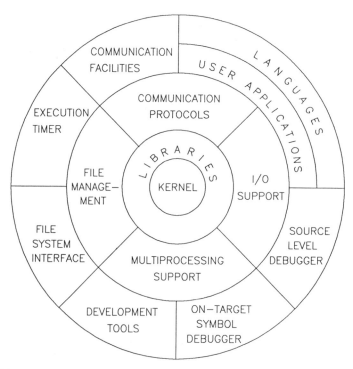

Figure 45.9 Structure of a real-time operating system, illustrating the central high-speed kernel and its interface to supporting services

- The facilities provided for acquiring status information regarding system performance, together with error handling structures and mechanisms.

- The facilities provided for inter-task communication, including task maintenance, synchronization and suspension, together with the support provided for communicating with external devices.

The key component of the operating system for real-time applications is the kernel, which provides critical and indispensable support, including interrupt handling, task scheduling and low-level communication with system hardware (Figure 45.9). Often the capabilities of the kernel invokes a clear dilemma for the system designer, as it must be both small and compact enough for high-speed operation and use in embedded applications, whilst also offering appropriate facilities and services for what is often complex and demanding application software. The casualty is usually the provision of safety and status structures, and consequently the responsibility for providing these is placed on the software engineer implementing the target code.

45.4.3 Application Languages

The process of developing application software may generally be decomposed into four distinct stages:

(1) The structure of the target software must first be elucidated to ensure an efficient and easily supportable implementation, which will also ensure convenient maintenance and future upgrade.

(2) Source code is created, usually on a host computer using a standard word processor or editor. This code must then be assembled or compiled.

(3) Object code is ported to RAM on the target system, possibly preceded by execution on a simulator if this is available. The application code is then executed and debugged.

(4) The final code must usually be blown into one or more EPROMs, and a stand-alone, bootable target system is then created.

The choice of application language has an obvious impact on the performance of the associated hardware, and the facilities provided by different languages should thus be considered with care. Assemblers and some high-level languages such as C, FORTH, PASCAL, Modula-2 and BASIC are now available for most types of single-board computer and microcontroller, but the facilities provided for multi-tasking, debugging and real-time control vary considerably, and should thus be carefully evaluated before a commitment is made to specific hardware. The C language in particular is widely available and supported, and its speed of operation, coupled with its widespread use and acceptance, often make this a prime choice, despite the lack of a multi-tasking implementation.

Key functions that must be provided or supported by the software include:

- Reliable and robust operation in electrically hostile environments, possibly with a degree of fault tolerance.

- Portable and easy to support, with efficient coding.

- Fast operation, both for arithmetic functions if data analysis is involved, together with I/O instructions and interrupt handling.

- Facilities for remote initialization and re-boot, possibly based on a 'watch-dog' feature, together with diskless operation.

- Suitable facilities for providing a convenient man–machine interface.

- Provisions for embedding in-line assembler code within the high- level language, which is an important requirement whenever time- critical tasks arise.

- Support for convenient I/O, including low-level byte/word read and write instructions for configuring and interacting with external hardware, together with the ability to specify and install device drivers, possibly provided by a third party.

Clearly the structure and facilities offered by high-level languages can greatly reduce development times and allow modifications to be made with ease, and if such languages provide the appropriate performance and facilities, this approach is to be preferred. Often such development work can be performed on a host computer to take advantage of its file storage and retrieval facilities, which are rarely necessary on target systems. Unfortunately, however, the ease and convenience of high-level languages are often either unavailable or inappropriate, and code must then be developed in assembly language, necessitating the use of assemblers that offer suitable test and debugging facilities. Many such assemblers are available commercially, together with integrated development packages that provide the designer with all the facilities necessary to develop, debug and port software to the stand-alone target.

Finally, brief mention should be made of the use of computer aided software engineering (CASE) tools, whose function is to allow significant improvements to the quality of applications software whilst also increasing the efficiency of its implementation, which consequently results in improved productivity and software integrity. At the same time, the costs and time scales associated with modifications, upgrades and support are reduced.

Although the use of structured or object-oriented code and the judicious application of comment fields and detailed documentation can greatly improve the ability to support the software, CASE tools allow the structure of the application code to be assessed at a more fundamental level. A wide variety of such tools, usually based on workstations or personal computers, provide varying levels of support for the major phases involved in the development of the software, including structured analysis, design, coding, testing, modelling and documentation. A wide range of useful information can be generated by such tools, which can provide invaluable assistance to the software engineer, including control flow graphs, process activation tables, resource allocations and state-transition diagrams.

45.5. A CASE STUDY

It is evident that the inclusion of distributed intelligence within an instrument or a data acquisition and control system can have a drastic influence on its performance, and to illustrate this point a practical implementation is now described, in which a number of microcontrollers are used to supervise the operation of a complex system of voltage supplies.

High-resolution electron spectrometers require large numbers of floating, highly-stable voltage supplies, ranging in value from a few volts to 300 V. Such supplies, which typically number between 10 and 50, are conventionally provided by passive resistor–divider networks, or more recently by digitally controlled supplies that rely on the provision of unidirectional parallel data, generated by an appropriate computer (Howell, 1985, 1987). Unfortunately, the corresponding cost and complexity of such systems, particularly when large numbers of supplies must be individually isolated, entails considerable expense and usually results in a system that is severely limited in its capabilities.

To alleviate such restrictions, a system of autonomous intelligent voltage supplies has been constructed (Howell et al., 1990), in which the operation of individual voltages is managed by an inexpensive Z8671 8-bit microcontroller, conveniently programmed via an on-board BASIC interpreter. Each supply receives serial information from a host computer in a pre-defined format, which is opto-isolated and interpreted by the local intelligence, which then writes the appropriate value to a DAC whose output is subsequently amplified to provide the required voltage. As the system is based on a serial bus the vast numbers of buffers and opto-isolators inherent in a parallel system are eliminated; in addition, bidirectional transfer of data is available, allowing the microcontroller to provide complex status information about the voltage supply (i.e. local temperature, output current, present operating mode, etc.), which is vital for large systems of voltage supplies.

Control software, executed directly from local EPROM, provides access to a wide range of exciting facilities which would usually require extensive intervention from the host computer. These facilities include the ability to download complex voltage profiles into the memory of one or more supplies, the automatic ramping of one or more voltages between pre-defined limits, synchronization of the update of voltage supplies, either with each other or with external events, and automatic calibration, repeated at pre-defined intervals. This last mode is particularly useful, as it allows DAC and amplifier offsets and linearities to be determined, ensuring that the microcontroller is able to set the output voltage of the supply to the exact value commanded by the user.

Local intelligence also provides another vital facility, in that it allows individual voltage supplies to be controlled manually via a digital potentiometer. The quadrature signals generated by this device are interpreted by the microcontroller, which then calculates the appropriate data to write to the DAC. Intelligent software algorithms allow a single digi-pot to provide coarse and fine control of an output voltage, and it is also possible to conveniently disable any potentiometer and to transmit the present manually set voltage to the host computer.

Although these facilities include all the options that are presently considered desirable and necessary, the incorporation of local intelligence ensures that future enhancements are rapidly and conveniently implemented by updating the control software, provided the supporting hardware has been designed with care. As a result, the flexibility and lifetime of the system is greatly increased, with obvious advantages.

45.6 CONCLUDING REMARKS

The distribution of intelligence around a complex data acquisition and control system can result in significant increases in the performance of that system, whilst also providing facilities that are difficult or impossible to attain using conventional techniques. This intelligence may be implemented in a number of distinct ways using a broad range of technologies, ranging from single integrated circuits to complete microcomputer systems, depending on the processing power and I/O facilities required. With increasingly complex and time-consuming demands made on remote instrumentation, the benefits of incorporating dedicated, distributed intelligence will become increasingly apparent, and it is evident that there will be considerable future interest in the development and application of intelligent instrumentation.

45.7 SOURCES OF FURTHER ADVICE

The reader is referred to the following textbooks and journals for further information on intelligent instruments.

Books

Barney, G. C. (1988) *Intelligent Instrumentation* Prentice-Hall, Hemel Hempstead.

Evans, W. A. (1988) *Trends in Instrumentation II* Adam Hilger, Bristol.

Hofmann, D. (ed.) (1986) *Intelligent Measurement* IMEKO, Budapest.

Ohba, R. (1992) *Intelligent Sensor Technology* John Wiley, Chichester.

Tran Tien Lang (1991) *Computerized Instrumentation* John Wiley, Chichester.

Journals

Intelligent Systems Engineering IEE, UK.

Measurement Science and Technology (formerly *Journal of Physics E: Scientific Instruments*) Institute of Physics, UK.

REFERENCES

Atkinson, J.K. (1987). 'Communication protocols in instrumentation', *J. Phys. E: Sci. Instrum.*, **20**, 484–91.

Brignell, J.E. (1984). 'Sensors within systems', *J. Phys. E: Sci. Instrum.*, **17**, 759–65.

Brignell, J.E. (1985). 'Interfacing solid state sensors with digital systems', *J. Phys. E: Sci. Instrum.*, **18**, 559–65.

Brignell, J.E. and Dorey, A.P. (1983). 'Sensors for microprocessor-based applications', *J. Phys. E: Sci. Instrum.*, **16**, 952–8.

CLiP (1989). *A Control Language for Intelligent Peripherals*, Arcom Control Systems Ltd and British Telecom Microprocessor Systems version 1.2.

Favennec J.-M. (1987) 'Smart sensors in industry', *J. Phys. E: Sci. Instrum.*, **20**, 1087–90.

Hamilton, T.D.S., Howell, S.K. and Turton, B.C.H. (1990). 'An Intelligent voltage/current scanner', *Meas. Sci. Technol.*, **1**, 371–2.

Howell, S.K. (1985). *The Real-time Optimisation of Electron Spectrometers*, Ph.D. Thesis, Victoria University of Manchester.

Howell, S.K. (1987). 'A system of computer-controlled voltage supplies', *J. Phys. E: Sci. Instrum.*, **20**, 288–93.

Howell, S.K., Hamilton, T.D.S. and Turton, B.C.H. (1989). 'Improvements to a system of computer-controlled voltage supplies', *J. Phys. E: Sci. Instrum.*, **22**, 994–7.

Howell, S.K., Hamilton, T.D.S. and Turton, B.C.H. (1990). 'A system of autonomous, intelligent voltage supplies based on a serial bus', *Meas. Sci. Technol.*, **1**, 17–23.

Khan, S. (1987). 'A single chip microcontroller-based portable multi-channel analyser'. *Nucl. Instrum. and Methods in Phys. Res.*, **A257**, 325–30.

Peatman, J.B. (1988). *Design with Microcontrollers'*, McGraw-Hill, New York.

Westbrook, M.H. (1989). 'Future developments in automotive sensors and their systems', *J. Phys. E: Sci. Instrum.*, **22**, 693–9.

Zhu, L. (1989). *Proc. 1st International Symposium on Measurement Technology and Intelligent Instrumentation'*, Wuhan, China Huazhong University of Science and Technology Press.

Chapter

46 F. W. UMBACH

Design of Extensive Measurement Systems

Editorial introduction

The preceding chapters often show the need for large systems of sensors. Today's designer's task will be more related to *extensive* measurement systems than for intensive considerations because extensive systems cannot be mass produced in instrument factories. Designers will need increasing systems understanding in the future. This chapter provides an overview of this need. Chapter 28 in Volume 2 is also relevant.

46.1 INTRODUCTION

In its simplest form a measurement system can consist of a stand-alone measurement device, such as a digital voltmeter or counter. However, a measurement system now usually consists of measurement elements which are often multisensor and, in combination with control systems, multi-actuator systems. A general functional scheme of such a modern measurement system is given in Figure 46.1. In a simple case Figure 46.1 gives the scheme of a system for the control of a production machine, where sensors measure the quality of the product and the actuators control machine parameters. In a complex case it gives the basic scheme of a launch site for missiles. Other examples of such *extensive* measurement systems are aircraft instrumentation, plant control, nuclear reactor instrumentation and train-traffic control.

Handbook of Measurement Science, Volume 3
Edited by P. H. Sydenham and R. Thorn
© 1992 John Wiley & Sons Ltd

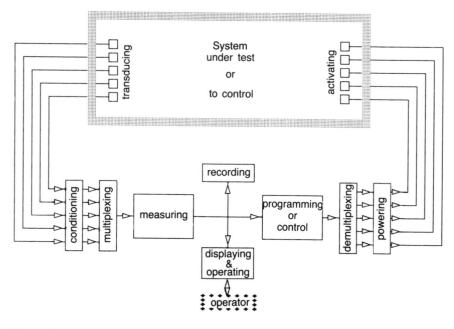

Figure 46.1   Generalized functional scheme of a measurement system

Very often these instrumentation systems work together with one or more operators, forming a *man–machine system*. The instrumentation system is then not a useful object but merely an environment for the operator. Consider, for example, a pilot in his cockpit.

In such cases the design of the instrumentation system is not only concerned with the design of the hardware and software for the technical part of the man–machine system, but also with the design of operator tasks. When problems occur with such complex systems, such as nuclear-power plant burn-out and aeroplane accidents then there are usually discussions as to whether the cause of the problem was of a technical nature or human failure.

46.2   FEATURES OF EXTENSIVE MEASUREMENT SYSTEMS

Complexity. Extensive measurement systems are complex systems in two respects. Firstly, complex in the sense that considerable hardware and software is involved with control, interface functions and protocols added to it. Secondly, complex in the sense that they are difficult to consider from both the technical and operator's viewpoint.

Modularity. Because of the extensive nature such systems have a *distributed* organization consisting of many organic parts interconnected by interfaces. This modularity has several advantages: some modules may be standard modules;

modules may be delivered by different suppliers; servicing can be done by replacement of modules; and the system can physically be split up. The degree of modularization also depends on whether it is a general-purpose or special-purpose system. In general-purpose systems there may exist additional variations of a basic system modularity for logistic advantages.

Difference between functional and physical organization. A division into modules does not always result in a physical organization that is a reflection of the functional structure of the system. This also contributes to the difficulty of comprehending such systems.

Interfaces. The consequence of dividing the system into modular parts is that these parts have to be interconnected for the system to work as a whole. Each part has to be joined through interfaces.

Hierarchical control. Because of the extensive nature and complexity of the system there is a need for system control functions. Because of the modular organization this control is often distributed in a hierarchical sense.

Operator–system interaction. When there is interaction between an operator and the system this interaction requires careful design. It is complicated by the opaque nature of the system, which means the operator may not always have the right perception of the system.

Man–machine interface. The operator communicates with the system by means of the man–machine interface. This includes both the means, such as displays and keyboards, and the protocol for the interaction. Design of both the operator–system interaction and the man–machine interface belong to the discipline of *ergonomics*.

Social acceptance. Once the system is realized and installed (or even better before that time) attention has to be given to ensuring that the people who will work with the system accept it. This means they accept, emotionally and rationally, working with the system (for example, the introduction of computers to the office). Existing social organizations are also often influenced by the introduction of the new system (for example, two- or three-person cockpit crew changes)

The design: from chaos to order. Because of the extensive and complex nature of these systems, and because of the often intensive interactions with the physical and organizational environment, it is not possible for a human to adequately comprehend a concept of the system in his mind. This problem already arises when the person commissioning a system often has great difficulty in specifying exactly what is wanted and in stating the restraints and relationships with the environment-including the operator–machine relationship needed.

This is especially so with totally new design systems. In the beginning there is only a glimpse of a concept and much vagueness. The design process, therefore, has to proceed in a structured way in order to arrive at a justified result. It is not acceptable to develop it by trial and error. A documentation scheme is needed that not only gives a clear insight into the results of each design step but also allows for communication between the people involved in the design. Gradually, as the design proceeds, this forms the basis for the final manual and technical descriptions.

46.3 SYSTEMATIC DESIGN

Relevance of systematic design

In manufacturing systems the direct costs of the design itself are a minor part of overall systems production costs. However, the influence of the design on the overall systems costs is tremendous (see Chapter 44). Inadequate design can increase the realization costs by factors of two or three.

The design determines the components and facilities to be purchased to fabricate the product and also the work that has to be done to realize the system in the production department. If, by bad design, the finished product contains unexpected factors for which unavoidable changes are incurred, total production cost can rise enormously.

To avoid these situations as much as possible the design has to be done very carefully. It should not be hurried. It should be checked, over and over again, for all of its aspects.

To ensure that the design process is carried out with appropriate safeguards, its process should be systematic, with embodied experience and knowledge being explicitly clear for all of its steps.

Design dimensions

In design three elements are involved that can be identified by the following questions (see Figure 46.2):

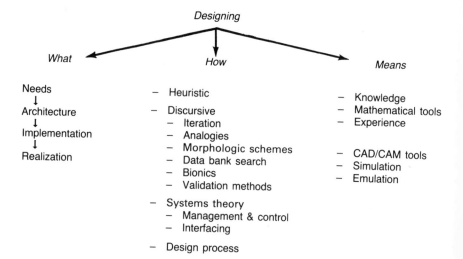

Figure 46.2   The dimensions of design

- *what* must be designed?
- *how* will it be designed?
- with which *means* will it be designed?

The first question *what* not only refers to what has to be produced, but also to what has to be done in the design process to arrive at design products, such as drawings and diagrams, that form the basis for production of the system. These elements are discussed in more detail in later sections.

46.4 SYSTEM DECOMPOSITION

Definitions

Systems consist of hardware, software and human operator tasks.

A *system* is 'a conceived set of elements with mutual relations and relations with the environment'. In general systems have different sizes and complexity and arise at different levels. Elements may be components, boards, devices or subsystems.

The *environment* of a system is 'that part of the total reality with which the system has relationships'.

A system can be divided into *subsystems*: 'A subsystem is a subset of the elements of a system together with the relationships between the elements of this subset.'

The *structure* of a system is 'the set of all relationships between the elements of a system'. Structure can be divided into *substructures*, also called *aspect systems*: 'An aspect system contains part of the relationships in a system.' It is useful to choose aspect systems based on different types of relationships. For example, in a measuring instrument power distribution can be seen as an aspect system, whereas the power supply can be seen as a subsystem. In a coronary care centre in a hospital it is possible to distinguish the staff, instrumental and information aspect systems. In a factory the economic, social and technical aspect system are easily distinguished.

Stratification

When considering a particular type of system, agreement is needed about the *stratification level* of the system along with definition of what kind the elements will be. For example, study of a measurement system implies consideration of a system consisting of measurement devices. For a measurement device the system elements are hardware boards and, maybe, software programs.

In cases where a system is formed by simply joining elements together the situation is relatively straightforward. In using *top-down* design it is not that simple. Consider the example of a measurement device. To arrive at the definition of the constituent subsystems, boards, etc., it has to be decided how the system functions will be grouped into meaningful clusters that have the characteristics to become a subsystem. Thus definition of subsystems, in general, is a process of first going into details and then subsequently integrating the findings into suitable parts.

Levels of abstraction

In linguistics a piece of language can be considered at three levels of abstraction:

- *pragmatics* is concerned with the effect required, the message;

- *semantics* is concerned with the meaning of words or symbols;

- *syntax* is the grammar, the construction of sentences.

These concepts can be applied to systems, in general. This is already commonplace in software engineering where there exists the concept of three forms of models:

- *systologic models* describe why and when an information system has to deliver information into a management system;

- *infologic models* describe the internal functional and operational structure of the information system;

- *datalogic models* describe the way the information system has to be realized.

If we apply these concepts to systems design we can translate them into the terms of (see also Figure 46.3):

	Language	Software engineering	Technical engineering
Pragmatics	Message	What for systologic model	Objectives, use Architecture
Semantics	Meanings of words, sentences Stylistics	Data processing Infologic model	Internal functions, processes Implementation
Syntaxis	Rules for forming sentences Grammar	Programming, data banks etc. Datalogic model	Means Realisation design
Real life system	Book	Automation system	Clock

Figure 46.3   Levels of abstraction, with examples

- *architecture*;

- *implementation*;

- *realization*.

These terms are explained in Section 46.5.

Processes

Dynamic systems are characterized by the existence of one or more processes. A *process* is defined as 'a set of activities by which entities are generated, assembled or whose attributes are transformed'. Activities are understood here as the performance of actions needed to achieve an effect.

According to Maarschalk (1971) two types of processes can be distinguished namely *flow systems* and *conditioning systems*. Flow systems are characterized by an input–output relationship of a relatively simple kind and a sequential nature in which time is an ever-present parameter. Conditioning systems on the other hand have no clear input–output relationship. They have several inputs from different points in the decision structure and several outputs. Time is not a parameter. One can merely speak of events. Its nature is rather more combinatorial than sequential and the structure may vary in time.

Functions

The term *function* has different meanings for different people. In organizational science and practice it has the meaning of 'the place in the organization'. Here the context, however, follows the technical interpretation defining a function as 'the specific office, oriented towards the total systems objective, of a certain element'.

Organisms

There exist dynamic systems of different nature, for example biological, organizational or socio-technical. Each have, at their own disposal, the feature of internal organization.

Such systems are called *organisms*. In organisms one can discern the following main functions (Paterson, 1973):

- *transformation*—producing output(s) out of input(s);

- *regeneration*—replacement of defect parts in some way or another;

- *support*—delivery of the means transformation need to fulfil its task;

- *adaptation* and *coordination*—two management functions. Adaptation means the change of the system's qualities in response to varying internal or external circumstances.

- *Coordination*—control of all the system's elements and relations to direct them to the common objectives.

46.5 TOP-DOWN DESIGN

46.5.1 The Design Scheme

Returning to the question 'what' has to be designed when designing systems, a detailed description can now be given.

Top-down design has two characteristics. Firstly, it means that the design starts at the highest level of abstraction, the architecture, gradually taking more shape as the design process proceeds. Secondly, it means that at each level the scope is most general in the beginning and becomes more detailed at the process provides elaboration. At the start of design the aggregation level is low, becoming more extensive as design proceeds. Figure 46.4 gives the elements needed for a complete top-down design.

In general the system or product to be designed relates to an external *need*. Each need is embedded in an environment that determines the specification of this need. Moreover there will be constraints upon the way the need shall be satisfied. For instance there might be a price limitation.

The first task is, therefore, to analyse the need in its context and with the limitations put upon the solution possibilities. This results in a set of *specifications* and a list of *constraints*. With the specifications and the constraints known the next phase can be started, in which an 'outside' design of the system is made. This does not cover the way in which the system will work, but how it will look to the outside world. This design is called the *architecture design*.

Architecture

The *architecture* (Blaauw, 1972) describes the system to be realized in terms of what the system will contribute to and what its effects will be for its environment. It consists of three components:

- *functional* architecture dealing with the functions that the system places at the disposal of the environment

- *operational* architecture dealing with the processes of the environment in which the system plays its part

- *organic, physical* architecture dealing with the physical nature the system will have.

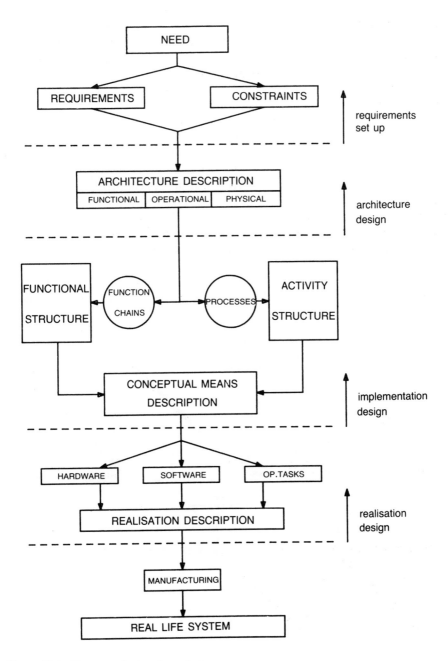

Figure 46.4   Elements of a top-down design

As an example, consider a clock. Will it be a clock with analog or digital display? Will it have only hours and minutes or also seconds, days or possibly more time data? Will it have alarm functions or programmable switches? How has the clock to be operated? How will the energy be reloaded? In what way can time and other date be updated? Is it a throw-away product or has it to be serviced? What will the clock look like—big or small, modern or old-fashioned? Will it be a hanging device or a standing one?

After the architecture design is finished it forms, together with the specifications, a preliminary reporting manual of the system.

Consider another example, that of a traffic light system at a road crossing. It is a system with traffic lights at the four corners, detection loops in the road surface and signalling boxes for pedestrians. Figure 46.5(a) gives a scheme of how the functional architecture may be conceived by an observer.

Driven by how the designer conceives that the system performs the observer concludes that there is a programming function controlling the sequencing of red and green lighting in order to secure a safe traffic flow across the crossing plus a signalling function allowing for cars and pedestrians to influence the sequencing in order that waiting time for them is minimized.

The operational architecture in this case describes how the traffic light system controls the traffic flow across the crossing. A manner of representing this is by means of a timing diagram as shown in Figure 46.5(b). In this figure a denotes

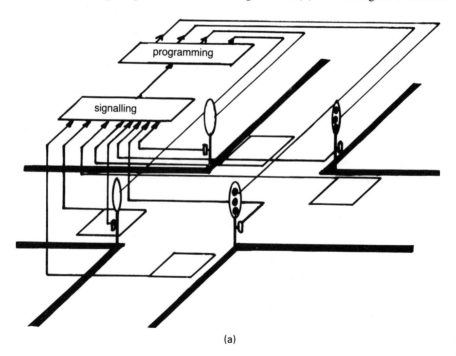

(a)

Figure 46.5   Aspects of a traffic light system: (a) functional architecture;

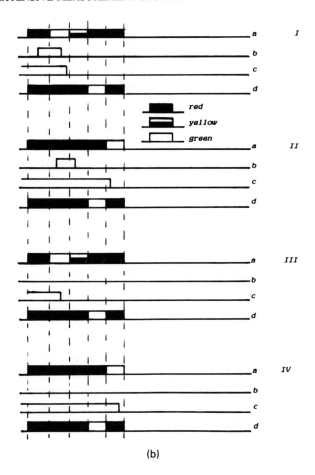

(b)

Figure 46.5 (*cont.*): (b) timing diagram;

the lights, *b* denotes the signalling push-button for the pedestrians, *c* denotes the signalling of a car by the detection loop and *d* denotes the lights for the pedestrians. The observer, by observation, can work out what happens and create a timing diagram, such as that of Figure 46.5(b) with its four phases, I–IV.

The physical architecture of the system to the observer consists of the lightpoles (a) with buttons (b) and the control-boxes (c) for the pedestrians, a cabinet containing the electronics and electrics and possibly the detection loops (d). This is shown in Figure 46.5(c).

Implementation

The *implementation design* deals with the internal functions that will perform the overall system goal and the subfunctions that are needed as a support to the main

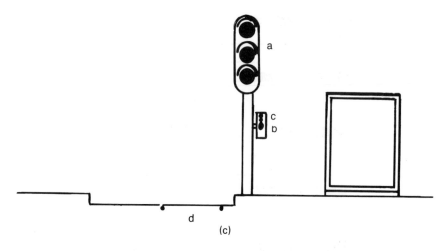

Figure 46.5 (*cont*.): (c) physical form

functions. This also deals with the internal processes that determine, by their activities, the dynamic behaviour of the system.

Whereas the architecture is an outside description of the system, the implementation design is a first description from the inside. It describes *how* the external goals will be reached. However, it is still expressed in abstract terms as functions and processes, not yet being in terms of hardware or software.

The overall result is a design of the internal functional and activity structure of the system with specifications for the conceptual constituent parts, that is, subsystems or components, either hardware, software or operator tasks and also interfaces. To these specifications also belong the constraints for the physical realization of the parts.

It is of importance here to recall what has been said about system decomposition in the design process. These conceptual constituent parts generally cannot be defined before the design has gone one step further in the reticulation process. That is usually needed to decide which functions should be put together to form these parts.

Again consider the clock example, now considering the internal functionality of the clock. This depends on the chosen architecture, for instance, whether it is of analog or digital form. Such questions also need to be answered as how to power the unit, how time rates will be realized, how the precision will be guaranteed, how settings will be effected.

In the case where a specific traffic light system has to be designed, the architecture description was a description aimed at the customers. The implementation design is the first phase in the design of the system's internal operation. Programming has to be designed in terms of control sequences (a conditioning system), the generation and operation of the sequences by the signals from the detection loops and pedestrian push-buttons. What will be the character of the output of the programming function? Will it be a stream of bytes to be translated into signals for the individual lights or will the output already consist of data for the various lights? How will the control signals be transformed to enable them to control the power for each light. These form the kind of decisions concerned with internal implementation.

Realization

In the third design phase, called the *realization design*, the formerly realized specifications of the conceptual constituent parts are transformed into materials which form the basis for production. It includes detailed hardware design down to such detail as component specification, software design down to program line statements and establishment of human task procedures.

In the example of the clock this phase includes the detailed design of the many different parts the clock will consist of including the housing, the colours, the ultimate physical appearance and more.

In the traffic light system example it includes the detailed design of the electronic and electric system parts and the writing of the programs that control the sequences of the lights. It also includes design of the physical parts of the system.

46.5.2   Systems Methodology

The *how* deals with the methods for system design, sometimes called *systems methodology*.

These methods are in general different to those of the architectural design, the implementation design or the realization design. The main distinction between methods is that *heuristic* or *intuitive* approaches are used for over-all designs development, whereas *discursive* approaches are applied once the design requirement has been reticulated down to the component small subsystems level.

The evolution process of systems development seems to be the only natural process. It consists of generating varieties of an existing form and confronting these varieties with the system constraints. That variety that copes best with these constraints is chosen. This iteration method is discursive and can also be used to design artificial (man-made) systems. Other discursive design methods include:

- searching for analogies either between systems in the same discipline or between systems in different disciplines. Analogy also offers the possibility of simulation;

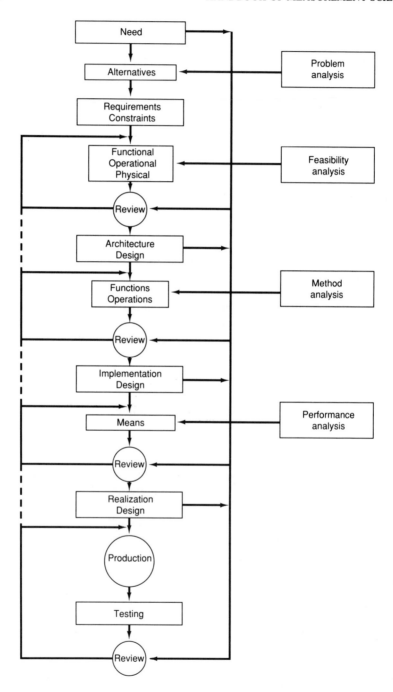

Figure 46.6   The design process

- morphologic schemes, for instance functions and possible realization methods for these functions, are placed in a matrix and the best way through the matrix is chosen;

- putting possible methods with their features in a data bank and, having a specific problem, searching this data bank for an adequate solution;

- the bionics approach wherein is sought solutions already given by natural systems;

- there also exist methods to enumerate how to select if there is a problem of choice and how to evaluate and validate a certain choice.

These methods relate closely to the science of decision making (Kaufmann, 1968), for design is essentially a continous process of making decisions given needs and constraints.

It is interesting to note that in software engineering several, more or less, formal methods have been developed, whereas these have not yet found the same level of application in hardware engineering.

Some of these methods are:

- structured analysis and design technique (SADT);

- information systems work and analysis change (ISAC);

- system matrix (SMX);

- Jackson system development (JSD).

In the body of systems theory only two subjects have fundamental studies on designing 'how'—management and control, and interfacing.

Design, as an activity, is not a straightforward process but comprises many loops of different sizes. Figure 46.6 illustrates this.

Note in this diagram the review loops which not only check the former specification, but also the specifications from earlier steps in the design process.

46.5.3 Design Means

At the 'means' level an enormous range of possibilities usually confronts the designer. This ranges from knowledge of and experience in the discipline, knowledge about systems theory and systems methodology, knowledge of mathematics, knowledge of economics, knowledge of fabrication, etc., to CAD/CAM design methods, simulation, emulation, realization design aids and testing design knowledge.

46.5.4   Concluding Remarks on Top-down Design

Using a recorded procedure for systems design guarantees, to a large extent, that
the design of the system will be done carefully and, therefore, that the risks in
producing it are minimized. Another benefit of this method is that it provides a
disciplined, structured way of thinking and designing, guiding the process from
a situation of 'chaos' through to 'order'.

Some special topics, such as design of system control, interfacing and man–
machine interaction are dealt with in more detail in following sections.

46.6   SYSTEMS CONTROL

Systems control will be performed by a *conditioning system* as defined in Section
46.4. In a special-purpose system having concentrated hardware this may be a
one-hierarchical level process. Often a design weakness is made in that this
conditioning process is carried out with the same hardware means, for example
a computer that is used for the carrying out of flow processes.

The time dependence of both types of processes is very different. In flow
processes having time as a running parameter actions are executed at regular time
intervals. Conditioning systems in contrast, usually operate on the basis of events.
When an event arises many operations have to be carried out in virtually real
time, leaving little spare capability to control the flow process.

Additionally the kind of hardware required is different for both types of processes. For these two reasons, in systems where time is a critical factor, a better
solution is to have the systems control separated from the conditioning system.

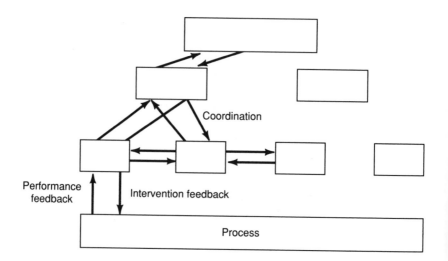

Figure 46.7   Pyramidal structure of hierarchical control

Extensive measuring systems with concentrated hardware are becoming less favoured. These systems now generally have a distributed configuration, either in the form of boards, modules, cabinets, racks or installations. In a distributed organization control is also distributed with mutual coupling by interfaces.

It is then recommended that control be organized hierarchically (Mesarovic, 1970). As a consequence control is structured in three hierarchical levels with distribution realized at the lower levels. This leads to a pyramidal structure as shown in Figure 46.7. In man–machine systems the highest, strategic level, is effected by the operator, whether or not assisted by artificial means. The modern approach is to bring this level into the machine by the introduction of artificial intelligence or use of expert systems.

46.7 INTERFACING

46.7.1 Definition

As said before, in a distributed system interfaces couple the various system parts together as shown in Figure 46.8. An interface may be defined therefore as *the means of establishing the functional relations between physically separated system parts*. It may be emphasized here that an interface is not just a physical, organic connection such as a cable, but a functional coupling by which a translation process maps data from one part into another part preserving both the combinational and sequential relations between parameters or measurands.

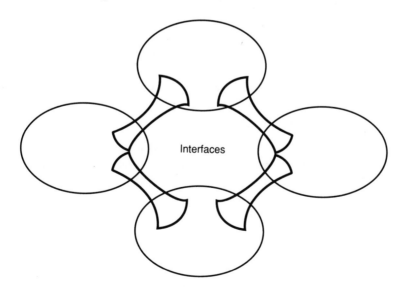

Figure 46.8 Interfacing several units

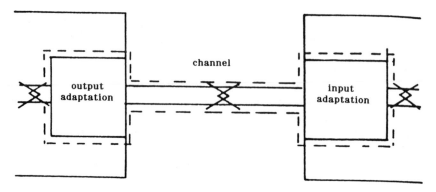

Figure 46.9   An interface channel between two units

The exchange of data between the system parts goes through a channel, a medium, with specific properties depending on the physical form of this channel, for example wire, optical fibre, radio waves. This necessitates converters between the parts and the channel as is shown in Figure 46.9.

There is also another way to consider interfaces, a way that is becoming more prevalent. In the foregoing interfaces are treated as connections to be established. However, it is also possible to consider the structure of all the interfaces in a system, an aspect system, seeing this as a communication system at the disposal of all system parts as the means to exchange their data. This communication system is then called a *network*. In this approach the system of interfaces is substantiated into a subsystem that provides for all necessary communication.

46.7.2   Interface Topologies

In Figure 46.10 basic topologies for interfaces are shown. These topologies may be mixed to obtain intermediate forms.

The *point-to-point* topology is the most simple and can be optimized for just the two parts involved.

In a *star* topology there often exists a hierarchical (master–slave) relation between the parts. The danger of these topologies is failure of the central part.

In *bus* topologies (also called *data-highway* or *party-line*) the bus consists of a contiguous medium, for instance a cable, with stations attached to it without breaking the path of the medium. This topology offers high reliability and allows for high data rates.

In the *daisy chain* topology (also called *ring*) the data passes all the connected parts and is picked up by the addressed part but may keep circling around if the addressed part is not (immediately) capable of handling the data.

More detail of instrument communications systems is given in Chapter 42.

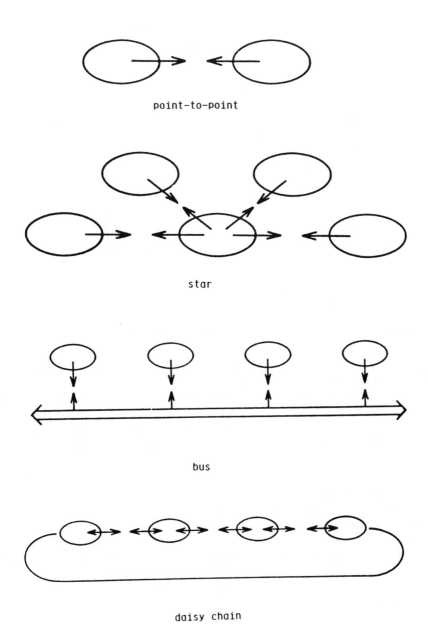

Figure 46.10   Interface topologies

46.7.3   Interface functions

To realize correct operation an interface, as depicted in Figure 46.9, needs some interface functions. The more complex the interface the more functions are needed. A general scheme is given in Figure 46.11.

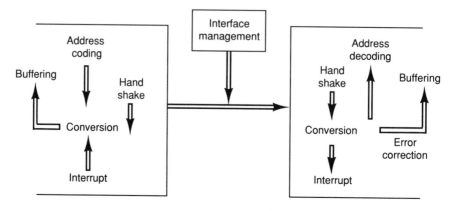

Figure 46.11   Interface functions

The *conversion* function adapts the data to the channel to suit the type of coding used.

Address coding is needed in a multiple element interface in order that the data reaches the appropriate destination.

Handshaking takes care of the required timing of data transport over the channels, including cases where there is irregular or uncertain delay in the channel.

The *interrupt function* allows normal communication to be stopped to facilitate sending of special messages, for instance, interface management messages, over the channel.

A *buffering function* may be necessary if the interface parts are not able to continuously receive data or if the channel is not always able to transmit data at the times they are delivered by a device.

The *error correction* function allows for correction of errors in the data caused by the channel itself.

Interface management is needed in complex types of interfaces to secure proper functioning of the communication. Operations performed by interface management may be: initialization of the interface, interrupt handling, congestion avoidance and so on.

Protocols

The description of a specific interface includes not only the description of the topology structure and the interface functions but also the physical nature, the signal properties and the protocol for the communication medium. To illustrate this consider some interfaces commonly used in measuring systems.

46.7.4 Measuring Interfaces

Off-line interfaces

Off-line interfaces are very common. For example, a sheet of paper with measurement results is an off-line interface between the measuring system and the operator or user. Other types are magnetic tapes, floppy disks and optically read media such as photographs.

Analog interfaces

Analog interfaces transport analog signals or time-discrete signals, such as pulse duration and pulse position signals, between two nodes. Analog interfaces are often point-to-point interfaces. A channel may, however, support various signal transmitters by the use of time or frequency-division multiplexing.

Time-division multiplexing, the commonest multiplexing form used in interfaces for instruments, allows each transmitter to occupy the channel for a certain amount of time in a regular sequence. This principle is used in combination with A/D converters.

Analog interfaces not using multiplexing may require a large amount of wiring: for instance in a chemical plant there are many kilometres of wire. In airplanes the weight of the copper wiring is substantial. In such cases multiplexing is difficult because of the dispersed allocation of the transmitters. Also the cost of additional modulators and demodulators that combine signals may not justify its use.

Analog interfaces normally have only a few interface functions. The conversion function adapts the original signal into a standard format such as in voltage or current form. In frequency-division multiplexing modulators are needed. In a time-multiplex interfacing synchronization must be provided between the multiplexing and the demultiplexing switches.

Because of the difficulties related to networking with analog interfaces there now exists a trend to convert, at the sensor, the analog signals into a digital data format for transmission through digital interfaces.

Serial digital interfaces

Digital interfaces are interfaces using coded binary signals. Serial digital interfaces transport the bits of the code one after the other. Well-known examples in the past were telegraph and telex communication, now largely replaced by facsimile systems.

Also commonly used is the *RS232 interface* for communication between computers and peripherals. Earlier examples used only two-wire interface connections but now there are additional wires for *handshaking*. In the two-wire case it is not easy to secure proper synchronization between the transmitter and receiver. In serial digital interfaces multiplexing can be used on the time-division basis. The various transmitters transmit a byte of their message in turn or, more commonly send a complete message at a time. In measuring situations a message consists of a header, the measurement unit, a sign, a number and an exponent. Interface functions relating to serial interfaces include handshaking, error correction and buffering.

In measuring systems the serial interfaces provide the simplest digital interfaces needing (where permissible) only two wires. They are suitable for long-distance transport of sensor data. Another use, in a mixed analog–digital form, is to convert sensor output voltages into equivalent frequencies (with a voltage-to-frequency converter) sending this binary, but not coded, signal over a two-wire interface. Instead of wires a radio link may form the transmission channel—for instance, as in communication with satellites. For high data rates coaxial cables are needed. In environments where there arises much electrical interference, fibre-optic cables are used.

Parallel digital interfaces

Parallel interfaces transport the digital bits of units in parallel and thus need multiple lines. One line is needed per bit of the unit. Additional lines are required for handshaking and special functions such as addressing. Complex parallel interfaces may have up to 64 lines. In measuring systems parallel interfaces are mostly used inside cabinets because of the amount of wiring involved. A special type of parallel interface, the IEEE-488 interface, is discussed below. Parallel interfaces may be implemented with many types of interface function. They are designed to offer a high data rate and reliable communication in a complex environment.

A common topology is the *bus form*, sometimes with a hierarchical interrupt scheme incorporated to allow for extensive interface management facilities. Multiplexing is accomplished by time division. An example of an industrial standard parallel interface is the VME bus.

46.7.5   The IEEE-488 Interface

General

This interface, also called HPIB (Hewlett Packard Interface Bus), General-Purpose Interface Bus (GPIB) and IEC-625 interface, is a specially developed interface for interconnection of measuring instruments. The international agreements are laid down in IEC publication 625-1 'Standard Interface Systems for programmable measuring equipment' (Part 2: Byte serial/bit parallel interface systems).

This document describes an interface for the interconnection of several instruments, programmable or non-programmable, for building up a measuring system. The mechanical, electrical and functional requirements of instruments for connection to this interface are defined. Standardized code and format conventions are laid down in IEC Publication 625-2. The organization of the IEC interface is based on a bus-line topology.

All instruments of a bus-line system are interconnected via a common set of lines, the 'bus' (Figure 46.12). Data transfer, via the IEC bus, is byte-serial and bit-parallel. One advantage of the IEC interface system (generally called the *IEC bus*) is that there are no cabling problems, and extension of the system is very simple. All instruments provided with facilities for the IEC-bus interface can be used together in one system, no matter where they were made or who made them.

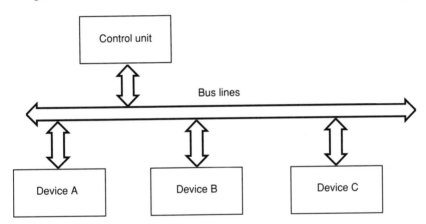

Figure 46.12   Schematic of IEC-bus interface system

Other characteristics of the IEC interface system are that instruments with different data rates can be coupled together without problems and that asynchronous data transfer is possible up to 1 Mbyte per second.

The flexibility of the internationally accepted IEC system is particularly useful in situations where systems varying from very basic to highly complex set-ups have to be used in quick succession.

IEC-bus-compatible instruments can be grouped into the following classes:

(1)   A *listener* device can receive data from other instruments. Examples are printers, display units, programmable power supplies and signal sources.

(2)   A *talker* device can send data to other instruments. Examples are voltmeters and counters with data output facilities, tape readers and data loggers.

(3) A *controller* device can control the information traffic on the bus lines. It determines which device has to send and which is to receive data. It can also send special commands and control signals. Each device has its own address and before data transfer can take place between a given pair of devices, they both have to be addressed by the controller. A controller is a digital computer, this term being used here to include minicomputers, microprocessors and microcontrollers.

Communication between the instruments of an IEC bus system is organized on well-defined rules. For example, only one device at a time can 'talk' (transmit data over the bus), while several can 'listen' (receive messages) simultaneously. The rate at which information is transmitted is automatically adapted to the speed of the instrument which is slowest when processing information.

Since a controller passes information (such as device-dependent programming instructions) regularly to the bus, it must have a 'talk' function in addition to its 'control' function. It generally also has a 'listen' function – in order to enable it to know that information is passing along the bus (and, of course, to receive information itself).

It is possible for a given device to act both as a talker and as a listener. Addition of the possibility of sending two special control messages, 'interface clear' (IFC) and 'remote enable' (REN) turns a control function into a 'system control' function. The device is then called a *system controller*. IFC and REN are general interface management messages. The structure of the IEC interface of a device determines whether the instrument can function as a talker, listener or (system) controller.

The IEC interface can also contain other functions. The 'handshake' functions play an important role in data-byte transfer control.

Other interface functions are:

- *Device clear function* initiates the instrument's device functions by the controller.

- *Device trigger function* triggers one or more instruments, or enables them to start a measurement simultaneously with the aid of bus commands.

- *Remote/local function* chooses between remote control and local (manual) control for the various device functions of an instrument.

- *Service request function*—an instrument informs the system controller that something has happened, requesting the system controller to take appropriate specific action asynchronously with respect to other bus operations.

- *Parallel poll function*—the instrument transmits status data via eight data lines, after the controller has asked for this information.

Bus lines

The 16 lines of the IEC bus can be subdivided into three groups, each with different functions:

- 8 lines are combined in the data bus; DIO 1–8;

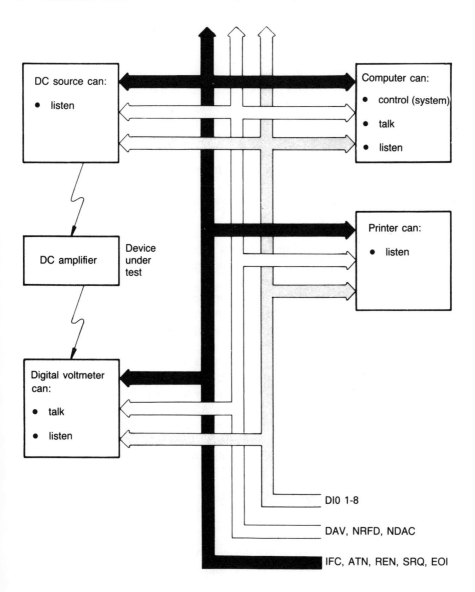

Figure 46.13   Bus lines for IEC-bus used to test a d.c. amplifier

- 3 lines are used for the data-byte transfer control; DAV, NRFD, NDAC (handshake lines);

- 5 lines are for general interface management messages; ATN, IFC, SRQ, REN, EOI.

The significance of these abbreviations is explained below. Figure 46.13 gives an impression of the structure of the bus. (Signals on the interface bus use negative logic with positive potential.)

The eight data input/output lines (DIO 1–8) are used for sending data bytes, addresses, programming instructions, status bytes and special bus commands. One data byte consists of eight parallel data bits. Data transfer on the IEC bus is byte-serial and bit-parallel and is also two-way and asynchronous. A complete message can consist of one or many bytes in series. The maximum transport rate is 1 Mbytes^{-1}. The IEC bus system has two transfer modes: data mode and command mode. One of the general interface management lines (ATN = attention) is for selection of the transfer mode. In the data mode (ATN = false), data bytes or programming instructions are sent via the data bus from an addressed device having a talk function to one or more addressed devices having a listen function.

Programming instructions are device-dependent and are selected by the equipment designer. These types of messages are used to program a certain addressed device. The code is usually sent from the controller, acting as a talker, to a single device addressed as a listener beforehand.

Data bytes are codes for status, display or measurement data. The meaning of each byte is device dependent. They are sent from an addressed talker to one or more addressed listeners, each of which must agree to the meaning of the codes they use for meaningful communication.

In the command mode (ATN = true) the data bus is used to address a device as a talker or a listener and to send special bus commands to program one or more devices. All of these messages are sent by the controller device.

Generally the international 7-bit ISO (or ASCII) code is used for data and messages. This code covers nearly all figures, letters and signs required in practice. The 8th bit is available for a parity check. Each transfer of a byte, over the bus, is obtained by activation of the three data-byte transfer control lines.

Handshaking

The data transfer handshake lines are:

- DAV: *Data valid.* From source device DAV = true indicates that the mess age on the data bus is correct and suitable for acceptance.

- NRFD: *Not ready for data.* From acceptor devices NRFD = false indicates that all instruments are ready to accept a new message.

- NDAC: *Not data accepted.* From acceptor devices. NDAC = false indicates that all listening devices have accepted the message.

The NRFD and NDAC lines are connected in wired-OR configuration. The message on the bus can thus only be 'high' (false) if all instrument outputs connected to those lines are 'high'. If just one of the instruments is not ready to accept data, then NRFD = 'low' (true). Similarly, NDAC = 'low' (true) if just one of the instruments has not yet accepted the data.

Suppose three devices of an IEC bus are addressed, one as a talker (source) and the other two as listeners (acceptors). The task is to send data bytes from the talker to the listener (data mode). Denote the output signals of the acceptor handshake function of listener (1) by nrfd(1) and ndac(1), and the output signals for the handshake lines of listeners (2) by nrfd(2) and ndac(2).

Figure 46.14 gives the handshake timing diagram for this case. This three-wire handshake is typical in IEC-bus use.

Interface management

The five lines of the general interface management group each have a specific function for communication between the controller and the other instruments.

Some of these functions have already been mentioned. The EOI line has two functions in combination with ATN. In the data mode (ATN = false), a talker can use EOI to indicate the end of a multiple-byte transfer (EOI = true). In the command mode (ATN = true) a controller uses EOI = true for the execution of a polling sequence.

An example

Now are given, with reference to Figure 46.13, the main steps involved in the operation of an automatic test system used, for example, to check the linearity of a d.c amplifier.

The d.c. source supplies a series of different d.c. voltages to the input of the amplifier. The digital voltmeter measures the output voltage of the amplifier, and the printer records the data.

The computer coordinates the whole process with the aid of a stored program. The measured values are fed into the computer from time to time, for storage or further processing.

The main steps involved in the process are as follows:

- The computer places the whole system in a initial state where all devices are unaddressed and capable of being put into the 'remote' state.

- The computer addresses itself and the DVM as a listener, in order to program the latter.

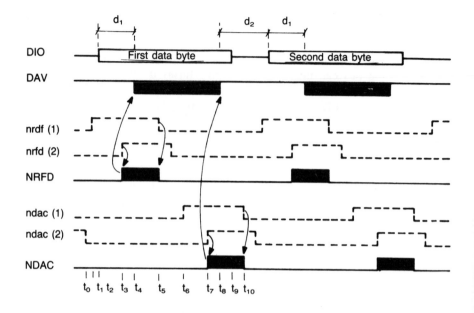

Figure 46.14   Timing of handshake signals for example use of IEC-bus—see text

- The computer sends one or more programming instructions to the DVM, for example, to set the range and the nature of the voltage to be measured.

- The computer unaddresses the DVM as a listener, and remains a talker.

- The computer addresses the d.c. source as a listener, in order to program the latter.

- The computer sends programming data to the d.c. source, which then supplies a certain voltage. The entire message consists of several bytes, transmitted in series.

- The computer unaddresses the d.c. source and addresses the DVM as a listener. The computer still remains as a talker.

- The DVM receives a start code from the computer, and carries out a measurement.

- The computer unaddresses the DVM as a listener, and addresses the DVM as a talker and the printer and itself as listeners.

- The measured data can now be sent from the DVM to the printer and the computer at the same time. The string transmitted consists of several bytes in series, having information about the type, unity, polarity an numeric value of the data.

● When the DVM has sent its last data byte the computer now unaddresses all active listeners (printer and computer) and addresses itself as a talker and the printer again as a listener. When the computer places the codes for CR (carriage return) and LF (line feed) on the bus, the printer will receive these codes and is now ready to print the next measured value on a new line.

● A full cycle has now been completed. It is repeated until all test values are completed.

A detailed exposition of this bus is given in Grimberg (1979)—many texts have covered its operation.

46.7.6 Networks

The world-wide growth of application of computer networks will undoubtedly impact upon the measuring discipline. As said in Section 46.7.1 networking is the other way of looking at interfaces. It has become a discipline in itself (Sinderen and Pras, 1987). Chapter 42 of this Handbook volume addresses communications for measurement systems.

46.8 MAN–MACHINE RELATION

People have varying relationships to technical measuring systems. Sometimes the technical system acts on people as, for instance do medical measuring systems. People can have a management relation to the system, for instance, if the task is servicing of the system. They can also be users of the technical system, for instance, by using a voltmeter or a counter. This complexity of the operator interaction can be very sophisticated, for instance between a pilot and the instruments. Interactive operation is studied by the science of *ergonomics or human engineering*. It is often of crucial importance in extensive measuring systems, because either the technical system is so complex that it needs a human operator, or the complexity of the technical measuring system is so great that the user, for example a pilot, can barely develop an adequate 'mental image' of the system and needs well-chosen 'help' from the system. If the design of the system has not given proper attention to the ergonomic factors it may lead to disasters, called *human failures*.

The design of the man–machine interaction needs knowledge of the psychological, the physiological and the organic properties of the human. Bosman, in Volume 2 of this Handbook, deals extensively with the display part of these needs (Bosman, 1983).

Use of this knowledge, and of the properties of the man–machine interface (Umbach, 1978), can avoid mistakes in design. However, the human is not a predictable machine and, therefore, does not always act as a machine does. Kalsbeek (1981) describes how to regard an operator as a human being and what the consequences are for the operator task development. Moreover, design of the man–machine system should be such that in cases of failure or disturbance the system declines in performance along a path of graceful and not of sudden degradation.

46.9 SOCIAL IMPACT OF MEASUREMENT AUTOMATION

Measurement automation and modernization can have great impact upon social groups. The introduction of electronic measuring systems in hospitals, for instance coronary care centres, has put the nurse behind an operator's desk instead at the bedside. This has strong influence on the patient's feeling of being in safe hands. Introduction of electrical measuring methods in chemical plants, with measuring data being easily transmitted to a central operator room, has changed the work of the operators who, before, went through the plant terrain to manually read the various meters. Introduction of advanced measuring and control instrumentation in aircraft made it possible to reduce the cockpit crew from three to two persons.

Often such a change gives rise to social upset and opposition. Introduction of changes such as these above influence the people involved in two ways. They change the task of individual persons and also have an impact upon the order of the technico-social system. For instance, if an automation exercise has an outcome that a team is reduced from ten to three people, the mutual task division will be different, the authority relationships alter, resulting in social interactions changing. This brings with it the possibility that graceful degradation features will be affected.

The consequences should be carefully investigated and mal-effects controlled by better system design and personnel awareness.

46.10 SOURCES OF FURTHER ADVICE

The reader is referred to the following textbooks and journals for further information on the design of extensive measurement systems.

Books

Blanchard, B. S. (1991) *Fundamentals of Systems Engineering* John Wiley, New York.

Mitchell, R. J. (1990) *Managing Complexity in Software Engineering* IEE, Stevenage.

Hubka, V. (1988) *Theory of Technical Systems* Springer-Verlag, Berlin.

Kerr, R. (1990) *Knowledge-Based Manufacturing Management* Addison-Wesley, Sydney.

Jamshidi, M. (1983) *Large Scale Systems* North-Holland, Amsterdam.

Journals

IEEE Transactions on Systems, Man and Cybernetics IEEE, USA.

Measurement IMEKO, UK.

REFERENCES

Blaauw, G.A. (1972). 'Computer architecture', *Electronische Rechenanlagen*, **14**(4).

Bosman, D. (1983). 'Human factors in display design', in *Handbook of Measurement Science*, Sydenham, P.H. (ed.), Vol. 2, Wiley, Chichester.

Grimberg, J.A.M. (1979). 'IEC Bus Interface', *Digital Instrument Course*, Part 4, Philips, The Netherlands.

Kalsbeek, J.W.H. (1981). 'The production of Behaviour and its Accompanying Stresses', in *Stress, Work Design and Productivity*, Corlett, E.N. and Richardson, J. (eds.), Wiley, Chichester.

Kaufmann, A. (1968). *The Science of Decision Making*, Weidenfeld and Nicolson, London.

Maarschalk, C.G.D. (1971). 'Aspect systems in a general model', *Annals of Systems Research*, Stenfert Kroeze, Leiden, The Netherlands.

Mesarovic, M.D. (1970). *Theory of Hierarchical, Multilevel Systems*, Academic Press, New York.

Paterson, T.T. (1973). *Management Theory*, translation M. de Jong, Samson, Alphen a.d. Rijn.

Sinderen, M.J. van and Pras, A., (1987). *An Introduction to Computer Networks*, Lecture notes, Twente University, The Netherlands.

Umbach, F.W. (1978). 'Interface research and design methodology', *Proc. Conference on 'The Operator Instrument Interface'*, Institute of Measurement and Control, Teesside.

Index

DESIGN AND MEASUREMENT IN ELECTRONIC ENGINEERING

Series Editors

D. V. Morgan,
*School of Electrical, Electronic
and Systems Engineering,
University of Wales
College of Cardiff,
Cardiff, UK*

H. L. Grubin,
*Scientific Research Associates Inc,
Glastonbury, Connecticut, USA*